LIZARD ECOLOGY

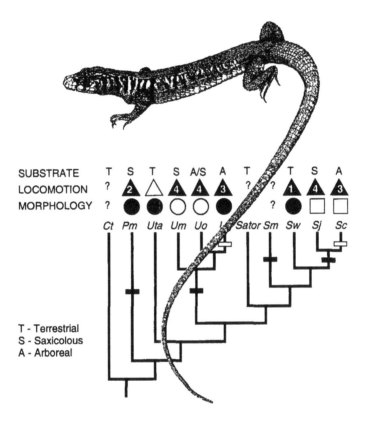

SUBSTRATE   T   S   T   S   A/S   A   T   T   S   A

LOCOMOTION

MORPHOLOGY

Ct   *Pm*   *Uta*   *Um*   *Uo*   *Ls*   *Sator*   *Sm*   *Sw*   *Sj*   *Sc*

T - Terrestrial
S - Saxicolous
A - Arboreal

Drawing of *Nucras tessellata* by Janet Young

LIZARD ECOLOGY

Historical and Experimental Perspectives

Edited by

Laurie J. Vitt and Eric R. Pianka

Princeton University Press, Princeton, New Jersey

Library of Congress Cataloging–in–Publication Data

Lizard ecology: historical and experimental perspectives /
edited by Laurie J. Vitt and Eric R. Pianka.
p. cm.
Papers from a symposium held at the annual meetings of the American
Society of Ichthyologists and Herpetologists and the Herpetologists' League
at the University of Texas at Austin in 1993.
Includes bibliographical references and indexes.
ISBN 0-691-03649-7
1. Lizards—Ecology—Congresses. I. Vitt, Laurie J. II. Pianka, Eric R.
III. American Society of Ichthyologists and Herpetologists.
IV. Herpetologists' League
QL666.L2L57 1994
597.95'045—dc20   93-46274

This book has been composed in New Caledonia font,
prepared in FrameMaker © on a Macintosh IIfx,
and designed by the editors.

2 4 6 8 10 9 7 5 3 1

CONTENTS

Part III. Evolutionary Ecology

Part IV. Population and Community Ecology

CONTRIBUTORS

ROBIN M. ANDREWS, Department of Biology, Virginia Polytechnic Institute & State University, Blacksburg, Virginia 24061 USA

ROBERT BARBAULT, Laboratoire d'Ecologie, Ecole Normale Supérieure, 46 rue d'Ulm, 75230 Paris Cedex 05 FRANCE

C. MICHAEL BULL, School of Biological Sciences, Flinders University, GPO Box 2100, Adelaide, SA, 5042 AUSTRALIA

TED J. CASE, Department of Biology, University of California at San Diego, La Jolla, California 92093 USA

JEAN CLOBERT, Laboratoire d'Ecologie, Ecole Normale Supérieure, 46 rue d'Ulm, 75230 Paris Cedex 05 FRANCE

WILLIAM E. COOPER, JR., Department of Biology, Indiana University-Purdue University, Fort Wayne, Indiana 46805 USA

ARTHUR E. DUNHAM, Department of Biology, University of Pennsylvania, Philadelphia, Pennsylvania 19104 USA

MICHELLE DE FRAIPONT, Laboratoire d'Ecologie, Ecole Normale Supérieure, 46 rue d'Ulm, 75230 Paris Cedex 05 FRANCE

THEODORE GARLAND, JR., Department of Zoology, Birge Hall, 430 Lincoln Drive, University of Wisconsin, Madison, Wisconsin 53706 USA

CRAIG GUYER, Department of Zoology & Wildlife Sciences, Auburn University, Auburn, Alabama 36849 USA

RAYMOND B. HUEY, Department of Zoology, University of Washington, Seattle, Washington 98195 USA

CRAIG D. JAMES, CSIRO Division of Wildlife & Ecology, P.O. Box 84, Lyneham, ACT, 2602 AUSTRALIA

JANE LECOMTE, Laboratoire d'Ecologie, Ecole Normale Supérieure, 46 rue d'Ulm, 75230 Paris Cedex 05 FRANCE

JONATHAN B. LOSOS, Department of Biology, Washington University, Campus Box 1137, St. Louis, Missouri 63130 USA

EMÍLIA P. MARTINS, Department of Genetics SK-50, University of Washington, Seattle, Washington 98195 USA

MANUEL MASSOT, Laboratoire d'Ecologie, Ecole Normale Supérieure, 46 rue d'Ulm, 75230 Paris Cedex 05 FRANCE

DONALD B. MILES, Department of Biological Sciences & Program in Conservation Biology, Ohio University, Athens, Ohio 45701 USA

PETER H. NIEWIAROWSKI, Leidy Labs, University of Pennsylvania, Philadelphia, Pennsylvania 19104 USA (Present Address: Savannah River Ecology Laboratory, Drawer E, Aiken, South Carolina 29802 USA)

KAREN L. OVERALL, Department of Zoology, University of Wisconsin, Madison, Wisconsin 53706 USA & Department of Clinical Studies, School of Veterinary Medicine, University of Pennsylvania, Philadelphia, Pennsylvania 19104 USA

A. STANLEY RAND, Smithsonian Tropical Research Institute, P. O. Box 2072, Balboa PANAMA

LIN SCHWARZKOPF, Department of Zoology, University of British Columbia, 6270 University Boulevard, Vancouver BC V6T 1Z4 CANADA

BARRY SINERVO, Department of Biology, Jordan Hall, Indiana University, Bloomington, Indiana 47405 USA

GABRIELE SORCI, Laboratoire d'Ecologie, Ecole Normale Supérieure, 46 rue d'Ulm, 75230 Paris Cedex 05 FRANCE

S. JOSEPH WRIGHT, Smithsonian Tropical Research Institute, Apartado 2072, Balboa PANAMA

INTRODUCTION AND ACKNOWLEDGMENTS

Laurie J. Vitt and Eric R. Pianka

This book follows in the footsteps of two other symposia on lizard ecology. The first, based on a conference held in 1965, was edited by the late W. W. Milstead (*Lizard Ecology: A Symposium*, University of Missouri Press, Columbia, published in 1967). That volume demonstrated to a broad audience that lizards are often extraordinarily good subjects for ecological studies: many species are abundant, easily observed and captured, low in mobility, and very hardy in captivity. Individuals can be marked and monitored for many years. Of course, such ecologically tractable animals have attracted the attention of some of our best ecologists. Key papers in this volume have provided an intellectual standard and a methodological and conceptual framework that has stimulated a great deal of subsequent research. In addition, most of the participants in this conference continued to contribute important research on lizard ecology long after the first lizard ecology symposium was held.

A second conference on lizard ecology was convened in December of 1980 at a symposium held during the annual meeting of the American Society of Zoologists in Seattle, Washington. Authors were chosen by three editors (Raymond B. Huey, Eric R. Pianka, and Thomas W. Schoener) and by three participants from the original 1965 symposium, William R. Dawson, Rodolfo Ruibal, and the late Donald W. Tinkle. Many authors in the second conference were students of participants in the original lizard ecology symposium. The resulting book, *Lizard Ecology: Studies of a Model Organism*, was published by Harvard University Press in 1983. This volume was dedicated to three pioneers of lizard ecology, Raymond B. Cowles, William W. Milstead, and Donald W. Tinkle. This book noted that lizards are replacing birds as the paradigmatic vertebrates for ecological study. Once again, participants continued to contribute important research long after the symposium was held.

After conferring with several of the participants in the first two symposia, Laurie Vitt and Eric Pianka invited the third generation of lizard ecologists to make their own contributions to the field. Many of the people are graduate students of participants from one of the earlier symposia. For example, Peter Niewiarowski was a graduate student of Arthur Dunham who was in turn a graduate student of Donald Tinkle. The third symposium, entitled "Lizard Ecology: The Third Generation," was held in Austin, Texas, where papers covering a wide range of topics were presented (14 of which are included in the present book). The symposium was followed by a panel discussion which included five members selected from both previous symposia (A. Stanley Rand,° Ted J. Case, Arthur E. Dunham, Raymond B. Huey, Eric R. Pianka) and Laurie J. Vitt. Topics covered were comprehensive, including problems of conservation and the future of funding for research on lizard ecology. The saurocentric answer to the homocentric question "What good are lizards?" was given: "What good are people?"

We have divided this volume into four sections, *Reproductive Ecology, Behavioral Ecology, Evolutionary Ecology,* and *Population and Community Ecology.* Headings of these sections alone suggest what we believe to be significant progress in the evolution of the study of lizard ecology: areas of research are converging as we continue to appreciate the complex and interactive nature of all aspects of lizard ecology. Modern phylogenetic systematics has revitalized comparative evolutionary biology. Understanding evolutionary aspects of ecology (e.g., life histories, behavior) no longer consists of simple discussions of adaptation, but now relies heavily on incorporation of phylogenetic history to sort out evolutionary pathways of characters under investigation. The importance of systematics has never been more appreciated. It must be recognized, however, that extant phylogenies are hypotheses based on the best data currently available. Ecological analyses based on these hypotheses will need to be revised as phylogenetic hypotheses become more robust. Consequently, conclusions based on these analyses are only as robust as the phylogenies. Nevertheless, this exciting advancement is revolutionizing lizard ecology.

Four pioneers in lizard ecology were selected by the editors to introduce each section: Arthur E. Dunham of the University of Pennsylvania, A. Stanley Rand of the Smithsonian Tropical Research Institute, Raymond B. Huey of the University of Washington, and Ted J. Case of the University of California. Their perspectives as well as the perspectives in each chapter illustrate the range of philosophical approaches to lizard ecology. This kind of

---

° The only person to attend all three lizard ecology symposia.

diversity is exactly what leads to major advances in science particularly when controversy over philosophical approaches results. Often viewed by the uninformed as indicating chaos, controversy causes investigators to more critically examine evidence supporting their ideas, resulting in major break-throughs amidst the debate.

The interval between this and the last Lizard Ecology symposium has witnessed an explosion of experimental approaches to lizard ecology ranging from controlled laboratory experiments (dating back much farther) to cre-ative field experiments such as reciprocal transplant "common garden" experiments. Lizards are particularly amenable to field manipulations—resource availability, habitat structure, population density, and even entire sections of food webs influencing lizards can be manipulated. These kinds of studies are in the mainstream of current ecology and their impact is helping shape theory as it applies to all organisms.

A major shortcoming in lizard ecology is the small number of species for which we have even minimal data. A vast majority of community and popu-lation-level field experiments have been performed on tropical anoline liz-ards. Most experimental studies of North American lizards have been conducted on a handful of species in a single family, Phrynosomatidae. An amazing number and diversity of studies have taken place within 10 km of a single field station in the Chiricahua Mountains of southeastern Arizona. We know a lot about desert lizards but very little about tropical or subtropical forest species. For a vast majority of lizard species in the world (most of which are tropical and either skinks or geckos), virtually nothing is known about their ecology. Exactly how general models based on the small number of studied species will be remains unknown. Even though nearly every lizard ecologist appreciates the value of detailed and long-term ecological studies (natural history studies), only a handful of these studies are currently in progress. The existence of a long-term population data base on *Sceloporus merriami*, for example, has fostered an incredible amount of collaborative research that can be interpreted within the context of known population dynamics of those populations. The relevance of these studies to large-bod-ied aquatic teiids such as *Crocodilurus* or subterranean microteiids such as *Bachia* is not so clear. Likewise, consequences of interactions among closely related and morphologically similar anoles may have little bearing on factors affecting structure of taxonomically and morphologically diverse lizard assemblages in lowland tropical forests of the Amazon region. Islands are much less complex than continental landscapes, which contain not only diverse topologies, vegetation, and climates, but also a hidden history of col-onization patterns that has influenced present day distributions. Challenges in this area will require extremely creative thinking as well as the collection of enormous amounts of baseline ecological data on all species.

On a different note, comparisons of the composition of participants in the three lizard ecology symposia to date, as well as an examination of published literature, indicate some positive trends as well. The number of women making major contributions to the field has increased enormously. Lizard Ecology I included no women scientists, Lizard Ecology II included two women, and Lizard Ecology III included four women speakers (with an additional three as coauthors). Likewise, the impact of non-North American scientists on lizard ecology is being felt not only through participation in the symposia, but also in terms of published research. We expect interactions among lizard ecologists from throughout the world to result in the data collection and creative thinking necessary to more firmly establish lizards as "model" organisms in ecology.

At least two critical reviews were obtained for each chapter after which authors revised their manuscripts. We express our sincere appreciation to the following colleagues as well as several reviewers wishing to remain anonymous: Roger Anderson, Royce Ballinger, Al Bennett, Gordon Burghardt, Ted Case, Dennis Claussen, Guarino Colli, Justin Congdon, Ben Dial, Ted Fleming, Anders Forsman, Stan Fox, Harry Greene, Tim Halliday, Ray Huey, Ross Kiester, Jon Losos, Alan Muth, Gary Packard, Bill Parker, Joe Pechmann, Barbara Rose, Doug Ruby, Joe Schall, Rich Seigel, Rick Shine, Carol Simon, Judy Stamps, Dick Tracy, and Kirk Winemiller. Special thanks to Royce Ballinger, Gad Perry, Pete Zani, and Janalee Caldwell, who read the entire manuscript.

We thank the American Society of Ichthyologists and Herpetologists and the Herpetologists' League for providing a forum for "Lizard Ecology: The Third Generation" at the 1993 annual meetings held at the University of Texas, Austin. David Canatella and Dean Hendrickson deserve special recognition for their efforts which led to tight coordination of the symposium. Finally, we acknowledge George Middendorf and the participants in the *Sceloporus* symposium for generating an atmosphere that contributed to the success of "Lizard Ecology: The Third Generation."

LIZARD ECOLOGY

PART I
REPRODUCTIVE ECOLOGY

Studies of lizard reproductive ecology have proliferated over the last fifteen years providing valuable data on such reproductive characteristics as reproductive mode, clutch size, clutch frequency, age at first reproduction, and offspring size for a wide variety of populations. Data are sufficient to allow analysis of geographic variation in reproductive characteristics and demography for only a few species with broad geographic distributions including *Sceloporus undulatus, S. graciosus, Uta stansburiana*, and *Urosaurus ornatus*. Improvement in quantitative techniques and additional data have resulted in reinterpretation by my student, Peter Niewiarowski, of conclusions reached earlier by Don Tinkle, Royce Ballinger, and me. Recent comparative analyses by Don Miles and me allow partitioning of observed phenotypic variation in reproductive and life-history characteristics of different species and populations into components attributable to phylogeny and those due to local population-specific effects such as adaptation to local selective environments. However, the number of species for which this type of analysis is possible remains small due to a relative paucity of high-quality data. Most data on reproductive characteristics of lizards are based on very short-term studies that render adequate quantification of within-population variation in these characters impossible. In addition, most studies provide data on only a subset of the reproductive characteristics, and available data are significantly biased taxonomically. Tropical and old world temperate species remain poorly represented in published studies and it is unlikely that funding agencies will appreciate the value of such time-intensive studies or would provide the needed support. For an understanding of lizard reproductive ecology to progress, deficiencies in available data must be corrected. Finally, only in very few cases do we know whether observed phenotypic differences in reproductive or life-history characteristics have a genetic basis or simply reflect plastic phenotypic responses to local environments. Some of these problems are being addressed in innovative reciprocal transplant or "common garden" experiments being performed by a number of workers, including Gary Ferguson, Larry Talent, Peter Niewiarowski, and Willem Roosenburg. These promise to improve our understanding of the degree to which observed phenotypic variation in reproductive and other life-history characteristics represent evolved differences among populations.

High-quality data on reproductive ecology of local populations are essential for progress in two related areas: understanding the evolution of life histories in lizards and refining life-history theory in general. A diversity of opinions exists concerning the most productive approaches to studying life-history evolution in lizards. Rick Shine and Lin Schwarzkopf have developed a provocative model in which they assume that survival costs are generally more important than energetic costs in explaining patterns of life-history variation in squamate reptiles. My students and I believe that understanding patterns of allocation of limited time and energy or nutrient resources and the fitness consequences of variation in allocation of such resources are critical to a predictive theory of life-history evolution. Pioneers in developing this viewpoint with respect to reptiles include Mike Hirshfield, Don Tinkle, and Justin Congdon. One approach that makes allocation decisions by individual organisms, as well as the fitness consequences of those decisions, explicit, was outlined by Bruce Grant, Karen Overall, and me. In this approach, life histories are viewed as a set of rules that specifies a time-ordered set of allocations conditioned on the environment an organism experiences and the physiological state of the individual. The set of life-history rules specify the allocation of available time into exhaustive and mutually exclusive activities (e.g., foraging, territorial defense, courtship), the allocation of net resources into a set of exhaustive and mutually exclusive functions (e.g., growth, maintenance, storage, reproduction, repair), and the allocation of those resources devoted to produce individual offspring. These allocations are subject to a set of population-specific trade-offs and constraints. Attendant upon each allocation decision made by an individual will be specific costs and benefits, both of which must be measured in the currency of expected lifetime reproductive success. Our model makes allocation decisions explicit and measures the costs and benefits of a particular allocation decision in the common currency of expected lifetime reproductive success of individuals making those decisions. This approach separates reproductive effort from the fitness costs of the effort and considers all potential costs and benefits. We believe energetic costs are of prime importance.

Carefully controlled experimental studies that allow manipulation of reproductive allocation using some of the creative techniques developed by Barry Sinervo are especially likely to increase understanding of selective consequences of variation in allocation, particularly if carried out in demographically well known natural populations. However, most allometric engineering manipulations are whole-organism manipulations that may alter factors other than allocation and, in that sense, are uncontrolled. Implications of such experiments for allocation theory must be interpreted with caution.

High-quality data on reproductive ecology of local populations are essential in attempts to develop a mechanistic, predictive model of local population response to environmental variation. Long-term studies of variation in reproductive characteristics of individuals in response to environmental variation provide essential data for the construction of individual-based, physiologically structured models of population dynamics. For these approaches to be successful, data on environmental factors influencing reproductive output and survival of individuals are essential.

In many lizard populations, variation in recruitment is a major factor influencing population dynamics. Efforts at understanding factors that affect recruitment to natural populations are currently hampered by a paucity of studies of environmental factors affecting egg and newborn survivorship in demographically well-known populations. Studies of thermal and hydric effects on egg survival, incubation period, size of emerging hatchlings, and their subsequent growth and survival such as those reported in this volume by Karen Overall for *Sceloporus merriami* are especially important to developing a mechanistic understanding of reptilian population dynamics.

Finally, fewer than a dozen high-quality, long-term, mark-recapture studies exist for natural lizard populations. These studies are unquestionably the only way of obtaining reliable data on within-population variation in reproductive performance of individuals and relating that variation to environmental variation. Progress in each of the above-mentioned areas will be slow until more investigators commit to this type of study. The paucity of data on taxa other than sceloporines seriously constrains our ability to develop theoretical models that will be widely applicable as well.

Arthur E. Dunham

CHAPTER 1
MEASURING TRADE-OFFS:
A REVIEW OF STUDIES OF COSTS
OF REPRODUCTION IN LIZARDS

Lin Schwarzkopf

The life history of an organism may be defined as a co-adapted complex of traits (Stearns 1976, 1989a; Ballinger 1983). If life-history traits are co-adapted, the concept of trade-offs is central to the understanding of life-history evolution. This is because the optimization of a single aspect of reproduction cannot occur except at the expense of other, related aspects. Thus, natural selection should produce phenotypes that are compromises between the costs and benefits of changing any particular character.

The concept of costs of reproduction first arose to address an apparent contradiction between theory and observation. All things being equal, it is assume that organisms producing more offspring should have higher reproductive fitness (i.e., they leave more descendants) than those that do not. Thus, natural selection should maximize lifetime production of offspring. Total lifetime production of offspring could be maximized by maturing early and devoting all resources to the production of one very large clutch of offspring (Cole 1954). Many organisms, however, do not seem to follow this strategy; instead, maturity is often delayed, organisms are iteroparous, and fecundity at any particular reproductive episode is lower than that which is physiologically possible. Williams (1966a) suggested that the solution to this apparent contradiction lies in the existence of costs of reproduction. These costs can be seen as the negative aspects, or disadvantages, of any particular level of reproductive expenditure and they are measured in terms of future reproductive output. An individual's expected future reproductive success (probable number of offspring produced during the rest of its life) may be defined as "residual reproductive value" (Fisher 1930; Williams 1966a; Pianka 1976). Costs of reproduction are decrements to residual reproductive value accruing from a given act of current reproduction (Fisher 1930; Williams 1966a). For instance, in iteroparous organisms, an increase in current reproductive expenditure may decrease the probability of survival of the reproducing individual or the amount of energy it has available for reproduction in the future. In this case, lifetime production of offspring may be maximized by a strategy of relatively low reproductive expenditure in any one reproductive episode. By keeping reproductive expenditure relatively low at any particular reproductive event, an organism may reproduce many times in its lifetime, thereby increasing lifetime reproductive success relative to an organism that devotes more to a single reproductive bout (e.g., Pianka and Parker 1975). In this view, reproductive output is determined by trade-offs between the costs to future reproduction arising from the benefits of present reproduction.

Two components to reproductive costs are "survival costs" and "fecundity costs" (Bell 1980). Survival costs reduce the survival of reproducing organisms. Fecundity costs may reduce an organism's ability to reproduce in the future in two ways. They may directly reduce energy stores that could otherwise be used for future reproduction (Stearns 1989a). I refer to this type of fecundity cost as a "direct fecundity cost." In organisms that grow after maturity, and in which fecundity is related to body size, fecundity will be reduced if growth rate is reduced (Williams 1966b). In this case, fecundity is reduced through an indirect effect of reproduction on growth rate, and I refer to this as an "indirect fecundity cost" or a "growth cost." Survival costs and fecundity costs may function separately, or they may be linked. For example, if energy stores are depleted by reproduction, reproducing individuals may need to forage more frequently, and hence may be exposed to increased risks of predation.

Measuring the Costs of Reproduction

There has been controversy in the literature concerning the best way to measure costs of reproduction. A variety of reviews found supporting evidence for the existence of costs of reproduction (Clutton-Brock 1984; Partridge and Harvey 1985; Reznick 1985; Bell and Koufopanou 1986; Stearns 1992). However, these studies also found several circumstances under which negative effects of reproduction do not occur or cannot be measured. These circumstances may be summarized as follows: (1) Positive correlations between survival and reproduction, or between past and future reproduction have often been observed in field studies of costs (Reznick 1985). Positive correlations are usually explained in terms of variation among individuals in resource acquisition abilities, i.e., if some individuals can acquire more resources than others, they may be better able to survive and reproduce, or they may be able to reproduce more frequently than individuals that are less able to acquire these resources (Clutton-Brock 1984; van Noordwijk and de Jong 1986). Positive correlations among phenotypic characters measured in the field can mask underlying negative genetic correlations that can be measured in the laboratory (Rose and Charlesworth 1981a, b), and are not evidence for the nonexistence of costs. (2) Reproduction may not be energetically costly. If energy is not a limiting resource, or if investment in reproduction can be made without a concurrent decrement in growth or energy stores, then there is no direct fecundity cost of reproduction (Stearns 1992).

Reznick (1985) and Bell and Koufopanou (1986) evaluated four different methods that have been used to measure costs of reproduction. (1) Phenotypic correlations are most commonly used to study costs. This type of study correlates some index of reproductive expenditure (e.g., clutch size, parental care) with some index of a cost, and can be done in the field (e.g., Reznick and Endler 1982). (2) Experimental manipulations are another way that costs can be studied. In this case, some aspect of reproduction is manipulated, and a cost-related response is monitored. Experimental manipulations have often been performed in the field, or in the laboratory under seminatural situations (Calow and Woollhead 1977; Marler and Moore 1988; Roitberg 1989). (3) Alternatively, genetic correlations may be used to study costs. These studies have used quantitative genetic designs to determine genetic correlations between reproductive expenditure and cost (e.g., Rose and Charlesworth 1981a; Reznick 1983). These studies are usually conducted in the laboratory under controlled conditions. (4) Finally, selection experiments may be used to measure costs. This type of study observes correlated responses to selection pressures on particular characters in the laboratory (Rose and Charlesworth 1981b).

Reznick (1985, 1992) strongly recommends that selection experiments be conducted to verify the basic assumption that observable variation in life-history characteristics has a genetic basis. Evolution cannot occur in response to variation with no genetic basis. Although selection experiments are critical to validate the basic assumption behind studies of costs, observational, and especially experimental field studies of phenotypic variation are necessary to study costs that are mediated by the environment (e.g., increased predation) and also to elucidate ways that organisms avoid costs (Reznick 1992; Stearns 1992). Various authors have stressed the need for field manipulations of reproductive allocation to understand environmental influences on costs (Pianka 1976; Partridge and Harvey 1985; Bell and Koufopanou 1986; Partridge 1992; Stearns 1992). In these experiments, the presence of sufficient genetic variation to allow evolutionary response to selection is assumed to exist. There is strong evidence from artificial selection (Maynard Smith 1978) and from selection experiments conducted under field conditions (e.g., Reznick et al. 1990) that this assumption is justified, although there are circumstances under which available variation is constrained by antagonistic pleiotropy and linkage disequilibrium (Stearns 1992).

Lizards as Model Organisms

Most field studies of costs of reproduction have been conducted using birds (reviewed by Lindén and Møller 1989). This concentration of studies on birds has limited the kinds of questions that researchers have asked to those prescribed by bird life histories. In particular, studies of costs in birds have measured the costs of parental care, and how these costs vary with fecundity. In certain ways, lizard life-history characteristics are more variable than those of birds. All birds are oviparous and most have some form of parental care (Blackburn and Evans 1986). Lizard species, however, span the range from oviparous to viviparous, and oviparous species may or may not have parental care (Shine 1988). Also, unlike birds, lizards may grow significantly after sexual maturity, so that there is the potential for indirect fecundity costs to be important in determining their reproductive output. Because of these differences in life-history characteristics between birds and lizards, studies of lizards allow us to address different questions about costs of reproduction.

Lizards make good subjects for the study of costs of reproduction because both survival costs and direct and indirect fecundity costs can influence reproductive investment in lizards. Many species store resources as fat over long periods (e.g., Derickson 1976), and therefore direct fecundity costs may be important determinants of reproductive investment. As mentioned above, many species show significant growth after reproduction (Ballinger 1983). In most species without fixed clutch sizes, clutch mass increases with body size (reviewed by Tinkle et al. 1970; Barbault 1976; Vitt and Congdon 1978; Vitt and Price 1982), so that indirect fecundity costs may influence reproductive allocation.

Another feature of lizard natural history that makes them suitable subjects for the study of costs, is that the ecology of juvenile lizards is similar to that of their parents. Lizards do not have larval phases during development as do many amphibians, fish, and invertebrates. It is possible to capture, mark, and determine the fate of offspring of particular individuals in the field, and therefore one can measure the fitness of particular traits or reproductive strategies (Sinervo 1990a; Brodie 1992).

Lizards have generally been considered too long-lived, and too large to be practical subjects for genetic correlation or selection studies of life-history characters. More recently, common-environment experiments have been used to determine the genetic contribution to variation in growth rates and life-history characteristics (Sinervo and Adolph 1989; Adolph and Porter 1993; Ferguson and Talent 1993). There have also been some interesting studies of selection on morphological and performance characters in juvenile snakes (Brodie 1992). Similar studies could be conducted using lizards as subjects. Studies of the genetic basis for life-history characteristics are limited in scope, as they must be confined to smaller bodied, short-lived species. However, it is critical to determine the extent of the genetic contribution to variation in life-history characters, because a variety of factors other than trade-offs among life-history characters can influence reproductive investment.

Factors Influencing Evolution of
Reproductive Investment

Of course, trade-offs among life-history characters are not the only factors determining reproductive output in lizards (Ballinger 1983). To a varying degree, we expect the evolution of characters to be constrained by history (Gould and Lewontin 1979). Although life-history characteristics vary among and within species (Tinkle 1969; Vitt and Congdon 1978; Vitt and Price 1982; Dunham and Miles 1985; Barbault 1988; Dunham et al. 1988a) there is a strong tendency for most species in any particular group (family, genus) to have similar life-history characteristics (Vitt and Congdon 1978; Vitt and Price 1982; Dunham et al. 1988a). Thus, phylogenetic lineage appears to have an important constraining influence on patterns of life-history characteristics (Stearns 1980; Ballinger 1983; Dunham and Miles 1985; Dunham et al. 1988a; Vitt 1992). However, despite constraints, lizard life-history characters can evolve (Adolph and Porter 1993). Variation in life-history characters not accounted for by phylogenetic lineage will include evolved responses to different environments. For example, species in lineages with small, fixed clutch sizes (e.g., geckos, anoles) would not be expected to have large clutches in any environment where they occur, but evolution of egg size and clutch frequency in these species in various environments is expected (e.g., Andrews 1979).

Low survival rates in populations of lizards are often associated with early maturity and a relatively high reproductive output, and the latter characteristics are assumed to be an evolved response to the former (e.g., Pianka 1970; Tinkle and Ballinger 1972; Ballinger 1973; Vinegar 1975b; Barbault 1976; Ballinger 1977; Ballinger and Congdon 1981; Tinkle and Dunham 1986; Jones et al. 1987a). Recently, Adolph and Porter (1993) have suggested that similar relationships among survival and reproductive output can be explained as environmentally induced variation in these characters. The length of the activity period (daily and seasonal), and therefore, the energy acquisition rate, may be influenced by temperature, and this may in turn determine important variables such as growth rate and total clutch mass produced per year (Adolph and Porter 1993). In addition, survival rate may be related to the length of the activity period (Marler and Moore 1991; Adolph and Porter 1993). In at least one lizard species, *Sceloporus undulatus*, a significant proportion of the variation in important life-history traits such as survival rate and total annual fecundity and egg mass can be explained by the length of the activity season (Adolph and Porter 1993). However, it appears that other important life-history characters such as size at maturity are not explained by the length of the activity season in some populations of *S. undulatus*, suggesting that evolution may have influenced these characters (Adolph and Porter 1993). Similarly, Ferguson and Talent (1993) found genetically based differences among populations of this species in growth rate, egg size, and age of maturity.

An understanding of patterns of life-history characteristics occurring among species of lizards requires knowledge of the phylogenetic background of the species being considered. In addition, experiments to determine environmental influences on these characteristics among populations within species are also necessary. However, the presence of constraints due to phylogeny or the environment does not mean that costs cannot be important determinants of individual reproductive output. Phylogeny and environment provide the bounds within which a cost structure may operate. It is important to be aware of the influence of these two factors on patterns of reproductive characteristics at the species level and above.

Behavioral and Morphological Changes Associated with Reproduction Likely To Be Costly in Lizards

Reproduction may reduce survival and/or energy stores in lizards in several ways. In both sexes of most species of lizards, obvious behavioral and morphological changes associated with reproduction increase mortality, either by increasing the probability that individuals will be detected and captured by predators, or by decreasing energy available for maintenance. If mortality is not increased due to these changes, energy stores may be depleted, and growth rate and future fecundity may be compromised. I have listed these changes in Table 1.1, and now discuss each in more detail.

Increased movement

In many territorial (Stamps 1983) and nonterritorial species (Kingsbury 1989), distance moved per unit time increases in reproductively active individuals, especially males. Increased movement is energetically costly. Moving animals may also be more susceptible to predation than stationary individuals, because movement is an important cue used by visual predators to detect prey (Curio 1976, pp. 88-90). In addition to increased movement, territorial displays such as push-ups and head-bobs may attract the attention of predators (Marler and Moore 1988, 1991).

Bright coloration

In some lizard taxa, males use brightly colored patches, marks and/or ornaments in sexual display (reviewed by Cooper and Greenberg 1992). Bright ornaments can disrupt cryptic coloration and may increase the probability of detection and capture of displaying lizards by predators (Greene 1988).

Table 1.1. Changes in behavior or morphological or physiological state associated with reproduction in lizards that may result in a cost of reproduction (i.e., a reduction in future ability to reproduce). P = behaviors likely to result in decreased survival due to increased predation, E = behaviors likely to be associated with decreased survival due to decreased energy stores.

Behavioral or Morphological Change	Cost Associated	Sex Affected
conspicuous coloration	survival (P)	males, females
decreased wariness	survival (P)	males, females
increased foraging after reproduction	survival (P)	males, females
reduced escape abilities	survival (P)	females
increased basking	survival (P/E) fecundity	females
increased movement	survival (P/E) fecundity	males
increased susceptibility to disease/stress	survival (E) fecundity	males, females
decreased food intake during reproduction	survival (E) fecundity	males, females

Decreased wariness

Aggression associated with territoriality during reproduction may decrease the probability that reproducing animals will flee in response to predators (Marler and Moore 1988). Also, several studies have documented an unwillingness of gravid females to flee approaching humans (Bauwens and Thoen 1981; Schwarzkopf and Shine 1992). Being less likely to flee could increase susceptibility to predation. However, unwillingness to move in response to the approach of potential predators may not mean that wariness is decreased. Instead, it may be due to a change in escape behavior from flight to crypsis in response to decreased escape abilities (see section on "Avoiding Costs") (Bauwens and Thoen 1981; Schwarzkopf and Shine 1992).

Reduced escape abilities

Reproducing female lizards must carry a burden of eggs or young, and several studies have shown that this decreases running speed (Shine 1980; Bauwens and Thoen 1981; Cooper et al. 1990; Sinervo et al. 1991) and/or endurance (Cooper et al. 1990; Sinervo et al. 1991). Reduced running speed may translate into increased predation if predator capture efficiency is relatively low (Vermeij 1982), and if running is the usual method used to escape predators. As viviparous species must carry clutches of eggs for longer periods than oviparous species, they may be especially susceptible to this cost

(Tinkle 1969). Also, aggressive behavior during territorial defense may take individuals farther from cover than they would otherwise venture, potentially reducing the probability of escaping predation (Lima and Dill 1990).

Increased basking

Females of some species of lizards bask more when they are gravid (reviewed by Shine 1980; Bauwens and Thoen 1981). Increased basking reduces gestation period in at least one species (Schwarzkopf and Shine 1991). If carrying offspring is expensive in terms of survival or energy, it should be advantageous to reduce the period that offspring are carried. Also, reduced gestation period and/or maternal control of incubation temperatures may be advantageous for offspring fitness (Sexton and Marion 1974; Beuchat 1988). However, increased basking may also be costly because females may be exposed to predators during basking (Shine 1980; Schwarzkopf and Shine 1991), and because maintaining high body temperatures is energetically expensive. In addition to increased basking while gravid, reproductive females may continue to bask more than other individuals even after giving birth (Schwarzkopf and Shine 1991). Also, males of several species of lizards bask during winter (Tinkle and Hadley 1973) or in very early spring (Schwarzkopf and Shine 1991). This behavior is thought to increase the rate of testicular development prior to mating (reviewed by Gregory 1982). Basking when few other individuals are active may expose them to high levels of predation, or may be energetically expensive.

Increased foraging after reproduction

In some species, reproduction decreases food intake (reviewed by Shine 1980; Rose 1982; Hailey et al. 1987), and this may reduce energy stores that would otherwise be available for maintenance (Barbault 1976; Jones and Ballinger 1987). Also, increased foraging activity after reproduction may increase mortality through increased risks of predation (Lima and Dill 1990). In species that skip opportunities for reproduction, predation risks due to increased foraging in nonreproductive seasons may impose significant survival costs.

Increased susceptibility to disease and parasites

Energy depletion due to reproduction could increase the incidence of disease and/or parasitism of reproducing individuals, and in turn increase their rate of mortality (Schall 1983).

Some of these changes have been specifically examined and shown to be costly in at least a few species of lizards. Others remain to be studied. In the following sections, I review the literature on studies of costs of reproduction in lizards and discuss areas that require further examination.

Documenting Costs of Reproduction in Lizards

Patterns of reproductive characteristics have been documented in a variety of species, and in some cases in different populations of the same species (Pianka 1970; Tinkle and Ballinger 1972; Ballinger 1973, 1977; Vinegar 1975b; Barbault 1976; Ballinger and Congdon 1981; Tinkle and Dunham 1986; Jones et al. 1987a). These studies were not specifically designed to measure costs of reproduction, but many suggest that trade-offs between reproduction and survival, or between reproduction and growth or energy storage, could have determined observed patterns. More recently, certain studies have focused directly on measuring these costs in lizards. I have tabulated the results of studies focusing directly on trade-offs in Table 1.2, and I will discuss the implications of the approaches used to measure costs in these studies and the results obtained.

To create Table 1.2, I collated several different types of studies. I included as many papers as possible that explicitly addressed questions about costs of reproduction in lizards, as well as several others that incidentally provide information on this topic. Two studies in Table 1.2 did not explicitly measure costs, but measured the influence of variations in food supply on reproductive variables in lizards (Ballinger 1977; Guyer 1988a,b). The responses of individuals to food supplementation or shortage provides insight into the nature of trade-offs in these species. I also included three studies that measured the "metabolic cost" of reproduction. These studies compared metabolic rates in reproducing and nonreproducing individuals, either in the field (Nagy 1983), or in the laboratory (Beuchat and Vleck 1990; DeMarco and Guillette 1992).

General trends in studies of costs

Studies of, or pertaining to, costs of reproduction have been conducted in ten species of Iguanine lizards in three families (Iguanidae, Phrynosomatidae, and Polychridae), six species of lizards in the family Scincidae, and one in the family Lacertidae (Table 1.2). The taxonomic distribution of these studies reflects the geographic distribution of researchers and the common lizard groups in the corresponding areas: iguanine lizards have been studied in North and South America, lacertids in Europe, and, with two exceptions, skinks have been studied in Australia. Compared to general studies of variation in life-history characteristics, which are very heavily concentrated on phrynosomatid lizards, quite a wide range of groups have been investigated. In spite of this, general statements about differences in the operation of costs in different groups are not possible because of the wide range of approaches that have been used to study a variety of different aspects of costs, on relatively few species. Nevertheless, because lineage determines

Table 1.2. A summary of studies of costs of reproduction in lizards. These studies are classified by taxonomic group, sex studied, type of trade-offs measured, direction of trade-off (-, 0, or +), location (field = [F] or the laboratory = [L]), and whether it was correlational (C, experimental (E), or a "natural experiment" (N). In addition, I have recorded whether any mechanisms for avoiding costs were noted.

Family Species	Sex	Trade-offs Survival vs. Reprod.	Trade-offs Fat Stores vs. Reprod.	Trade-offs Growth vs. Reprod.	Study Type	Costs Avoided?	Authors
Iguanidae							
Amblyrhynchus cristatus	F	-(e)			F, C	skip	Laurie 1990
Phrynosomatidae							
Holbrookia maculata	F		-		F, N	1 clutch	Jones & Ballinger 1987
Sceloporus undulatus	F	-	?				Jones et al. 1987
Sceloporus jarrovi	M	-			F, E	↓aggress.	Marler & Moore 1988, 1991
Sceloporus jarrovi	F		-(m)		L, E		Beuchat & Vleck 1990; DeMarco & Guillette 1992
Sceloporus occidentalis	F	-(p)			L/F, C		Sinervo et al. 1991
Sceloporus virgatus	F	-(s)		-	F, N		Vinegar 1975
Urosaurus ornatus	F		-	-	F, N		Ballinger 1977
Uta stansburiana	F		-		F, C		Nagy 1983
	M	-		-			
Polychridae							
Norops humilis (mainland species)	F	-		0	F, E		Guyer 1988a,b
	M	-		-			

Lacertidae						
Lacerta vivipara	F	0		F, E (m/f)	change escape behav.	Bauwens & Thoen 1981
Scincidae						
Eumeces laticeps	F	- (p)		L/F, C	↓activity	Cooper et al. 1990
Eulamprus tympanum	F	- (p)	0 (fi)	F, N	skip, change escape behav.	Schwarzkopf 1993
Anolis maccoyi	F	- (p)	0 (fi)	L, E (m/f)		Shine 1980
Niveoscincus° coventryi	F	- (p)	0 (fi)			
Pseudemoia° entrecasteauxii	F	- (p)	0 (fi)			
Eulamprus† tympanum	F	- (p)	0 (fi)			

(e) = environmentally mediated cost (happens only in "bad" years); (m) = "metabolic cost"; (p) = performance reduced, may or may not be related to a cost in the field; (s) = survival of nonreproducing females was 13% higher than reproducing females; (c) = measured body condition; (fi) = measured food intake; (m/f) = males were compared to gravid females; skip = skip opportunities for reproduction; 1 clutch = single clutch produced per season; ↓aggress = low rates of aggressive behavior; change escape behav. = modified escape behavior while gravid; ↓activity = reduced activity above ground while gravid; ° = as *Leiolopisma*; † = as *Sphenomorphus*.

many life-history characters in lizards, costs should be similarly influenced. Therefore, future studies of costs should continue to examine and compare as wide a range of lizard groups as possible.

Most studies of costs of reproduction in lizards have used females as subjects (Table 1.2). This is probably because, compared to males, it is simpler to determine when females are reproducing and to measure fecundity in females. There is a general impression that males suffer high survival costs of reproduction, at least in territorial species, and two of the studies examining males have found that reproduction does indeed reduce survival (Guyer 1988a; Marler and Moore 1989, 1991). Also, both these studies showed that energy is depleted in reproducing males, and Guyer (1988b) showed that growth rate is reduced in reproducing males. Reduced energy for future reproduction and reduced growth could constitute a cost of reproduction in males.

There have been three basic approaches to the study of costs of reproduction in lizards. (1) Some studies have attempted to directly measure the influence of reproduction on survival, growth, and/or energy stores. These studies have correlated measures of fecundity or reproductive frequency with direct measures of survival, growth, and/or future reproductive output (Vinegar 1975b; Jones and Ballinger 1987; Hasegawa 1990; Laurie 1990; Schwarzkopf 1993). Alternatively, reproductive investment has been experimentally manipulated to measure the influence on survival, growth, and/or future reproductive output (Marler and Moore 1988, 1989, 1991; Landwer pers. comm.). (2) Some studies have examined the influence of reproduction on measures of performance thought to be correlated with survival (Shine 1980; Cooper et al. 1990; Sinervo et al. 1991). A few studies have measured the susceptibility of reproducing individuals to predation in captivity (Shine 1980; Schwarzkopf and Shine 1992). (3) Several studies have measured the metabolic or respiratory cost of reproduction by comparing reproducing and nonreproducing individuals, or by comparing the same individuals before and after reproducing (Nagy 1983; Beuchat and Vleck 1990; Guillette and DeMarco 1992; DeMarco 1993). These three approaches measure different aspects of costs.

The first approach is the most useful for determining if reproduction is costly in nature, and for determining when and how costs may be important. Patterns of mortality or energy depletion in relation to reproduction observed using this approach may suggest mechanisms through which costs operate. For example, reproducing females of some species have a high probability of dying during winter (Jones and Ballinger 1987), while in other species, higher mortality of gravid females occurs during midsummer (Schwarzkopf 1993). High overwinter mortality is thought to be due to depletion of energy resources that would otherwise be available for reproduction, while high mortality in summer suggests that the probability of predation on reproducing females is increased. Experimental manipulations increase the natural range of variation in reproductive output observed in a population, and therefore increase the probability of detecting the presence of costs (Partridge and Harvey 1985). As with correlational studies, manipulations of reproductive investment can suggest possible mechanisms for the operation of costs. For example, in an experiment manipulating both reproductive behavior and food intake, Marler and Moore (1991) showed that decreased survival associated with reproduction in male *S. jarrovi* was due to reduced food intake rather than increased predation.

The second approach commonly used in studies of costs in lizards examines the operation of specific factors that are a potential source of mortality or energy loss during reproduction. This type of study is most useful to elucidate mechanisms through which costs may operate. For example, various studies have found that sprint speed and/or endurance is reduced in gravid female lizards (Shine 1980; Bauwens and Thoen 1981; Cooper et al. 1990; Sinervo et al. 1991). These studies suggest that increased mortality of gravid females may be due to increased predation, as gravid animals are less able to escape predators. Because little is known about the operation of costs, this type of approach would be best if associated with a field study demonstrating that mortality of gravid females is increased during reproduction. For example, decreased performance is not necessarily correlated with increased predation (Bauwens and Thoen 1981; Schwarzkopf and Shine 1992). Gravid females may compensate for reduced performance by altering escape behavior (Bauwens and Thoen 1981; Cooper et al. 1990; Schwarzkopf and Shine 1992).

The third approach measures one specific aspect of costs: the amount of energy devoted to reproduction. This approach does not usually determine whether there are costs in the sense defined in this paper, that is, whether there is a reduction in survival or future reproductive output as a function of the energy invested in reproduction. However, these studies show that the energy expenditure of reproducing individuals is greater than that of nonreproducing individuals (Nagy 1983; Beuchat and Vleck 1990; DeMarco 1993), and this could constitute a cost of reproduction if energy intake does not compensate for reproductive expenditure, and if energy stores are required for reproduction.

In general, trade-offs between reproduction and survival, fat stores and growth were negative in the studies of lizards compiled here, supporting the hypothesis that costs are important determinants of life-history characteristics of lizards. In the following sections, I discuss each trade-off separately, how it is measured, and what implication this has for the interpretation of the result. In addition, I discuss cases in which trade-offs were not observed and reasons why trade-offs may not be observed under particular circumstances.

Survival costs

Of the ten studies listed in Table 1.2 that addressed survival costs, eight measured survival under field conditions. Only one such study concluded that survival was not reduced, under at least some conditions, due to reproduction (Bauwens and Thoen 1981). The predominance of studies showing reduced survival is interesting, because as many as half of the correlational studies of other taxonomic groups found positive correlations between reproduction and survival (reviewed by Stearns 1992). The apparent ease with which negative correlations between reproduction and survival can be measured in lizards suggests that variance in energy-acquisition ability among individual lizards may be less than factors influencing energy intake in all lizards in a population (van Noordwijk and de Jong 1986). All four studies that measured performance variables found that the performance of reproducing individuals was reduced. As mentioned above, reduced performance is not necessarily correlated with reduced survival, but it does suggest that survival costs may also be important in these species.

There can be strong environmental effects on the expression of the phenotypic trade-off in lizards. Two of the studies in Table 1.2 found that survival of reproducing females was reduced only when environmental conditions were poor (Laurie 1990; Schwarzkopf 1993). Cold and rainy weather, reduced food availability, and increased predation may all have contributed to the increased mortality observed in these two studies. Both species used for these studies, *Amblyrhynchus cristatus* and *Eulamprus tympanum*, are long-lived, and therefore have a high probability of survival when environmental conditions are suitable. In comparison, short-lived species show a negative relationship between reproductive frequency and survival even under "benign" conditions of plentiful food and few predators (Jones and Ballinger 1987). This pattern suggests that long-lived species have relatively low reproductive output per season, which increases the probability of survival to the following season. It may be difficult to observe costs in long-lived species, unless environmental conditions impose an extra stress on reproducing individuals. On the other hand, short-lived species may reproduce at close to maximum levels, risking high survival costs. In this case, observing a survival cost without manipulating reproductive investment should be more likely. Experiments that increase reproductive investment by long-lived species are required to determine whether high rates of survival are caused by lowered reproductive output in these species.

Direct fecundity costs

Nine studies listed in Table 1.2 measured some aspect of direct fecundity costs. Of these, four directly measured the influence of reproduction on fat stores, two measured metabolic costs of reproduction, one measured both of these, one examined condition after reproduction (an indirect measure of fat storage), and one measured changes in food intake during reproduction. Most of these measures document a reduction in energy (fat) stores caused by reproduction, but most do not determine whether the ability to reproduce in the future is impaired by this reduction in energy. Both steps are required to demonstrate that reproduction is costly in the sense defined here. If stored energy does not contribute much to reproduction, a reduction in fat stores may not constitute a cost of reproduction (Stearns 1980).

A continuum of energy utilization strategies was suggested for birds by Drent and Daan (1980). Using an economic analogy, they suggested that species that use previously stored energy for breeding rely on "capital investment" to determine reproductive allocation. Drent and Daan called these species "capital breeders." Alternatively, some species do not rely on fat stores for breeding: they have no "capital," and so adjust reproductive allocation according to food availability at the time of breeding. These species rely on "current net daily income" to determine reproductive output, and are referred to as "income breeders." Species intermediate on this continuum may rely on fat stores for their first clutch of the season, but after the first clutch may switch to income breeding. This concept has not been explicitly applied to lizards, but may be useful because it would aid in identifying those species that are likely to be influenced by direct fecundity costs. These costs are important in capital breeders, but do not apply to income breeders (Stearns 1992).

In short-lived species, "saving up" resources for future reproduction is not a good strategy, and energy not used in maintenance should go directly into reproduction. For example, relatively short-lived mainland anoles do not get fat in response to increased food availability; instead, egg production increases in females and males grow more rapidly (Guyer 1988b). In contrast, long-lived island anoles do not increase reproduction much in response to increased food intake, but do store fat (Licht 1974; Rose 1982).

Even in capital breeders, a reduction in fat storage incurred during breeding may not constitute a fecundity cost if energy stores can be replenished quickly after breeding. Fat stores were not significantly reduced in the period after breeding in female E. tympanum (Schwarzkopf 1993), and this was probably due to increased feeding by postparturient females relative to nongravid individuals (Schwarzkopf unpubl.).

In some reptiles, females reduce food intake, or completely stop feeding during reproduction, and this may be costly in terms of growth or future reproduction (reviewed by Bull and Shine 1979). One study in Table 1.2 measured food intake during reproduction and found no reduction in food intake as a result of breeding in four species of skinks (Shine 1980), but several other studies have shown that food intake is reduced in some species of gravid skinks (Hailey et al. 1987; Schwarzkopf unpubl.). The reason that females might eat less during reproduction and the consequences of this have not been studied. Changes in the behavior of gravid females, such as increased basking and decreased activity, may be related to reduced food intake and are worthy of further attention.

Compared to females, little is known of how males devote food and stored energy resources to reproduction. In males, it is more difficult to calculate the fecundity cost of reduced fat stores or growth. The amount of energy required for sperm production is much less than that required for egg production (Nagy 1983). In spite of this, energy expenditure on territorial behavior or mate searching adds to the energy expenditure related to reproduction in males, so that it often equals or exceeds the expenditure of females (Marler and Moore 1988). Fat cycles in males suggest that in many lizards energy is stored from the previous year for use in reproduction (reviewed by Gregory 1982; Vitt and Cooper 1985), but in an extreme "income breeder" such as *Norops humilis*, supplemented food is channelled directly into growth, and fat stores are not increased in fed males (Guyer 1988b).

Indirect fecundity costs

A significant proportion (15%–50% of maximum body length, mean = 30%) of bodily growth occurs after maturity in lizards (data from Appendix 1 of Andrews 1982a). Reproduction can significantly reduce growth rates of lizards, and this can have an important negative influence on future fecundity because there is usually a positive relationship between body size and fecundity in lizards. Six of the studies listed in Table 1.2 measured the influence of reproduction on growth. Four studies reported a reduction in growth rate related to reproduction, at least in one sex, while the others found no influence of reproduction on growth. One study suggested that reproduction may reduce growth rate in males but not in females, because when provided with supplemental food, males increased growth rate, but females did not (Guyer 1988b).

Comparing reproducing with nonreproducing females, Vinegar (1975b) found a reduction in growth of female S. *virgatus* due to reproduction. However, she calculated that this reduction was not truly costly, because lifetime reproductive output was greater in females that reproduced in both years of life than in those that skipped reproduction in their first year, and consequently grew larger. In contrast, early maturity does result in lowered fitness in S. *undulatus* because of reductions in growth rate (Tinkle and Ballinger 1972). As most of the studies in Table 1.2 did not determine lifetime reproductive output in the species they studied, most could not make a comparable calculation of the real cost of reductions in growth rate. Two studies of species that routinely skip reproductive seasons showed that growth rate is significantly reduced in years when reproduction occurs (Hasegawa 1990; Schwarzkopf 1993), suggesting that trade-offs between growth and reproduction may influence lifetime reproductive output in these species.

Although growth and reproduction may be interchangeable uses for stored energy in the model of costs, growth rate and reproduction may not be equivalent in their response to variations in energy availability (Tuomi et al. 1983). For example, female lizards infected with malaria reduce reproductive allocation by approximately 14% relative to uninfected lizards, but only reduce growth rate by approximately 4% (Schall 1983). This pattern of allocation is surprising, because infected lizards have reduced survival compared to uninfected lizards, and therefore have little to gain by growing. Similarly, male lizards infected with malaria had significantly smaller testes than uninfected males, but showed only a small reduction in growth rate (Schall 1983). Such results may indicate that there has been selection for variable reproductive allocation in response to variable energy stores (infected lizards have lowered energy reserves as well as lower reproductive allocation), but not for variable growth rate. In a long-lived species, reducing reproductive allocation in response to a short-term reduction in resource availability is advantageous, as long as survival is unaffected or increased and especially if growth rate is increased (Bull and Shine 1979). Alternatively, it may simply be less costly to grow than to reproduce under stress of disease, although this may not be an effective "strategy."

Avoiding costs

Six studies in Table 1.2 suggested ways that lizards may avoid costs of reproduction. Low reproductive frequency was the most commonly cited way that lizards may avoid costs. Lizards may produce a single clutch per season instead of multiple clutches (Jones and Ballinger 1987), or they may skip reproductive seasons. Lizards with a tendency to forgo reproduction were all long-lived, and were more likely to skip reproduction at smaller body sizes (Hasegawa 1990; Laurie 1990; Schwarzkopf 1993). The reasons that lizards may skip opportunities for reproduction are not known, but there are three plausible explanations for this life-history trait. (1) Energy stores may be so depleted after reproduction that females do not have sufficient energy to form a clutch of eggs in the following season without reducing their probability of survival (Seigel and Ford 1987). In this scenario, larger females have a greater capacity to store resources, and therefore can reproduce more frequently than smaller individuals. (2) Females that skip reproduction grow significantly more than those females that reproduce (Hasegawa

1990; Schwarzkopf 1993). Smaller females grow more rapidly than larger females, and so stand to gain more in terms of growth by skipping reproduction. Skipping may be especially advantageous if the relationship between body size and clutch size is very shallow (offspring number is low) and substantial amounts of growth are necessary to increase clutch size by relatively few offspring (Schwarzkopf 1992). (3) Smaller females may be more susceptible to predation than larger females, and if reproduction also increases susceptibility to predation, then smaller females take a greater risk when reproducing than larger females, and so may avoid reproducing (Bull and Shine 1979).

Assuming that both reproduction and growth can vary in response to increased food, experiments in which food is provided to individual females should help to clarify which cost, or combination of costs, may cause skipping. If reproductive rate increases more than growth rate in response to supplemental food, then energy is probably limiting reproduction. Alternatively, if growth rate is more responsive to increased food than reproductive rate, then females may be skipping reproduction to enhance growth.

Low levels of aggression in males was another mechanism of avoiding costs identified in Table 1.2. Although aggression is thought to increase reproductive success in male *Sceloporus*, levels of aggression are relatively low in field populations of *S. jarrovi* (Marler and Moore 1988). Marler and Moore (1988, 1991) manipulated natural levels of aggression using testosterone implants, and in this way were able to demonstrate that individuals with higher levels of aggression than those normally seen in the field had an increased probability of mortality. They concluded that low levels of aggression allow males to avoid high costs of reproduction.

Another way that reproducing lizards may avoid costs is by reducing activity. Females compromised by heavy burdens of eggs and reduced sprint speed and endurance may be able to avoid increased predation by refraining from above-ground activity (Cooper et al. 1990). Similarly, gravid females with reduced performance modify escape behavior, and this may reduce predation rate (Bauwens and Thoen 1981; Cooper et al. 1990; Schwarzkopf and Shine 1992). These plausible suggestions need to be verified by experiments that compare predation rates in reproducing and nonreproducing females. General changes in activity levels associated with reproduction may cause a correlated set of changes in behavior, some of which are beneficial, and others which are costly. For instance, avoiding predation may be a beneficial result of these changes, but reduced foraging ability may be costly (Ford pers. comm.).

Future Directions in Studies of Costs of Reproduction in Lizards

Examination of Table 1.2 revealed four basic questions that need to be answered if costs of reproduction are to be well understood in lizards.

1. How do reproductive trade-offs, phylogenetic history, and environmental influences interact to determine aspects of reproductive characteristics?

Clearly, phylogenetic lineage determines many aspects of the life histories of lizards (Vitt and Congdon 1978; Dunham et al. 1988a). Therefore, important trade-offs should differ among lizards with different phylogenetic backgrounds. For instance, in short-lived species with high reproductive output, fecundity trade-offs may not influence reproductive characteristics very much (Shine and Schwarzkopf 1992). Many studies of reproductive trade-offs have examined phrynosomatid lizards, limiting the phylogenetic scope of conclusions that can be drawn about trade-offs in different groups. In spite of the logistic difficulties associated with conducting long-term studies, more studies of costs in long-lived lizards with low reproductive output are required.

Environmental factors may account for some variation observed in life-history characteristics of lizards (Adolph and Porter 1993). Suites of characters that appear to be coadapted complexes may simply be correlated responses to an external influence. However, adaptation does account for some variation among populations. To distinguish environmental effects from adaptive responses to particular environments, it is necessary to do common-environment or reciprocal-transplant experiments (Sinervo and Adolph 1989; Ferguson and Talent 1993). In part, varied responses of the life-history characters of different populations to similar laboratory environments will reveal differences in trade-offs (Ferguson and Talent 1993).

2. How have costs influenced reproductive investment in males?

Much is known about factors influencing reproductive success in males of other vertebrates (Clutton-Brock 1988), but little is known about these factors in male lizards, especially in nonterritorial species. As a corollary to this, we know little about costs of reproduction in males. However, costs may be important determinants of various aspects of the life history of males, just as they are in females. For instance, males of many lizard species appear to have lower survival rates than females (reviewed by Turner 1977; Tinkle and Ballinger 1972; Stewart 1985; Tinkle and Dunham 1986; Strijbosch and Creemers 1988; Hasegawa 1990; Laurie 1990), and this is often attributed to differences in reproductive costs in the two sexes (e.g., Tinkle and Dunham 1986; Snell et al. 1988). It is not really clear, however, whether these differences in survival between the sexes are due to costs of reproduction or to other factors, such as differences in body size (Laurie 1990) or activity levels (Stewart 1985). Also, several studies compiled here have documented a trade-off between growth and reproduction in males of some lizard species, but since the fecundity advantage of large size is difficult to assess in males, it is hard to determine the cost of this trade-off (Nagy 1983; Guyer 1988b).

It is difficult to observe mating in lizards, and, as there are no examples of male parental care in squamate reptiles (Shine 1988), it is difficult to associate males of these species with their offspring. However, with the advent of genetic techniques of determining paternity, it should be possible to measure reproductive variables such as fecundity and mating success in males. Once factors determining reproductive success are known, it should be possible to measure the cost of high reproductive output in terms of survival, growth, and future reproductive output in males.

3. How should costs of reproduction be measured in the field?

The studies collected in Table 1.2 used a range of approaches to address questions about costs of reproduction in lizards, and different approaches are required to understand costs and how they work. Relatively few of the studies tabulated here, however, directly measured the influence of reproduction on survival, future reproduction, and/or growth in field populations of lizards. More such studies on various taxonomic groups and in various environments are required to flesh out the answer to question 1 above. Field studies in which some measure of reproductive output is correlated with measures of survival, growth, and future reproduction are instructive in this respect (e.g., Jones and Ballinger 1987), as are studies comparing reproducing and nonreproducing individuals in populations in which individuals reproduce only once every two or three years (e.g., Hasegawa 1990; Schwarzkopf 1993). Experiments manipulating reproductive allocation in the field should be performed so that variation in reproductive characteristics can be increased above naturally occurring levels, and the consequences of increased reproductive investment can be measured directly (Partridge and Harvey 1985).

Two relatively new techniques allow us to manipulate reproductive output in both sexes. (1) Follicle size alteration and ablation (Sinervo 1990a) can be used to alter levels of reproductive investment in females. (2) Hormone manipulation (Fox 1983; Marler and Moore 1988; Sinervo this volume) permits the researcher to influence hormonally controlled behaviors, such as aggression and thermoregulation, associated with reproduction in both males and females. These techniques can be used in conjunction with traditional methods of measuring demography (mark-recapture, radio-tracking) to determine the influence of changes in reproductive allocation on individuals and their offspring.

Alternatively, environmental conditions can be manipulated. Costs of reproduction are often only evident under conditions of stress (Laurie 1990; Schwarzkopf 1993). Experiments manipulating field conditions such as predator density (Reznick et al. 1990), or food availability (Guyer 1988a,b) can be used to observe the influence of changes in environmental conditions on reproductive allocation. However, it may be difficult to interpret responses to environmental manipulations if the form of intraindividual trade-offs is not well understood (Reznick 1983).

4. How do life-history characteristics respond to selection in lizards?

A variety of studies have shown that selection can influence life-history characters (reviewed by Stearns 1992). Responses to selection can be constrained, however, in some important ways (Reznick 1992). Therefore, it is important to demonstrate that characters thought to be subject to selection have a genetic basis. More common-environment experiments of the type conducted by Ferguson and Talent (1993) are required to determine the genetic contribution to variation in life-history characteristics. It is probably not possible to conduct artificial selection experiments on life-history characters on relatively large, long-lived animals such as lizards. However, if laboratory studies suggest that there is a genetic basis for particular characters, then selection experiments on phenotypic characters in the field could be performed to measure genetic trade-offs in lizards.

Summary

Risks and/or energy allocation associated with reproduction may reduce an organism's ability to reproduce in the future. This negative influence has been called the "cost" of reproduction. Trade-offs between advantages and disadvantages of particular levels of reproductive expenditure may affect the life histories of iteroparous organisms by influencing age at maturity, survival rate, litter size, and reproductive frequency. Costs may be divided into two groups: survival costs and fecundity costs. Survival costs are those that influence future reproduction by reducing the probability of survival of reproducing organisms. Fecundity costs influence energy available for future reproduction and may be direct or indirect. "Direct fecundity costs" reduce energy available for future reproduction by reducing long-term energy stores that are used for reproduction. "Indirect fecundity costs" reduce growth rates, indirectly influencing future reproduction through body size/ clutch size relationships in organisms in which these relationships occur.

There has been controversy in the literature about the best way to measure costs of reproduction. Costs may be measured using phenotypic correlations, manipulative experiments, genetic correlations, and selection experiments. Most authors agree that studies are needed to establish that there is a genetic basis for selection to act upon reproductive attributes. In addition, it is important to determine what effect changing environments have on costs in the field. Experimental manipulation of reproductive investment in the field is the most powerful method of examining environmentally mediated costs (related to predation and resource limitation) and ways that organisms avoid costs in the field.

Lizards rarely have parental care, but they may be viviparous or oviparous, and therefore they provide us with a variety of opportunities for examining questions about costs of reproduction that cannot be examined using birds, the organisms most commonly used to study costs. As lizards may be iteroparous, may store energy for use in future reproduction, and may live for several activity seasons, they make excellent model organisms for the study of costs of reproduction. Lizards can easily be studied under field situations, and reproductive investment can be manipulated in the field to answer questions about costs of reproduction.

In addition to costs of reproduction, several other factors may determine reproductive output in lizards. Both phylogenetic history and environmental effects explain some of the variation observed in life-history characteristics of lizards. There is a strong tendency for life-history characters to be similar among lizards of similar evolutionary backgrounds, even when they occur in very different environments. Also, recent work suggests that daylength, or activity time available to lizards in different environments, explains some of the variability in life histories among populations of the same species. Although these two variables are important determinants of differences in life histories among and within species of lizards, some variation in life-history characters is not explained by phylogeny or environmental influences. This suggests that microevolutionary responses to local selection pressures also determine aspects of life histories in lizards.

Within the context of phylogenetic and environmental influences, a variety of changes in behavior and physiology of reproducing lizards occur, and these may result in decreased survival or decreased energy available for future reproduction and growth. Increased mortality may be the result of increased predation on reproducing animals, or it may occur because energy stores are too depleted for maintenance, or for other activities requiring energy. Decreased energy available for reproduction may occur because stored resources that would otherwise be available are depleted by reproduction, or because growth is inhibited by reproduction, and therefore body size is decreased.

In this chapter, I summarized the results of 16 studies of costs of reproduction in lizards. This summary allowed me to identify four questions that require further investigation: (1) How do reproductive trade-offs, phylogenetic history, and environmental influences interact to determine aspects of reproductive characteristics? (2) How have costs influenced reproductive investment in males? (3) How should costs of reproduction be measured in the field? (4) How do life-history characteristics respond to selection in lizards?

Costs of reproduction have been shown to influence survival and reproductive output in certain populations of iguanid, polychrid, phrynosomatid, scincid, and lacertid lizards (Table 1.2). More experimental studies across a wider taxonomic range, and in both sexes of lizards are required to better illuminate the nature of costs of reproduction in these organisms. Recently developed techniques and approaches will allow us to experimentally determine how these costs operate, and to determine their influence in shaping the life-history characters that we observe today.

Acknowledgments

I would like to thank Eric Pianka and Laurie Vitt for inviting me to participate in this symposium. M. J. Caley, J.N.M Smith, R. Shine, and N. Ford made valuable comments that greatly improved the manuscript. While conducting this research I was supported by a Natural Sciences and Engineering Council of Canada Postdoctoral Research Fellowship.

CHAPTER 2
UNDERSTANDING GEOGRAPHIC
LIFE-HISTORY VARIATION IN LIZARDS

Peter H. Niewiarowski

Species distributed over broad geographic ranges often display extensive variation in life-history traits such as age at maturity, growth rate, and age-specific schedules of fecundity and survivorship. Variation in life-history traits has received tremendous theoretical and empirical attention (see Stearns 1976, 1977, 1992; Roff 1992 for reviews). Studies of intraspecific geographic variation have played an important role in identifying potential ecological sources of variation, and in providing hypotheses concerning the evolution of life histories. However, even in systems where a substantial amount of comparative data have been compiled, an understanding of the ecological and evolutionary significance of variation among populations frequently remains elusive.

Understanding the ecological and evolutionary significance of life-history variation is a convenient shorthand for interest in a number of questions fundamental to the study of life-history variation, including (1) How do various aspects of the environment affect the expression of life-history phenotypes? (2) What are the ecological and evolutionary implications of phenotypic plasticity (or lack thereof) in the expression of life-history traits? (3) How have the particular patterns of allocation of time and energy that determine life-history variation evolved?

The Eastern Fence Lizard, *Sceloporus undulatus*, is an excellent model species whose geographic life-history variation has been extensively studied. An impressive empirical data base has been compiled (e.g., see Tinkle and Ballinger 1972; Tinkle and Dunham 1986; Dunham et al. 1988a; Gillis and Ballinger 1992 for reviews), and it has been the subject of several "tests" of theory (e.g., Stearns 1980; Stearns and Crandall 1981; Grant and Porter 1992; Adolph and Porter 1993), yet we understand very little about the ecological and evolutionary significance of variation observed among populations (Tinkle and Dunham 1986; Gillis and Ballinger 1992; Ferguson and Talent 1993). A number of key problems have not yet been adequately resolved, including, but not limited to (1) determining the extent to which phenotypic variation among populations reflects genetic differentiation (due to local adaptation), (2) identifying ecological sources of variation, and (3) testing hypotheses about the evolution of life histories within the context of a population level phylogeny. The unresolved nature of these issues is not unique to the *S. undulatus* system.

Below, what is currently known about geographic life-history variation in
S. undulatus is summarized. The summary is not intended as an exhaustive
description of the specifics of life-history variation in this lizard species, nor
as a source of new hypotheses to explain the observed variation; synthetic
studies of these sorts can be found elsewhere (Ferguson et al. 1980; Ballinger et al. 1981; Tinkle and Dunham 1986; Gillis and Ballinger 1992).
Rather, I summarize some of the major patterns of geographic variation
observed, and consider how several studies have attempted to explain those
patterns. My review is primarily intended to describe our current understanding of the processes that have generated geographic life-history variation in this system rather than be a critique of previous work.

Given the extent of comparative data on ecological and life-history variation in this species, I believe we are in a position to make significant
advances in answering questions about geographic life-history variation. In
order to do so, I argue that future research should focus on experimental
and phylogenetic methods that allow evaluation of genetic and environmental influences on life-history variation within a phylogenetic framework.
Emphasis must shift from determining if observed patterns of variation are
consistent with life-history theory, to testing assumptions and predictions of
theory. Towards that end, I will outline a general approach stressing experimental and phylogenetic methods that I believe will allow us to build on the
uncommonly extensive *S. undulatus* database. Although most of my discussion focuses on *S. undulatus*, there are general implications for understanding geographic life-history variation within any species.

Patterns of Variation and Hypotheses

Tinkle and Ballinger (1972) and Tinkle (1972) initiated comparative demographic work on *S. undulatus* because they were impressed by its extensive
geographic range and life-history variation. They suggested a number of
non-mutually exclusive hypotheses to explain differences among five widely
distributed populations studied. Although they acknowledged that phenotypic life-history variation among populations could be attributed to environmentally induced variation and/or genetic differentiation, they assumed
instead, that differences among most populations in traits like growth rate,
age at maturity, fecundity, and longevity were probably genetically based,
brought about by adaptive evolution in each environment. While the empirical scope of their work, in terms of its comparative data base, was impressive, it also made an equally important contribution to generating
hypotheses that implicated various ecological factors as sources of selection
in the evolution of differences among populations.

These pioneering efforts (Tinkle and Ballinger 1972; Tinkle 1972) were followed by a number of studies extending comparative demographic data from six populations (Crenshaw 1955) to the current 14 (Vinegar 1975c; Ferguson et al. 1980; Ballinger et al. 1981; Tinkle and Dunham 1986; Jones and Ballinger 1987; Gillis and Ballinger 1992; Niewiarowski unpubl.). Furthermore, many of these studies have attempted to interpret broad patterns of geographic variation in life-history traits in light of the accumulating data.

From the outset, comparative demographic studies of *S. undulatus* have recognized a distinction between "proximal" (environmental) and "evolutionary" (genetic) sources of geographic variation in life-history traits (Tinkle 1972; Tinkle and Ballinger 1972). Proximal sources induce variation among phenotypes arising from the same genotype, and evolutionary sources imply that genetic differences among phenotypes are responsible for observed phenotypic variation. Many kinds of ecological or environmental factors have been implicated in producing phenotypic variation in life-history traits. For example, differences in individual growth rates among populations of many reptiles have often been attributed to differences in food availability or temperature (Cagle 1946; Gibbons 1967; Andrews 1976; Dunham 1978; Ballinger and Congdon 1980). Tinkle and Ballinger (1972) suggested that low food availability and strong thermal constraints on activity could explain relatively slow growth and delayed maturity of *S. undulatus* hatchlings from Ohio compared to those from Texas. They contrasted the hypothesis implicating "proximal" sources of variation (i.e., food and thermal environment) in growth rate and age at maturity between lizards from Ohio and Texas with an "evolutionary" hypothesis explaining the relatively slow growth and delayed maturity of lizards from South Carolina compared to Texas (Tinkle and Ballinger 1972). Growth rates of lizards from South Carolina, they suggested, were not directly limited (relative to Texas) by food availability or thermal environment. Instead, it was hypothesized that lizards in South Carolina that allocated energy to predator avoidance, at the expense of growth, experienced higher adult survivorship rates and greater lifetime reproductive success.

The above distinction, between environmental and genetic sources of variation in life-history traits, is conceptually simple and important, especially in the context of describing specific life-history traits as adaptations to local environments (Ferguson et al. 1980; Berven 1982; Ferguson and Talent 1993; Niewiarowski and Roosenburg 1993). Consider the above example in which the hypothesized distinction is between slow growth rates in Ohio relative to Texas due to low food availability or unfavorable thermal environments, versus slow growth in South Carolina relative to Texas arising from the hypothetically higher relative fitness of South Carolina lizards that divert energy away from growth in favor of predator escape. In the first case (Ohio vs. Texas), slow growth results directly from food limitation or thermal constraints; in the second case (South Carolina vs. Texas), slow growth is a consequence of an evolutionarily derived alternative energy allocation strategy.

Overall, Tinkle and Ballinger (1972) and Tinkle (1972) suggested many hypotheses, proximal and evolutionary, to account for variation among populations in growth rate, age at maturity, clutch size, body size, and other traits. Subsequent work by Ferguson et al. (1980), Ferguson and Brockman (1980), and Ballinger et al. (1981) made important contributions by (1) extending the number of populations studied, (2) suggesting that there were latitudinal and regional trends in the variation possibly associated with "selective regimes," and (3) by calling attention to the implicit assumption required by hypotheses of local adaptation; i.e., genetic differences between populations. Ferguson and Brockman's (1980) common garden laboratory growth experiment provided the first evidence to suggest that genetic differences might indeed contribute to differences in growth rates of hatchlings and, by extension, other life-history traits of lizards from populations in Texas, Kansas, South Carolina, and Utah.

Ferguson et al. (1980) suggested that differences in habitat type and latitudinal clines might account for similarities in life-history traits among groups of populations. They proposed a simple habitat classification scheme based on qualitative structural differences between grassland, canyonland, and woodland habitats. Rankings of populations for life-history traits appeared to be related to habitat type and also to latitude within habitat types. For example, the four lowest-ranking populations (lizards with delayed maturity, large adult body size, and high longevity) were from northern portions of the species' range. Populations from grassland habitats tended to rank relatively high (early maturity, small adult body size, low longevity), while populations from canyonland habitats tended to rank relatively low. Furthermore, differences between the rankings of northern versus southern populations were reversed between grassland habitats and woodland habitats.

Ballinger et al. (1981) reconsidered Ferguson et al.'s (1980) habitat hypothesis and noted that apparent habitat and latitudinal trends might arise from specific selective environments associated with habitat types, consistent with the local adaptation hypothesis of Tinkle and Ballinger (1972). Tinkle and Dunham (1986) more formally tested the hypothesis that habitat type (grassland, canyonland, woodland) as defined by Ferguson et al. (1980) and supported by Ballinger et al. (1981), could account for differences in life-history characteristics among species. They reasoned that if habitat type could account for differences among populations, then the life-history characteristics of individuals from any given population should resemble most closely the characteristics of individuals from other populations of the same habitat type. They constructed a Primm network based on the standardized rankings of the life-history variables used in Ferguson et al.'s (1980) analysis. While populations from the same habitat type were frequently in close proximity to one another in the Primm network (consistent with the habitat hypothesis), there were several notable exceptions of populations from different habitats that were more similar to one another than they were to populations from the same habitat (Tinkle and Dunham 1986). The Primm network was similarly used to test the hypothesis that differences among populations could be explained by recency of common ancestry reflected by taxonomic relationships ("phylogenetic" hypothesis). Once again, there were notable exceptions to the expectation that populations of the same subspecies should cluster together, leading Tinkle and Dunham (1986) to reject the general explanatory power of both the habitat and phylogenetic hypotheses.

Gillis and Ballinger (1992) recently presented reproductive data from another population (Eastern Colorado) and used a principal components analysis (PCA) to reexamine relationships among reproductive life-history characters for populations studied to date. They argued, based on the apparent clustering of populations in a space defined by the first three principle components explaining variation in reproductive characters, that Tinkle and Dunham (1986) prematurely rejected the habitat hypothesis first suggested by Ferguson et al. (1980) and later refined by Ballinger et al. (1981). To evaluate this argument, however, it is important to consider the limitations of cluster analysis: ad hoc clustering procedures (visual or quantitative) must be viewed cautiously because such procedures are often highly dependent upon the expected number of clusters and distance algorithms employed (Everitt 1974; SAS Institute 1989).

To demonstrate how the a priori choice of number of clusters and distance algorithm can affect interpretation of pattern, I reanalyzed the data using PCA and ran a quantitative cluster analysis on the multivariate data set. In addition to using the data set on 12 populations analyzed by Gillis and Ballinger (1992), I used data from a population of S. undulatus I am currently studying (New Jersey) to increase the sample size (Table 2.1, Fig. 2.1). Sceloporus undulatus in New Jersey grow slowly (Niewiarowski 1992; Niewiarowski and Roosenburg 1993), are large and old at reproductive maturity, and produce two intermediate-sized clutches of large eggs relative to other populations studied to date. Relative positions of populations in the original multivariate space were not changed much by the addition of the New Jesey population (Fig. 2.2; Table 2.2). Consider differences in clustering of S. undulatus populations when an average distance linkage versus a complete distance linkage algorithm is used (Fig. 2.3). More profound variation in clustering patterns is obtained when either the number of expected clusters or both the distance algorithm and cluster number are varied. In light of the sensitivity of a population's cluster membership to the linkage

Table 2.1. Reproductive traits of *S. undulatus* from a population in New Jersey (Niewiarowski unpubl.). Values of all traits except age at maturity (α), minimum mature SVL (MSVL), and average female SVL (ASVL) were estimated exclusively from samples of females collected during the active seasons of 1986 and 1988. Estimates of α, MSVL, and ASVL were based on collections and estimates from a mark-recapture study using standard techniques (1986–1992; Niewiarowski unpubl. data).

Trait	n	Mean	2 Standard Errors
Egg Wet Mass (g)	30	0.3572	0.0223
Egg Dry Mass (g)	15	0.1749	0.0112
Clutch Size (# of eggs)	30	8.87	0.7655
Clutch Frequency (per/year)*		2	–
Age at Maturity (Months)*		20	–
Minimum Mature SVL (mm)*		60	–
RCM	30	0.3012	0.0238
Average Female SVL (mm)*		73	–

* qualitative estimates based on mark-recapture study.

algorithm employed and/or expected number of clusters, it is unclear how (1) the results of a particular analysis can be used to test the phylogenetic, habitat, or other hypotheses, or (2) one set of choices for cluster number and linkage algorithm can be selected over another. However, one fairly consistent trend is interesting: clusters obtained from different combinations of distance algorithms and expected cluster number using the *S. undulatus* data set typically contain mixtures of habitat types and subspecies. This last result is consistent with the conclusions of Tinkle and Dunham (1986).

Although multivariate techniques like those employed by Gillis and Ballinger (1992) may often be useful as exploratory tools because they can suggest testable hypotheses, in the case of the *S. undulatus* system, the analysis did not reveal any new associations between populations not already identified by the Primm network (Tinkle and Dunham 1986). The multivariate analysis actually reinforces the conclusions of Tinkle and Dunham (1986), but rejection of the habitat hypothesis or phylogenetic hypothesis should not be confused with the view that aspects of specific habitats and the phylogenetic history of *S. undulatus* are unimportant in explaining geographic variation. On the contrary, it is clear only that the two hypotheses, as currently formulated, cannot explain the observed geographic variation (Grant and Porter 1992). This point is much more important than it may seem to casual observation because it emphasizes our lack of understanding of the specific combination of proximal and evolutionary factors contributing to life-history variation among populations of *S. undulatus*.

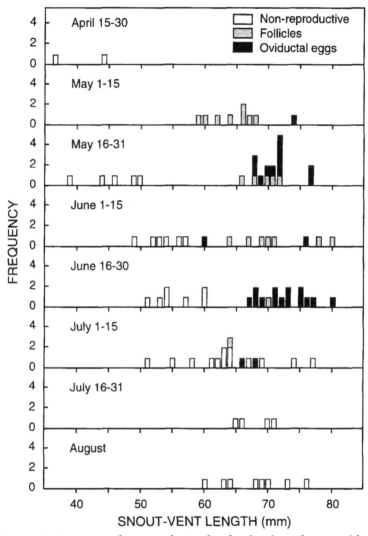

Figure 2.1. Frequency of nonreproductive females, females with ovarian follicles, and females with oviductal eggs collected during the activity seasons of 1986 and 1988 in New Jersey ($N = 95$).

Recent experimental work with *S. undulatus* suggests that a combination of environmental and genetic sources of variation is required to explain differences in life histories across the species' range (Ferguson and Talent 1993; Niewiarowski and Roosenburg 1993). For example, *S. undulatus* from Utah grew more rapidly than *S. undulatus* from Oklahoma in a common garden laboratory environment (ad libitum food and water, relaxed thermal constraints) suggesting that differences in growth rates between lizards from these two populations were genetically based (Ferguson and Talent 1993). However, growth rates of free-ranging Utah lizards were lower than growth rates of free-ranging Oklahoma lizards, implying that some aspect(s) of the Utah environment resulted in reduced phenotypic expression of a greater genetic potential for growth. Lizards from Utah and Oklahoma also exhibited differences in size and age at maturity, clutch size, and egg mass under the common garden laboratory conditions.

A reciprocal transplant of *S. undulatus* similarly implicated both population-specific (genetic) and environmental sources of variation in explaining differences in hatchling growth rates observed between Nebraska and New Jersey (Niewiarowski and Roosenburg 1993). Relatively slow-growing *S. undulatus* from New Jersey did not grow faster when transplanted to Nebraska, but *S. undulatus* from Nebraska transplanted to New Jersey grew at half the rate observed in Nebraska. This result is consistent with the hypothesis that both environmental differences between New Jersey and Nebraska, as well as genetic differences, explain observed phenotypic differences between New Jersey and Nebraska in the growth rates of free-ranging hatchlings.

Results, like those described above, implicating both environmental and genetic sources of variation in several life-history traits do not appear to be a unique aspect of *S. undulatus*. For example, common garden laboratory experiments by Sinervo and Adolph (1989) demonstrated significant familial and thermal environmental effects on growth rates in *Sceloporus occidentalis* and *Sceloporus graciosus*. Similarly, Sinervo (1990a) showed that laboratory thermal environments and familial effects contributed to variation in growth rates of *S. occidentalis* from two populations, and that the relative contribution of estimated genetic effects within each population was drastically different. Thus, studies of *S. undulatus* and other lizards that have attempted to estimate the contributions of genetic and environmental sources of variation in growth rates and other life-history traits suggest that both sources are important. Furthermore, the relative importance of either source may vary for different traits, populations, and species.

Experimental work with *S. undulatus* further suggests that differences in life-history phenotypes among populations might be partially explained by the plastic response of a common life-history genotype in different environments. In other words, geographic variation could be a result of phenotypic plasticity (Bradshaw 1965). While hypotheses of phenotypic plasticity versus local adaptation (achieved through genetic differentiation among populations) may be mutually exclusive, hypotheses of phenotypic plasticity and adaptive evolution (in general) are not, because plasticity itself may be adaptive (Via and Lande 1985; Newman 1988, 1992). Therefore, a distinction

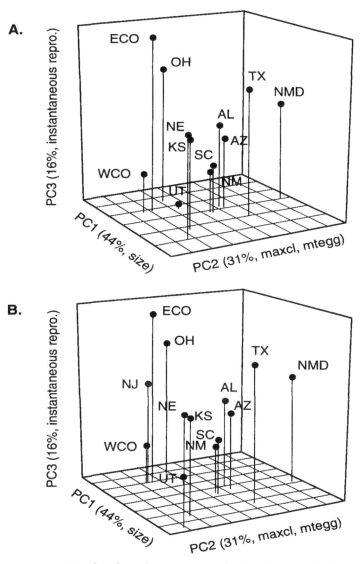

Figure 2.2. (A). Plot of population scores on the first three standardized principal components of reproductive characters from Table 2.2 (see Table 2 in Gillis and Ballinger 1992). Labels refer to location of populations: AL = Alabama; AZ = Sunflower, Arizona; ECO = eastern Colorado; KS = Belvue, Kansas; NE = Arthur, Nebraska; NM = Pines Altos, New Mexico; NMD = New Mexico desert; OH = Ohio; SC = South Carolina; TX = Eldorado, Texas; UT = Washington County, Utah; WCO = western Colorado. (B). Same as in A, but with the New Jersey (NJ) population added.

Table 2.2. Standardized PCA loadings for Figure 2.2

	Loadings for Fig. 2.2a		Loadings for Fig. 2.2b	
Population	PC1	PC2	PC1	PC2
AZ	0.0217	0.6303	-0.1634	0.7004
NM	-0.1389	0.1890	-0.2767	0.2353
UT	0.6993	-0.3291	0.6090	-0.0801
WCO	0.9256	-1.1574	0.9853	-0.8319
ECO	0.8774	-0.9282	0.9974	-0.6699
NE	-1.9205	-1.1349	-1.7855	-1.5156
KS	-1.2497	-0.8136	-1.1851	-1.0183
TX	-1.0000	0.8297	-1.1713	0.6535
NMD	0.0931	2.2456	-0.3331	2.2927
OH	1.5718	-0.3789	1.5674	0.0666
SC	-0.1930	0.2313	-0.3278	0.2608
AL	0.3132	0.6161	0.1334	0.7319
NJ	–	–	0.9503	-0.8254

between environmental and genetic sources of phenotypic life-history variation is incomplete without a consideration of the evolutionary significance of
plasticity (Newman 1988, 1989; Grant and Dunham 1990; Reznick 1990;
Newman 1992; Grant and Porter 1992; Bernardo 1993; Adolph and Porter
1993).

In summary, comparative demographic study of *S. undulatus* has
revealed patterns of variation that have not been adequately explained by
simple hypotheses (Tinkle and Dunham 1986; Gillis and Ballinger 1992).
This probably reflects, in part, our ignorance of the relative contributions
made by environmental and genetic sources of variation among populations
of *S. undulatus* in general. Experimental studies have suggested that both
environmental and genetic sources of variation are important in explaining
phenotypic differences among populations, and that the degree of phenotypic plasticity of some traits may also vary among populations (Ferguson
and Brockman 1980; Ferguson and Talent 1993; Niewiarowski and Roosenburg 1993).

Future Directions

Given the relatively large comparative data base, the focus of future
research should shift from determining whether new observations or new
analyses are consistent with previously suggested adaptive hypotheses (e.g.,
Gillis and Ballinger 1992) to testing proposed hypotheses and the assump-

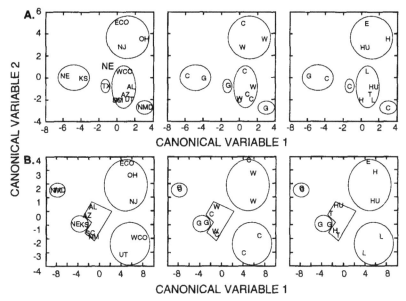

Figure 2.3. (A). Plots of population clusters based on scores from the first three standardized principal components in Fig. 2.2A. using an average linkage algorithm (Proc Cluster, SAS Institute 1989). Population clusters are plotted in a space defined by a canonical discriminant analysis on the principal component scores (Proc Candisc, SAS Institute 1989). Clustering preceded canonical discriminant analysis, which was only used to reduce the three PC's to two variables so that clustering could be easily visualized. No significance should be attached to the shape of clusters in any of the plots. Labels in left panels are location (as in Fig. 2.2). In center panels, C = Canyonland; G = Grassland; W = Woodland. In right panels, labels refer to subspecies as specified in Gillis and Ballinger (1992): C = *S. u. consobrinus*; E = *S. u. erythrocheilus*; G = *S. u. garmani*; H = *S. u. hyacinthinus*; HU = *S. u. hyacinthinus* × *S. u. undulatus* hybrid; L = *S. u. elongatus*; T = *S. u. tristichus*. (B). Same as in A but a complete linkage algorithm was used in the cluster analysis.

tions they require. As a specific example, revealing the relative contributions of population-specific (genetic) and environmental sources of phenotypic differences among populations should precede the proposal of new adaptive hypotheses that assume that differences between populations are the result of genetic divergence due to natural selection (Ferguson et al. 1980; Ferguson and Talent 1993; Niewiarowski and Roosenburg 1993). I describe below an approach with two components that simultaneously complement comparative data already collected and allow tests of hypotheses previously proposed. The first component stresses experimental methods that can reveal how environmental and genetic sources of variation in energy allocation contribute to geographic life-history variation. The second component stresses the testing of adaptive hypotheses using an independently derived population phylogeny.

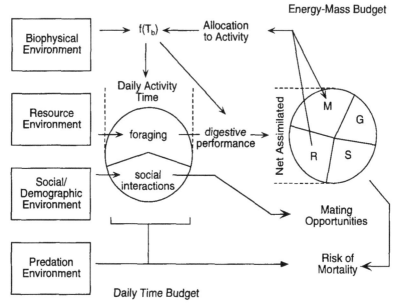

Figure 2.4. Allocation model of Dunham et al. (1989) linking the physiological and biophysical ecology of individuals to the life-history phenotype, through the allocation of time and energy. M = metabolism, G = growth, R = reproduction, S = storage, T_b = body temperature.

Life-history phenotypes, allocation rules,
and experimental methods

Dunham et al. (1989) presented a conceptual model (hereafter, "allocation model") linking study of physiological ecology and life histories of individuals with population biology (Fig. 2.4). The model views a life history as a heritable set of rules that, given the physiological state of the individual and the constraints imposed by its environment, determine allocations of time and energy. Life-history characters such as growth rate, age at maturity, adult body size, and fecundity are manifestations of the time-ordered (i.e., age-dependent) sequence of allocations made over an individual's lifetime. An important distinction is made, therefore, between the life-history phenotype and the allocation rules, which, together with the environment and physiological state of the individual, ultimately determine its life-history phenotype. This distinction focuses life-history theory on explaining the evolution of allocation rules rather than on particular phenotypic trait values, such as "delayed maturity" or "high fecundity." The allocation model also directly incorporates phenotypic plasticity as a potential consequence of context dependent (i.e., physiological state and operative environments; see below) allocation processes (see Houston and McNamara 1992, for a quantitative model implementing some of these ideas).

Three main sources of variation in life-history phenotypes are identified by this model: (1) allocation rules, (2) sets of operative environments, and (3) the physiological state of the individual (Dunham 1993). The first two sources represent genetic and environmental sources of variation, respectively, in life-history phenotypes. The third source is an umbrella for a variety of factors that affect allocation decisions. Examples include the physiological condition of an individual in terms of its energy stores and the functional relationship between temperature and physiological processes. Determining the role of each of the three sources above and their interaction in explaining life-history variation among populations, represents a significant challenge, but one that is central to understanding, for example, how different rules evolve or the significance of plasticity in the expression of a set of rules in different environments.

Within the conceptual framework of Dunham's allocation model, phenotypic differences among populations of S. *undulatus* can be interpreted as (1) differences in the rules that specify allocations of time and energy, (2) differences in how a single set of rules is expressed in different environments, (3) differences in physiological status, or (4) some combination of the three. Consider the first two sources of variation among populations (allocation rules, environmental effects on expression). The phenotypic response of genetically identical individuals (in life history or other traits) over a range of environments describes a reaction norm (Stearns 1989b). For a population made up of genetically heterogeneous individuals, a family of trajectories is expected (Stearns and Koella 1986), and we could consider the average phenotypic response of individuals to a range of environments as the population level analog of a reaction norm. Therefore, phenotypic differences between populations of S. *undulatus* can be interpreted in two ways: as (1) different points along a single population-specific norm of reaction, or (2) different points on two distinct population-specific norms of reaction. Different population-specific reaction norms would be the result of different sets of "rules" specifying the age-dependent allocation of time and energy individuals make given a set of operative environments and physiological status. But, because life-history phenotypes (e.g., growth rate, age, and size at maturity) result from the interaction of the rules with the environment (and physiological status) in which they are expressed, phenotypes arising from a single set of rules could alternatively vary because of differences in environments and physiological status.

We know very little about how the third source, variation in the physiological states of individuals, contributes to variation in life-history phenotypes. This is largely an artifact of a demographic approach to the study of life-history variation, which has led us to focus on population or species averages at the expense of ignoring interindividual variation. Indeed, interindividual variation in physiology has itself only recently received attention (Bennett 1987a; Garland and Adolph 1991). Nevertheless, several studies suggest that physiological variation potentially contributes to phenotypic life-history variation in a variety of ways (Sinervo and Adolph 1989; Sinervo 1990a; Dunham 1993; Beaupre 1993; Beaupre et al. 1993). Needless to say, lack of data in this area represents an important area of future research with respect to implementing the approach described herein.

Experimental approaches utilizing common garden laboratory techniques (e.g., Ferguson et al. 1980; Sinervo and Adolph 1989; Sinervo 1990a; Ferguson and Talent 1993) and reciprocal transplant techniques (e.g., Antonovics and Primack 1982; Reznick and Bryga 1987; Lacey 1988; Bernardo 1993; Niewiarowski and Roosenburg 1993) are well suited to identifying genetic and environmental sources of variation in life-history phenotypes among populations. In spirit and practice, such experiments can be used in a way closely related to one advocated by Ballinger (1983), in that the relative contributions of different sources of life-history variation (genetic and evolutionary) can be estimated statistically. However, rather than focus on the magnitude of variation explained by contributing sources, I suggest the approach be used to sort populations of S. undulatus into groups that have distinguishable sets of life-history rules as indicated by population-specific phenotypic responses across a range of experimental environments.

Operationally, we would distinguish between two sources of variation in average life-history phenotypes among populations: population-specific and environmental (Fig. 2.5). Population-specific sources include genetic differences in life-history rules, and environmental sources include differences in sets of operative environments (sensu Dunham et al. 1989). Notice that this dichotomy implicitly assumes variation in physiological status and allocation rules among individuals within a population is relatively less important than variation in rules and operative environments among populations. For the purposes of distinguishing among populations, this may not be an unreasonable assumption if (1) variation among individuals within a population in life-history phenotypes is much less than variation in averages among populations (e.g., Grossberg 1988), and (2) ranges of within-population variation do not overlap greatly. Using a multipopulation and environment transplant or common garden experiment with S. undulatus, the set of life-history phenotypes observed under each combination of source population and environmental treatments could be compared with MANOVA. Significant heterogeneity among populations in the multivariate phenotypic responses (life-history phenotype vector) across the range of environments used would be consistent with the hypothesis that populations differed with respect to

A. Sources of Variation: Allocation Model (Dunham *et al.* 1989)

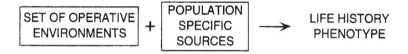

B. Sources of Variation: Transplant or Common Garden Experiment

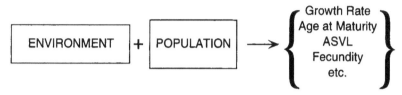

Figure 2.5. (A). Summary of two main sources of variation in life-history phenotypes specified by the Allocation model in Fig. 2.4. (B). Sources of variation in life-history phenotypes from common garden or reciprocal transplant experiments corresponding to sources in Fig. 2.5A. Note that the life-history phenotype is a set of traits including, but not limited to, those pictured.

their life-history rules (Fig. 2.6). Multiple comparisons testing the significance of source population and source population by environment interactions could then be used to group populations with indistinguishable phenotypic responses.

Groups of populations with distinct sets of allocation rules, as inferred from responses to transplantation or common garden experiments, could form the basis for constructing hypotheses concerning the evolution of population differences. However, evaluating the role of natural selection in producing current differences in particular cases (i.e., testing hypotheses of the adaptive evolution of differences in allocation rules) would remain a formidable challenge. This would involve evaluating the fitness consequences of alternative allocation rules in a given environment. For example, consider inferred differences in allocation rules between hatchling *S. undulatus* from Nebraska and New Jersey manifested as a lack of plasticity in growth rates of New Jersey lizards (Niewiarowski and Roosenburg 1993). On what basis could we establish the potential role of natural selection in producing differences in allocation strategies between these two populations? Is fixed allocation to growth (via direct or indirect mechanisms) favored in New Jersey? Flexibility in allocation to growth by Nebraska hatchlings appears to allow individuals to realize potential fitness advantages resulting from rapid growth (e.g., size-dependence of survival, early maturity, large clutch size, and increased clutch frequency) when conditions are favorable (Jones and Ballinger 1987). These same potential fitness advantages may be absent in

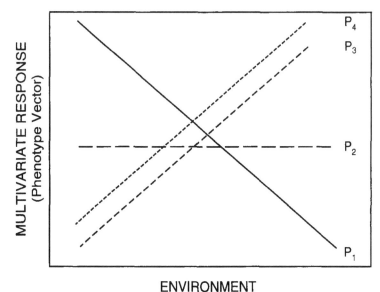

ENVIRONMENT

Figure 2.6. Hypothetical multivariate population-specific response to transplant or common garden experiment. P_i is population, and trajectories represent the average set of life-history phenotypes analyzed with a MANOVA. In this example, P_3 and P_4 are indistinguishable, though significantly different from P_2 and P_1, which are also significantly different from each other, implying three distinct sets of allocation rules.

New Jersey, or may seldom arise if the appropriate kind of environmental variation is rare. At least one potential fitness advantage of rapid hatchling growth appears to be absent in New Jersey: hatchlings with high growth rates or large body size did not have a higher probability of fall or overwinter survival in two years in which it was measured (Niewiarowski 1992). However, other potential advantages or disadvantages, and their temporal variation, are currently unknown.

The response of a single population in transplant and common garden experiments (the multivariate response profile) would also ultimately provide an important source of information concerning how to specify the rules which determine allocations of time and energy. However, as noted above, different population-specific responses across a range of environments potentially include two sources of variation: differences in allocation rules and differences in physiological states of individuals. The potential confounding of these two sources is a nontrivial obstacle which will impede rapid progress in our ability to specify the life-history rules as a time-ordered sequence of context specific (environment and physiology) allocation decisions and in understanding their evolution (Dunham 1993). Presently, how differences in physiological states among individuals contribute to variation in life-history phenotypes is not well understood but could be explored using individual-based physiologically structured models (see Dunham 1993 for an example).

Historical hypotheses and phylogenetic methods

Even when differences in allocation rules between populations can be unambiguously demonstrated and related to putative current habitat-specific sources of selection, absence of population-level phylogenetic information can confound interpretation of the historical role of natural selection in producing the differences. This problem can be viewed as a specific example (trait being considered is the set of allocation rules) of the general problem of evaluating adaptational hypotheses within an historical framework, for which cladistic methods may provide significant insight (Coddington 1988; Baum and Larson 1991; Miles and Dunham 1992, 1993; but see Lauder et al. 1993).

The demonstration that a particular set of allocation rules results in allocation patterns that confer higher fitness relative to other sets of rules in a given environment, suggests that the set is adaptive in the environment under consideration. This inference focuses on the "current utility" of the trait (sensu Gould and Vrba 1982; see Baum and Larson 1991 for a discussion) and may be satisfactory for hypotheses in a variety of contexts. Nevertheless, the inference implicitly ignores the evolutionary history of the trait (allocation rules). Phylogenetic analyses of adaptation at the species and supraspecies levels are common in comparative studies (e.g., Brooks and McClennan 1991; Harvey and Pagel 1991) because the importance of history is widely appreciated. In contrast, an acknowledgment of the potential influence of history on trait distributions among populations is largely absent. But populations, like species and higher taxa, also have an evolutionary history (Avise 1989), and ignoring that history can affect inferences about the processes responsible for observed patterns of trait variation. Lack of population-level phylogenetic perspective in studies of intraspecific geographic variation is no longer excusable, given widely-used molecular techniques designed to resolve phylogenetic relationships among populations (Avise et al. 1987).

Baum and Larson (1991) developed an approach that could be applied at the population level to add a historical perspective to interpreting geographic life-history variation (see Davis and Nixon 1992 for a discussion of appropriate terminal taxa in phylogenetic analyses). Consider the distribution of sets of allocation rules (a single, multistate trait) and "current" selective environments superimposed on the phylogeny of a hypothetical taxon, where the nodes represent populations (Fig. 2.7). The hypothesis that rule

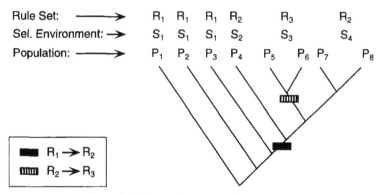

Figure 2.7. Hypothetical phylogeny that shows the rule set, selective environment, and population designation for each terminal node. Character (set of
allocation rules) transitions are shown on the tree. The pattern of evolution,
given the independently derived phylogeny, suggests that rule set two (R_2) is
not an adaptation to selective environment four (S_4), because it is a primitive
trait, first associated with selective environment two (S_2). See text for details.

set R_2 is an adaptation to selective environment S_4 may be falsified if it can
be shown that rule set R_2 is a primitive trait, having evolved in response
(perhaps) to a change in selective environments from S_1 to S_2 in population
P_4. A test like this is subject to all of the assumptions employed by cladistic
methods involved, including the use of parsimony in assigning changes in
traits to specific nodes (Maddison et al. 1984; Coddington 1988; Baum and
Larson 1991). It is sobering to consider how much we need to know about
populations and their environments in such an analysis. Clearly, phylogenetically controlled tests of this sort will require an ability to specify "selective
environments" and at least identify distinct sets of life-history rules for each
terminal node of the phylogeny. In spite of the difficulties involved, such
tests have been tractable in examining adaptational hypotheses about morphological evolution and may provide one more means of evaluating hypotheses about the evolution of allocation rules.

Conclusion

In light of extensive comparative life-history data in S. *undulatus* and the
relative lack of explanatory success of simple hypotheses, I advocate an
approach emphasizing experimental and phylogenetic techniques to understand geographic life-history variation. I have used S. *undulatus* as an example, but the approach is a general one which builds on comparative data
available for many species.

Although the potential to significantly advance understanding of the
causes and consequences of life-history variation is great with a multifaceted

approach like the one described here, theoretical and logistical difficulties involved cannot be underestimated. Building a life-history theory based on Dunham's allocation model will require, in any particular case, "discovering" allocation rules. Quantifying the phenotypic responses of individuals to transplant experiments, common garden experiments, and other manipulative approaches represents only one step in the ability to specify the rules of allocation. A causal link between the pertinent environmental variables and allocation rules must also be established. It is not enough to know, for example, that lizards transplanted from Nebraska to New Jersey change allocations to growth. We must also establish which aspects of these two different environments are responsible for the changes in allocations. The challenges are significant, but this should not be seen as a unique feature of the allocation model or phylogenetic techniques. Rather, it is a consequence of the depth of understanding sought.

Acknowledgments

Ideas presented in this manuscript arose out of regular discussions with Steve Beaupre, Justin Congdon, Art Dunham, and Greg Haenel. Art Dunham provided critical comments on the manuscript, especially with respect to interpreting his allocation model. Justin Congdon, Laurie Vitt, Eric Pianka, and three anonymous reviewers made many helpful comments on the manuscript. Any errors in presentation that remain are my own. I would like to thank Gary Ferguson and Larry Talent for sending proofs of their manuscript before it went to press. Steve Jones provided some demographic data for the New Jersey population in 1986. Finally, I thank Carol Murphy for her unwavering support and encouragement.

CHAPTER 3
LIZARD EGG ENVIRONMENTS

Karen L. Overall

Studies of lizard population biology have focused largely on posthatching or postparturition stages. In many lizard populations, variation in survival of eggs or embryos may represent the largest contribution to variation in annual recruitment. To understand population fluctuations or factors that affect population dynamics of animals, it is critical to study factors that influence survival of eggs or embryos. In short-lived animals like small lizards, population dynamics appear to be particularly sensitive to fluctuations in annual recruitment and survival of egg and juvenile stages compared to longer-lived animals in which survival of adults may have greater impact on population dynamics (Congdon et al. 1993a,b). For oviparous lizards, recruitment can be affected in the early stages by (1) the number of eggs laid; (2) the number of eggs that survive; and (3) the number of hatchlings that survive to reproduce. Factors that affect any one of these stages are germane to the understanding of population dynamics and to the study of life-history evolution. I discuss two experiments designed to elucidate the effect of the nest environment on egg survivorship and hatchling growth for the canyon lizard, *Sceloporus merriami*. The importance of the egg and juvenile stages is supported by Andrews (this volume) for *Anolis* and has been discussed for other species elsewhere (Muth 1980; Tinkle et al. 1981).

Only one of three factors affecting recruitment listed above has been studied with any thoroughness: the number of eggs produced by females. Two approaches have been taken. The first involves autopsy data from populations for which population parameters may or may not have been known; the second involves laboratory studies of eggs for which no population-level data are available. Both approaches are flawed, the second particularly so since it lacks a biological context in which to interpret the results.

Because clutch size varies as a function of female body size in most lizard species (Dunham et al. 1988a; Barbault and Mou 1988), clutch-size data obtained by autopsy are used in regression models relating clutch size to female body length. In combination with estimates of the number of sexually mature females in a population, the number of eggs produced by that population at any time can be estimated (Dunham et al. 1988b). The best data for demographic analyses of population dynamics are mortality and fecundity data obtained from long-term, mark-recapture studies on the population of interest; however, it is difficult to capture and mark all hatchlings. Even if all hatchlings could be censused, it is difficult to know the number of eggs placed into the environment and, therefore, overall egg survivorship.

Also, data obtained by autopsy may not accurately represent the mass, water content, or linear dimensions of the egg at oviposition.

Laboratory studies, rather than studies utilizing regression models based on autopsy data, can potentially provide data to estimate egg production better and to assay factors affecting egg survivorship. Some studies provide valuable data on population-specific variation in clutch size (Ferguson et al. 1982; Ferguson and Snell 1986; Muth 1980), suggesting that interindividual variation in clutch size and number may be important to variation in recruitment and, hence, to variation in population dynamics. Unfortunately, the populations for which most laboratory data are available are not ones for which concomitant, population-level life history or demographic data have been collected. Long-term demographic studies have demonstrated considerable variation among populations within a species in important life-history phenotypes including, but not restricted to, age at maturity, clutch frequency, clutch number, average clutch size, age at first reproduction, average cohort generation time, preferred body temperature, and size of home range or territory (e.g., Niewiarowski, this volume; review in Dunham et al. 1988b).

Finally, because most squamate eggs are ectohydric, the nest environment can influence another factor that affects recruitment: the number of eggs that survive to hatching. Laboratory studies have shown that both the thermal and hydric environments can influence egg survival in several species (Gutzke and Packard 1987; Kobayashi et al. 1983; Muth 1980; Packard and Packard 1988; Packard et al. 1977; Sexton and Marion 1974; Snell and Tracy 1985; Tracy and Snell 1985) and the transport of minerals (Packard et al. 1992; Phillips et al. 1990). To evaluate the biological value of laboratory data for effects of thermal and hydric environments on parameters affecting eggs and hatchlings, the context of the demographic environment is crucial. One needs to ascertain that laboratory conditions bracket and mimic those found in the field, that care of the animals in the lab is adequate to keep them healthy, and that animals resulting from the laboratory manipulations of eggs are representative of the range of those found in the field. Achieving this requires access to relevant population-level data. Most laboratory studies are conducted using animals for which no demographic, population-level data exist. In the absence of demographic data, the value of laboratory data is restricted to setting boundary conditions that may or may not be biologically relevant. Regardless, such laboratory data do suggest areas for further work and may suggest testable hypotheses about physiological ecology of eggs and nest site selection.

Neither autopsy nor laboratory data are suitable to evaluate possible effects of age of females, female physical condition, female behavioral status, clutch size, and clutch frequency, all of which may affect egg and hatchling size and survival. This renders them of limited use in generating mechanistic hypotheses about causes of population variation.

Integrating data obtained on factors affecting number of eggs laid, egg survival, and hatchling survival into a broader scheme of population-level processes, though difficult, is not impossible. One approach to integrating such data into a predictive model of population dynamics was provided by Dunham et al. (1989) and Dunham (1993). In this approach, a life history is viewed as a set of rules that specifies three classes of age-specific allocations that depend on physiological state and on environmental conditions and are subject to a set of population-specific trade-offs and constraints (see Fig. 3 in Dunham 1993 or Fig. 2.4 in Niewiarowski this volume). Female age-specific survivorship influences recruitment through the number of eggs laid. The definition of a life history as a set of rules provides a framework for investigating age-specific allocation decisions, constraints, and trade-offs that occur due to variation in the set of operative (sensu Dunham et al. 1989; Dunham 1993) environments (e.g., resource, biophysical, social/demographic, and predation environments) within which every animal must simultaneously function. Variation in these environments may influence individual time-energy budgets over time frames within which the animal fulfills social and nutritional obligations, influencing the amount of resources available for allocation to growth, storage, maintenance, and reproduction. In any year, any of the four operative environments could be an overwhelming determinant of reproductive production. All animals in a population are subject to rules determining allocation decisions; the summation of their individual responses to operative environmental conditions over time will yield population-specific mortality and fecundity schedules (i.e., the resultant life-history phenotypes). The extent to which life-history phenotypes vary within a population represents the variability in the individual response to the operative environments given the set of rules (life-history).

The above type of heuristic model (Dunham et al. 1989; Dunham 1993) must include operative environments of eggs. For eggs, operative environments must include the biophysical, predation, and resource environments. The biophysical environment is critical for lizards because thermal and hydric environments during incubation affect developmental period, survivorship, and offspring size at hatching (Christian and Tracy 1981; Cunningham and Huene 1938; Fitch and Fitch 1967; Gordon 1960; Muth 1980; Packard et al. 1980; Snell and Tracy 1985; Tracy 1980; Tracy and Snell 1985; Van Damme et al. 1992; Werner 1988). Effects of the biophysical environment may be mitigated by the egg-resource environment, which is determined by the female and fixed in oviparous lizards, except for water exchange, at oviposition. Egg development and survival may be influenced by factors controlled by the female: amount of yolk, amount of water in the albumin/allantois, and egg size and shape. Her choice of nest site determines the biophysical environment to which the egg is exposed. Egg shape (dimension) is less variable within a clutch than within a species (Overall unpubl.), while shell structure (size and distribution of pores) appears to be relatively constant within a species (Packard et al. 1982; Sexton et al. 1979), as does energy density of eggs and yolk (Dunham 1981; Ricklefs and Cullen 1973; Troyer 1983). Egg size, amount of yolk, amount of albumin, and allantoic storage may all interact with the local biophysical environment to influence water transport (Packard and Packard 1987). At oviposition, the amount of energy available for allocation to growth, storage, and maintenance of the embryo and newly emerged hatchling is fixed. Thus, the allocation model for eggs in a given population will generally be simpler than that for adults. The interaction of the biophysical, maternal (i.e., resource), and predation environments determines whether or not eggs survive. Summed over the population of all eggs laid, the response determines the second factor affecting recruitment, survivorship of eggs.

Effects of incubation environment on the size of resultant hatchlings have been reported in other studies (Gutzke and Packard 1985, 1987; Miller et al. 1987; Morris et al. 1983; Plummer and Snell 1988; Rand 1972; Snell and Tracy 1985; Tracy and Snell 1985; Van Damme et al. 1992; Webb and Cooper-Preston 1989). Two critical pieces of information are generally missing from such studies. Data on size distributions of newly hatched animals in natural populations are rarely reported so no context exists in which to evaluate the potential importance of hatching size for the individual lizard or for resultant population-level effects (Andrews, this volume; Miller et al. 1987; Sinervo 1990b; Sinervo and Adolph 1990; Sinervo and Huey 1990). No study to date has reported long-term effects on growth and/or survivorship of individual hatchlings of different masses or lengths (Sinervo 1990b; but see Webb et al. 1987). Here, I provide data on the effects of incubation environment on growth and on variable incubation environments on hatchling survivorship and development.

I discuss two experiments using the canyon lizard, *S. merriami*, from two populations in Big Bend National Park, Texas. The first experiment focuses on the effect of the thermal environment and elucidates the importance of constant and variable conditions for egg survivorship, stage of death, developmental period, and size of resultant hatchling. For this experiment only eggs from female *S. merriami* resident from one population, Boquillas, are used. The second experiment focuses on the effects of the hydric environment on emergent hatchling size and subsequent growth rate in hatchling *S. merriami* from both populations.

The System

This study was conducted in the Chihuahuan Desert in Big Bend National Park, Texas. Canyon lizards, *S. merriami*, are patchily abundant throughout the park. Two populations were chosen for investigation: Grapevine Hills (elevation 1036 m; average annual precipitation = 35.6 cm [range: 17.2–55.6 cm]) and Boquillas (elevation 560 m; average annual precipitation = 24.1 cm [range 12.4–38.0 cm]). Habitat at Grapevine Hills is formed by a weathering laccolith. Soil is primarily cobbly loam to 12 cm, cobbly clay loam, and exposed bedrock with slopes of 25%–40% (Cochran and Rives 1985). Dominant vegetation at this site includes lecheguilla (*Agave lecheguilla*), creosote (*Larrea tridentata*), ocotillo (*Fouquieria splendens*), sotol (*Dasylirion leiophyllum*), and Texas persimmon (*Diospyros texana*). Vegetation cover is approximately 33%.

Boquillas is xeric, set in a limestone canyon with east and west facing slopes. Vegetation cover at Boquillas is approximately 15%. Dominant vegetation is cactus (*Opuntia rufida*, *O. leptocaulus*, and *O. phaeacantha*), lecheguilla (*Agave lecheguilla*), candallia (*Euphorbia antisyphillitica*), yucca, (*Yucca rostrata* and *Y. torreyi*), false agave (*Hechtia scariosa*), leather stem (*Jatropha dioica*), and acacia (*Acacia constricta*). Soil is representative of the Upton-Nickel association, consisting of gravelly loam to 8–13 cm and sandy loam, with slopes of 1%–6%. Permeability and available water capacity are similar at both sites; moist bulk density is higher at Grapevine Hills than at Boquillas due to higher clay content of the soil (18%–35% vs. 5%–15%) (Cochran and Rives 1985).

Sceloporus merriami are small, 3–5 g saxicolous, heliothermic sceloporine phrynosomatid (sensu Frost and Etheridge 1989) lizards. This is numerically the dominant lizard species at both sites (Dunham 1978, 1980). Populations of *S. merriami* have been the subject of an intensive 20-year demographic study (Dunham 1978, 1980, 1981, 1982, 1983, 1993). These populations have been the focus of studies elucidating the effects of the biophysical environment on habitat use (Grant 1990; Grant and Dunham 1988, 1990), male social system (Ruby and Dunham 1987a), mating system (Overall and Dunham unpubl.), locomotor performance (Huey and Dunham 1987; Huey et al. 1990), physiological ecology (Beaupre et al. 1993a,b), and theoretical approaches for allocation strategies (Dunham et al. 1989; Dunham 1993). The egg stage has been understudied in this and many other demographic and long-term studies. The *S. merriami* system is an ideal one in which to investigate questions about eggs since demographic and biophysical environments are well studied, providing the field data necessary to validate and provide a context for laboratory data.

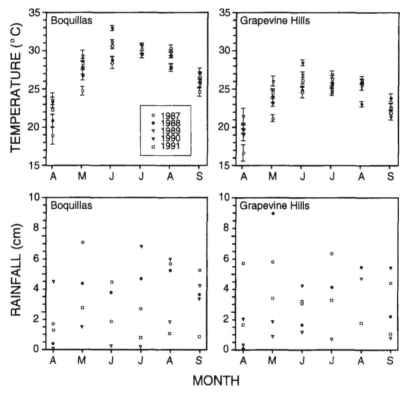

Figure 3.1. [upper]. Summary of monthly average temperature variation for the Boquillas and Grapevine Hills study localities during the lizard active seasons of the years 1987–1991. [lower]. Summary of total monthly rainfall for the lizard active seasons for the Boquillas and Grapevine Hills study localities during the years 1987–1991. Symbols are same as in upper panels.

Characterization of the Soil Moisture and Thermal Environments on Female Territories

Rainfall and temperature in the Chihuahuan Desert are highly variable within and between years. Monthly temperature and rainfall data from U.S. National Park Weather Stations near each study site are summarized for the lizard active seasons of 1987–1991 in Fig. 3.1. Boquillas is considerably hotter and drier, on average, than is Grapevine Hills, although Grapevine Hills may be more adversely affected by drought conditions. During the period of this study, rainfall was considerably below long-term averages for both sites. These fluctuations affect insect prey abundance for insectivorous lizards (Dunham 1978). Between-year variation in rainfall patterns affects soil moisture and temperature; the latter measurements were made annually at both Grapevine Hills and Boquillas sites.

Two areas known to encompass female territories were chosen at each site for intense biophysical monitoring. Using two Campbell CR7X data loggers, soil temperatures were measured every minute and averaged every 15 minutes for 7–10 days a minimum of three times per summer. Soil moisture was measured daily throughout summer. Both sites were monitored either within the same weekly period or simultaneously. The same territories were used each year to allow comparison of measurements within sites between years. Twelve years of intensive focal-observation data on mapped sites indicate that the majority of female *S. merriami* oviposit within their territories. Occasional females make egg-laying journeys. These are periods of high mortality (Ruby and Dunham 1987a; Dunham and Overall unpubl.). Thermocouple probes were placed at depths of 0 cm (surface), 2.5 cm, 5 cm, 10 cm, and, if possible, 20 cm, in the following biophysical microhabitats: open (= Open), under a rock (= Rock), under a bush (= Bush), and in the interface of bush and rock (= Bush/Rock). Microhabitats were opportunistically chosen in areas encompassing the range of habitats used by the females on each territory. The 20-cm probe was not placed in most locations because the bedrock base of the granitic laccolith of Grapevine Hills and the Cretaceous limestone cliffs at Boquillas prohibited this. Representative data from the Open and Rock microhabitats at Boquillas for June 1988 are presented in Figures 3.2 and 3.3. Data from the same time period from all four microhabitats (Open, Rock, Bush, and Bush/Rock) and both localities are summarized in Table 3.1.

Soil water potential may be measured using thermocouple psychrometry or tensiometry methods. Thermocouple psychrometry is better for dry soils, while tensiometry techniques work better for wetter soils (Campbell and Gardner 1971; Savage et al. 1981). The former is accurate to within 50 kPa (Campbell et al. 1973). Soil moisture profiles and drying curves were measured at both sites using both techniques (Overall unpubl.). Here I present representative data obtained from tensiometry. At 5 different female territories at each site, tensiometers (Soilmoisture, Santa Barbara, CA) were placed to a depth of 10 cm in the Open and Bush/Rock microhabitats. Tensiometers were not placed under rocks, because they could not be consistently, physically positioned there without destroying the habitat. Tensiometers were maintained at the same location for each year of the study, allowing within site, between year comparison. These data are presented in Figure 3.4.

At both sites, temperature fluctuations were largely damped at a depth of 10 cm in the Bush/Rock and Rock microhabitats. This is consistent with the depth that female lizards are generally thought to lay their eggs:

Lizard Ecology

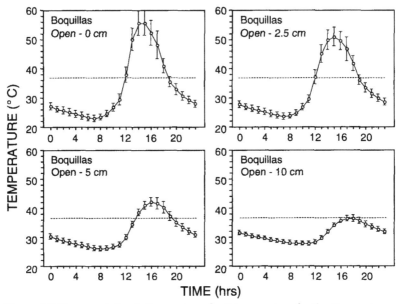

Figure 3.2. Representative daily marches of soil temperature for the Open microhabitat at the Boquillas study locality during June, 1988. Line represents 37°C.

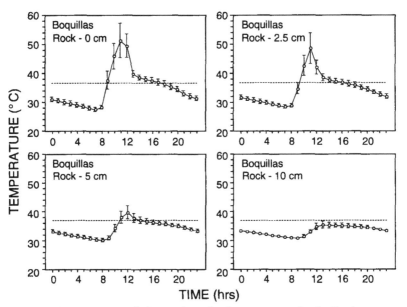

Figure 3.3. Representative daily marches of soil temperature for the Rock microhabitat at the Boquillas study locality during June, 1988. Line represents 37°C.

Table 3.1. Representative soil temperature profile data for Boquillas and Grapevine Hills in June 1988 for each potential oviposition microhabitat. Also shown are mean number of hours per day that each microhabitat depth combination experienced temperatures above 37°C. *CI* is 95% confidence interval.

Locality Microhabitat	Depth [cm]	Mean Temperature [°C ± 95% *CI*]	Hours ≥ 37°C
BOQUILLAS			
Open	0.0	33.5 ± 4.7	7.0
	2.5	33.0 ± 4.0	7.0
	5.0	32.1 ± 2.4	5.0
	10.0	31.5 ± 1.4	3.0
Rock	0.0	31.5 ± 2.9	9.0
	2.5	30.7 ± 2.3	8.0
	5.0	29.4 ± 1.3	6.0
	10.0	29.1 ± 68	0.5
Bush	0.0	35.3 ± 2.8	6.0
	2.5	34.7 ± 2.1	4.0
	5.0	34.0 ± 1.2	0.5
	10.0	33.4 ± 0.7	0.0
Bush/Rock	0.0	35.3 ± 4.4	8.0
	2.5	34.0 ± 2.9	7.0
	5.0	33.4 ± 2.1	6.0
	10.0	32.9 ± 1.2	3.0
GRAPEVINE HILLS			
Open	0.0	36.5 ± 5.5	10.5
	2.5	36.4 ± 4.6	10.0
	5.0	35.4 ± 3.1	9.5
	10.0	34.8 ± 1.9	9.0
Rock	0.0	31.7 ± 0.9	0.0
	2.5	31.3 ± 0.7	0.0
	5.0	30.4 ± 0.4	0.0
	10.0	29.8 ± 0.2	0.0
Bush	0.0	35.9 ± 3.9	10.5
	2.5	35.4 ± 3.0	10.0
	5.0	35.1 ± 2.6	8.0
	10.0	33.9 ± 1.0	6.0
Bush/Rock	0.0	29.7 ± 1.1	0.0
	2.5	30.5 ± 0.9	0.0
	5.0	30.1 ± 0.6	0.0
	10.0	30.0 ± 0.5	0.0

1–1.5 body lengths or 5–7.5 cm for *S. merriami* (Muth 1977, 1980; Rand 1972; Rand and Greene 1982). The critical thermal maximum (CTM) for these lizards is 41°C (Grant 1990) and their preferred body temperature is 33°C (Grant and Dunham 1988). Another experiment determined that constant incubation at 37°C is uniformly lethal to *S. merriami* eggs; the last 2/3 of incubation is more temperature sensitive. This is discussed elsewhere (Overall unpubl.). Extended exposure to temperatures higher than 37°C should kill eggs earlier and faster. Given the length of time the field soil temperature exceeds known lethal temperature, the data indicate that egg survivorship can occur only in the microhabitats of Rock and Bush/Rock. These soil temperature measurements and the magnitude of their diurnal fluctuations are repeatable within and between years. During the summer, there is no depth to which a female canyon lizard is competent to dig in the Bush and Open microhabitats that will not result in the death of her eggs. The incubation microhabitat that best damps temperature fluctuations while minimizing alterations in soil moisture is rock. This is consistent with observations of *S. merriami* nests. *Sceloporus merriami* eggs are extremely hard to find; females become secretive at oviposition. Despite excavation of large tracts only seven nests were located. One, in a Bush/Rock interface, contained decayed eggs; two nests were identified by observing gravid females digging under slabs of rock; two were identified by observing nascent hatchlings emerging from under large rocks; two were identified by observing patch-nosed snakes (*Salvadora deserticola*), predators on lizard eggs, that were digging under rocks. Areas extensively explored by the snakes were screened and emergent hatchlings caught.

In the years of this study, any sampled microhabitat on any female's territory met minimal hydric conditions for egg incubation (Fig. 3.4). Eggs of *S. merriami* require a minimum soil-moisture potential of –2850 kPa to develop, although mortality experienced under these conditions is in excess of 80% (Overall unpubl.). I cannot say absolutely that availability of hydrically acceptable oviposition sites did not limit female territorial distribution, but during this study, conducted during a drought, this seemed unlikely. Small-scale microhabitat distribution and variation in soil wetness within a female's territory is more difficult to define. These differences may be important, given the results on egg survivorship, developmental period, and growth rate discussed in this chapter.

Accession and Maintenance of Gravid Females

Gravid females were captured by noosing on areas near, but not part of, the gridded Boquillas and Grapevine Hills long-term study sites. These sites have been maintained for 14 and 20 years, respectively, and have not been

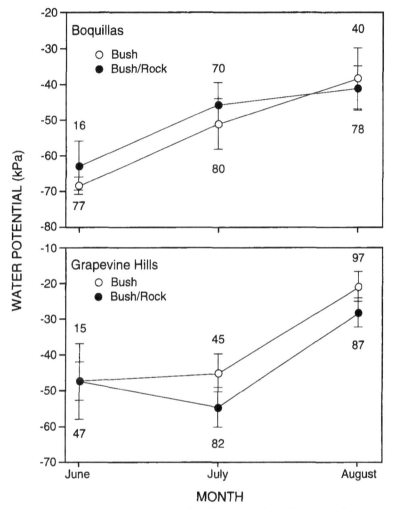

Figure 3.4. Summary of tensiometry data for the Bush and Bush/Rock micro-habitats at both study localities. Numbers are sample sizes.

subject to experimental manipulation for the last 14 years. Females were returned to the lab (Barker House) the same day, where they were weighed to the nearest 0.01 g, measured (snout-to-vent length [SVL], tail length, and hind limb length) to the nearest mm, assigned a unique toe number, palpated to determine gravidity status (small, medium, or large eggs and number of eggs), and individually housed.

Cages were constructed from food-grade containers with netting glued into lids. Each cage was provided with an oviposition site (a petri dish with

Whatman filter paper kept wet with nonchlorinated water), a white, non-chlorine-bleached paper towel on the bottom, and a small petri dish filled with rocks and nonchlorinated water for drinking and humidity. Toweling was chosen as the cage substrate instead of wet sphagnum or sand for two reasons. Pilot studies indicated that these lizards develop eye infections, have impaired hunting ability, stop eating, and lose condition when kept on sphagnum or sand. In every species for which such effects have been investigated for their impact on *in utero* development, negative effects have been reported. Some of these effects include reduction in ability to survive later in life and may not be noted at birth. One goal of this study was to evaluate effects of egg environment on survivorship, growth, and performance of hatchlings; therefore, I minimized uncontrolled risks, such as those associated with maternal debility. Because these females were returned to their territories, debility or injury was unacceptable. Finally, because very early exposure to thermal and hydric microenvironments may influence future development, it is essential to get freshly laid eggs to insure that artifacts are not being studied. The cage environment provided insures no hidden eggs.

All lizards were fed daily ad libitum, usually four to seven 2-week old, calcium dusted, tarsi-free, crickets (Fluker's Cricket Farm). Crickets were fed dog biscuits, carrots, apples, potatoes, and a protein-based dog and cat vitamin and mineral supplement. Lizard cages were changed and wiped, and water dishes cleaned and changed every other day or more often, if needed. Lizards were maintained on 14 h light / 31°–34° C: 10 h dark / 26°–28° C cycle using all solar wavelength Vita-lights. Cage position on shelves was rotated randomly twice daily. These conditions simulated those in the wild as closely as possible for 200 lizards. When hatchlings emerged, they were kept under conditions identical to those of the adults except that they were fed pinhead-sized crickets ad libitum. Gravid females were weighed and palpated minimally every other day. All changes in mass and gravidity status were noted. When gravid females decreased food intake and tapered their mass gain; had eggs that had become turgid, were visible through the lateral abdominal wall, and had moved into the pelvic canal; and had ligamentous loosening in the pelvic canal determined by palpation, they were induced with oxytocin to lay their eggs. Few lizards in this study laid naturally. Those that did usually oviposited late at night or early in the morning. Animals ovipositing naturally between the hours of 0800–1900 h were usually observed, and their eggs were freshly obtained. The oxytocin method was developed because eggs naturally laid very late or very early were often desiccated or damaged by crickets when found. By careful monitoring, oviposition could be induced at the time the females were ready to lay. Oxytocin at induction dosages will not affect unprimed tissue; induced births are common in other species. This method also provided excellent data on relative clutch mass (RCM), clutch mass, and individual egg mass and dimension critical to this study. Induction of oviposition also provided information on order of egg laying; effects of egg order on egg dimensions have some interesting implications for allocation theory.

Females were weighed to the nearest 0.01 g after oviposition, returned to clean cages, and fed for three days. Thereafter, they were toe-clipped and returned to their territories.

Handling and Maintenance of Eggs

Number 3 blasting sand was chosen as the incubation substrate because it is most similar to the natural substrate and possesses a consistent, fine grain, allowing comparison between treatments. Because air-filled pore space is non-conducting, coarse-textured materials (i.e., vermiculite) conduct water less freely than one of finer texture (i.e., sand) (Marshall and Holmes 1979), hence, oxygen exchange may differ (Ackerman et al. 1985). Gas diffusivities are lower in vermiculite than in sand because of the micellar structure of the former (Carter et al. 1986). When sand and vermiculite were compared in experiments using turtle (*Chelydra serpentina*) eggs, only hatchling size was affected by substrate; at any temperature and soil moisture, carcass dry mass was greater for eggs incubated in sand (Packard et al. 1987). No data are available to assess the relevance of resultant size distributions in natural populations. No comparable, experimental data exist for any lizard species. If size is influenced by substrate, substrates that mimic natural conditions are preferred.

Each bin held four eggs. All eggs were marked by color-coded flags that identified the female and clutch of origin and by charts that identified treatment. Bins were constructed from round, food-grade containers containing 600 g of oven-dried number 3 blasting sand (approximately 10 cm deep) and a specified amount of non-chlorinated water. Sand was measured using a Sartorius Scientific Balance and a Mettler pan balance (0.01 g). Dividers separating compartments for each of the four eggs were constructed of interlocking aluminum screens that fit snugly from the bottom of the sand to the inside of the lid. In this manner, any lizard that hatched could be contained in its own quadrant. Bins were numbered and color coded as to thermal and hydric treatments.

Eggs were weighed to the nearest 0.0001 g using a Sartorius Scientific Balance, measured in length and width to the nearest 0.01 mm using Mitutoyo digital calipers, and noted as to order of oviposition, if known. They were then carefully "planted" so that the top of the egg was at least 4 cm below the surface of the sand in the bins. Sand in each incubation bin was maintained at a predetermined experimental soil moisture. Eggs were

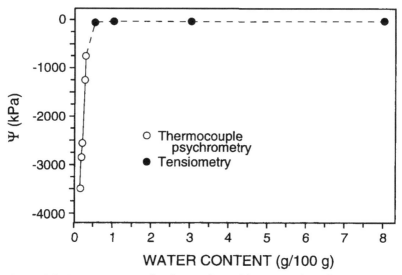

Figure 3.5. Saturation curve for the number 3 blasting sand used in all incubation experiments. Solid line indicates determinations carried out with thermocouple psychrometry. Dashed line indicates determinations carried out using tensiometers.

randomly assigned to treatments, controlling for maternal and clutch effects. Bins were then replaced in the temperature cabinet to which they had been assigned.

Data from field-measured temperature and soil moisture profiles were used to select and validate the range of soil moisture and temperatures used in the incubation experiments and to define boundary conditions for eggs in the natural nest and in laboratory experiments. All egg incubation experiments were then maintained at specific soil moistures (water potentials) gravimetrically so that eggs were minimally disturbed. The hydric treatments were constructed on a gravimetric basis and converted to a water potential (kPa) using a Wescor Model H-33T microvolt meter, a Decagon SC 10 thermocouple psychrometer (Campbell and Gardner 1971; Campbell et al. 1973; Savage et al. 1981), and a measured water retention curve for number 3 blasting sand (Fig. 3.5). For the first experiment, eggs were incubated at a subset of water potentials used for the second experiment (8.0 g H_2O / 100 g no. 3 blasting sand = –17 kPa; 1.0 g H_2O / 100 g no. 3 blasting sand = –34 kPa; 0.225 g H_2O / 100 g no. 3 blasting sand = –1250 kPa). These water potentials were previously determined to produce high egg survivorship. A second experiment was designed to test effects of the hydric environment on growth. In that experiment, water potentials used ranged from -2850 kPa to –17 kPa.

Egg bins were prepared gravimetrically on a 1.0 g water per 100 g of sand basis. All egg bins were maintained gravimetrically using the added masses of the eggs, plus the original wet mass. Since eggs absorbed water during the study, each treatment dried slightly as the study progress. Based on interim egg weights and initial total water mass of treatment, the boundary conditions of each treatment were bracketed. All treatments were prepared, thoroughly mixed, and maintained prior to planting any eggs. Egg containers were removed from temperature cabinets at least every other day only for as long as was necessary to plant another egg or to adjust water content. Although the containers were fairly airtight, some evaporation did occur. Evaporated water (0.00–0.09 g for 0.225 g/100 g treatment; 0.10–0.23 g for 8.0 g/100 g treatment) was replaced using a Terumo 27 gauge 1 cc syringe. The water was uniformly sprayed on the surface of the sand. Water was not injected into the soil for two reasons. First, there was the risk of hitting an egg or emerging hatchling. Second, most of the evaporative phase shift was probably taking place at the soil surface rather than deep in the soil. By reconstituting the mass lost to evaporation at the surface, I took advantage of the natural evaporative phase shifts that occur when any moist medium, whether an egg bin or a natural nest, contacts an air boundary (Hillel 1971). This technique, especially since round containers were used, also avoided unmeasurable and uncontrollable local pockets or gradients of moisture available only to some eggs, which is an inherent risk in techniques where water is injected into the soil or when vermiculite is used (Marshall and Holmes 1979).

Experiments

Variable temperature experiment

This experiment examined the relationship between variable versus constant incubation temperature and egg survivorship, stage at death, incubation period, and size of hatchlings. Only eggs of female S. merriami from Boquillas were used for this experiment. Soil moistures used were –17 kPa, -34 kPa, and –1250 kPa. A previous experiment determined that constant incubation at 37°C is uniformly lethal to S. merriami eggs. Soil-temperature data taken from female territories at Boquillas (Fig. 3.1; Table 3.1) indicate that the Rock and Bush/Rock incubation microhabitats are amenable to egg survivorship at depths of 5–10 cm, but that at these depths in these environments the soil temperature may increase to 37°C or above for as long as 0.5–6 h/day (Table 3.1). Average soil temperature in such microhabitats was between 31° and 34°C. Eggs were divided into two groups: those incubated for this experiment at 31°C and those incubated at 34°C (baseline

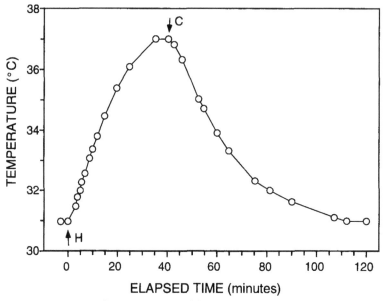

Figure 3.6. Average heating/cooling calibration curve for egg incubation bins used to regulate the time eggs were exposed to 37°C in the variable temperature experiment. Arrow marked H indicates the point at which heating began. Arrow marked C indicates the point at which cooling began (37°C).

temperatures). Each of these groups was then divided into three treatments: incubation at a constant baseline temperature (31° or 34°C), incubation at one of the baseline temperatures with exposure to 37°C for 1 h/day, and incubation at one of the baseline temperatures with exposure to 37°C for 3 h/day. The heating/cooling curve for the incubation bins used in this experiment (Fig. 3.6) permitted calculation of the total amount of time that the egg bin had to be exposed to increased temperatures to insure that the eggs were exposed to 37°C for one and three hours, respectively. Exposure to increased temperature was conducted at the same time every day, between 1300 and 1700, representing the period when this would occur in the field. The 37°C temperature exposure was maintained using a constant-temperature water bath; the remainder of the time the eggs were incubated in Percival temperature cabinets calibrated to either 31°C or 34°C. Temperatures in the Percival temperature cabinets were monitored using a Campbell CR 21X data logger and did not vary more than 0.2°C. Bins were relocated twice daily within the temperature cabinets. The experiment continued until all eggs hatched or were determined to be dead.

Growth experiment

Eggs of female *S. merriami* from Grapevine Hills and Boquillas were incubated at approximately 31°C in hydric environments ranging from –2850 kPa (0.175 g water/100 g no. 3 blasting sand) to –17 kPa (8.0 g water/100 g no 3 blasting sand). Hatchlings were maintained as described above and fed ad libitum. If animals were kept more than a few months they were moved into 10 gallon terraria with sand and rocky substrates taken from their natal sites, cover, heat lamps, water dishes, and a small plant. Animals were weighed every other day for the first few months and then weekly. During each weighing period tail, hind limb, and snout vent length were measured.

Results

Variable temperature experiment

The main effect of incubation temperature was on developmental time. Interaction terms indicate that soil-water potential has a negligible effect on incubation period and a slight, nonsignificant effect on other factors discussed within temperature treatments. The latter effect may be nonsignificant because of the small sample size. The number of days from oviposition to hatching averaged 37 ($2 SE = 0.4$) days at 31°C and 32 ($2 SE = 0.6$) days at 34°C at constant incubation temperature (Table 3.2). This approximately 5–day difference is statistically significant ($P < 0.001$). Exposing eggs incubated at these baseline temperatures to 37°C for either one or three hours resulted in a decrease in incubation period from that found in the corresponding constant temperature treatments. The incubation period for eggs incubated at a baseline temperature of 31°C and exposed to 37°C for 1 h/day decreased to 34 ($2 SE = 0.8$) days; that for eggs incubated at 31°C and exposed to 37°C for 3 h/day also decreased to 34 ($2 SE = 0.7$) days. No significant difference was noted in incubation period between eggs incubated at 31°C and exposed to 37°C for 1 h/day and those incubated at 31°C and exposed to 37°C for 3 h/day ($P > 0.05$). The decrease in incubation period in both variable temperature treatments when compared with that of the constant treatments was statistically significant ($P < 0.001$).

An analogous result was obtained for eggs incubated at a baseline temperature of 34°C and exposed to 37°C for either 1 or 3 h/day (Table 3.2). In this case, the magnitude of the temperature effect was much smaller: the incubation period for eggs exposed to either variable temperature treatment decreased by about 1 day over that of eggs incubated at a constant temperature of 34°C ($P < 0.01$). Although statistically significant, a decrease in incubation period of about 1 day is of doubtful ecological significance. No significant difference in incubation period was found between eggs

Table 3.2. Results of variable temperature experiment. $T_{Baseline}$ is the baseline temperature for incubation. Data for incubation period, hatchling mass, and hatchling SVL are presented as mean (n, 2 SE). Data on survival to hatching are presented as proportion surviving (n, number surviving). Stage at death was scored according to the following ordinal scale: 1 = no visible development, 2 = some CNS development, 3 = organogenesis, 4 = fully developed but not pipped, and 5 = pipped or partially out of shell. It was not possible to determine stage at death in all cases.

	Hours at 37°C		
$T_{Baseline}$ = 31°C	0	1	3
Incubation Period (d)	37.28 (76, 0.39)	34.13 (10, 0.80)	33.67 (6, 0.66)
Hatchling Mass (g)	0.37 (76, 0.02)	0.44 (10, 0.06)	0.41 (6, 0.04)
Hatchling SVL (mm)	22.3 (76, 0.22)	22.4 (10, 0.50)	22.2 (6, 0.56)
Survival to Hatching	0.79 (96, 76)	0.67 (15,10)	0.56 (11, 6)
Stage at Death	2.17 (30, 0.60)	1.67 (5,1.2)	1.20 (5,0.40)
$T_{Baseline}$ = 34°C	0	1	3
Incubation Period (d)	32.30 (37, 0.56)	31.60 (10, 0.72)	31.28 (6, 0.63)
Hatchling Mass (g)	0.36 (37, 0.02)	0.45 (10, 0.10)	0.32 (6, 0.08)
Hatchling SVL (mm)	22.0 (37, 0.22)	22.0 (10, 0.01)	21.4 (6, 1.54)
Survival to Hatching	0.54 (69, 37)	0.50 (20, 10)	0.30 (20, 6)
Stage at Death	2.69 (30, 0.42)	2.79 (10, 0.56)	2.33 (14, 0.61)

incubated at 34°C and exposed to 37°C for 1 h/day and those incubated at 34°C and exposed to 37°C for 3 h/day ($P > 0.05$). This finding suggested that 34°C is at the edge of the range of tolerance for *S. merriami* eggs.

Survival to hatching of eggs incubated at a constant temperature of 34°C (0.53) was significantly lower ($G[1, adj] = 9.83$; $P < 0.05$) than that (0.79) of eggs incubated at a constant 31°C (Table 3.2). For eggs raised at 34°C and then exposed to 37°C for 1 hour per day, no significant decrement in egg survivorship was seen, but the sample size is small. Increasing exposure of eggs to 37°C to three hours per day drastically decreases survival to hatching of eggs otherwise incubated at 34°C from 53% to 30% (Table 3.2), providing additional evidence for that eggs incubated at 34°C may be at the edge of tolerance of the thermal environment. Unlike results for eggs incubated at 34°C, eggs incubated at 31°C experience drastic reductions in survivorship for any exposure period to 37°C (Table 3.2). Regardless, in each treatment, more eggs hatch proportionally if the remainder of the incubation is conducted at 31°C than at 34°C.

Significant effects of the incubation temperature regime on the body mass of emerging hatchlings were found (Table 3.2: $F_{5,140} = 3.31$; $P < 0.0075$). Body sizes of hatchlings produced in this experiment were all within the range of hatchling masses and SVLs produced in natural populations. Body mass of hatchlings emerging from eggs incubated at a constant temperature of 31°C averaged 0.37 (2 SE = 0.02) g, while that of hatchlings emerging from eggs incubated at a baseline temperature of 31°C and exposed to 37°C for 1 h/day averaged 0.44 (2 SE = 0.06) g. Average body mass of hatchlings emerging from eggs incubated at a baseline temperature of 31°C and exposed to 37°C for 3 h/day was 0.41 (2 SE = 0.04) g. Multiple comparison tests (GT^2 tests) revealed that average body mass of hatchlings emerging from both variable incubation temperature treatments was significantly greater (P < 0.05) than that of hatchlings emerging from constant incubation temperature. Average body mass of hatchlings emerging from the two variable incubation temperature treatments (baseline incubation temperature = 31°C and exposure to 37°C for either 1 or 3 h/day) did not differ significantly (GT^2 test; P > 0.05). Similar results were obtained from hatchlings emerging from eggs incubated at a baseline temperature of 34°C. Body mass of hatchlings emerging from eggs incubated at a constant temperature of 34°C averaged 0.36 (2 SE = 0.02) g and while those of hatchlings emerging from eggs incubated at a baseline temperature of 34°C and exposed to 37°C for 1 h/day averaged 0.45 (2 SE = 0.10) g. Average body mass of hatchlings emerging from eggs incubated at a baseline temperature of 34°C and exposed to 37°C for 3 h/day was 0.32 (2 SE = 0.04) g. GT^2 tests revealed that average body mass of hatchlings emerging from eggs incubated at a baseline temperature of 34°C and exposed to 37°C for 1 h/day was significantly greater (P < 0.05) than those of hatchlings emerging from the constant incubation-temperature treatment. Average body mass of hatchlings emerging from two variable incubation-temperature treatments (baseline incubation temperature = 34°C and exposure to 37°C for either 1 or 3 h/day) were also significantly different (GT^2 test; P < 0.05).

Although significant variation among temperature treatments in average body mass of emerging hatchlings existed, no significant differences in SVL of those hatchlings among any temperature treatments were found ($F_{5,140}$ = 1.16; P > 0.33).

Dead embryos could be visually staged roughly using the criteria presented in Table 3.2. Eggs incubated at 34°C versus 31°C died at later stages (Mann-Whitney U test, P < 0.05). In fact, the only eggs to die fully developed and pipped occurred in the 34°C treatment. When otherwise incubated at 31°C, eggs that die do so at early stages; increasing their exposure to higher temperatures may kill them earlier in development, but sample sizes do not warrant statistical comparison. No significant difference in stage at death among temperature treatments for eggs incubated at a baseline temperature of 34°C was noted (Mann-Whitney U tests, all P > 0.05).

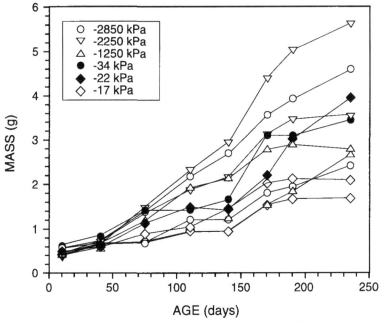

Figure 3.7. Growth trajectories for individual lizards raised under common garden (identical) conditions following incubation at 31°C and variable soil-water
potentials.

Growth experiment

Hatchlings grew faster at drier treatments (Fig. 3.7). Hatchlings from eggs
incubated at very wet treatments (–22 to –17 kPa) grew extremely
slowly, particularly in later months. At 10 months of age some *S. merriami*
hatchlings had not yet reached minimum size for reproduction (Fig. 3.7).
Hatchlings from dry treatments grew most quickly to day 200. Early
growth patterns are more complex and will be discussed elsewhere
(Overall unpubl.).

Conclusions

In *S. merriami*, effects of the hydric environment are probably less important for egg survival than those of the thermal environment, but not for all
parameters that may affect an animal's ability to succeed in a given demographic environment. Factors that affect age-specific size may be critical for
attainment of mates, avoidance of predation, and over-winter survival. Such
factors have been asserted to be important for other lizard populations, but
few data exist that allow evaluation of the extent to which this may be the

case (Ferguson and Fox 1984; Vince and Chinn 1971; Walls and Altig 1986). In *S. merriami*, minimum age and SVL for reproduction have been established, as have field active growth rates (Dunham 1978, 1981). These data are invaluable for providing a reference to test the importance of the nest environment. Experimental results indicate that growth rates through the first year of life to reproductive maturity are profoundly affected by the hydric environment, while the thermal environment determines overall survivorship. Females good at choosing appropriate nest sites should accrue a fitness advantage. Soil-moisture conditions necessary to produce animals with the best growth and survivorship rates were present on every territory measured during this study. This result underscores the importance of understanding the demographic context of any system to interpret the relevance of any life-history phenotype data produced by laboratory studies.

The egg environment also influences the stage at which eggs die. Patterns discussed in here suggest two things. First, an incubation temperature of 34°C appears to be close to edge of the range of tolerance for survivorship of *S. merriami* eggs. Given that eggs are exposed to the internal female environment until at least stage 30 (DuFaure and Hubert 1961), this should not be surprising for a lizard whose preferred body temperature is 33°C. Second, the importance of a nest environment that damps out temperature fluctuations at the higher end of the survivability spectrum is clear: hatchlings die at later developmental stages at warmer temperature, and even slight increases may have a profound effect.

Effects of incubation in variable temperature environments have interesting ramifications for "size" or condition (Sinervo 1990b) indices: hatchlings resulting from treatments using prolonged exposure to high temperature weigh significantly less than those reared from constant temperature treatments or those reared from treatments involving short exposure to extremes. This effect on mass is not accompanied by an effect on linear dimensions. This pattern holds for either initial incubation temperature, suggesting that selection for allocation strategies influenced by minor thermal variation has occurred. Such findings indicate a more complex and cautious story than suggested by current studies involving allometric engineering.

Findings above have direct application to conservation biology. The effect of the egg environment on growth has been underappreciated and seldom documented in a manner that allows evaluation of its importance to population recruitment. These are critical issues when choosing habitats to preserve and when breeding animals in captivity for reintroduction.

A note on the possible complication posed by the process of egg retention is warranted. Further development with retention is most common in species living at high elevations (Braña et al. 1991; Tinkle and Gibbons 1977). Regardless, no difference in degree of development in animals that oviposited in the lab after a relatively short versus a long stay has been noted for species in which this has been examined (Braña et al. 1991), nor was this the case in my study, which occurred at low or warm elevations. Intrauterine development appears to end at approximately the same stage (30; DuFaure and Hubert 1961), which is correlated with the end of the phase of primary differentiation (Guillette 1982; Shine 1983). No effect of development stage at oviposition, which is narrowly clustered within species and populations, on egg mass has been apparent (Braña et al. 1991).

Finally, data presented in this paper are less important than the approach demonstrated and the theory behind the approach. There are problems with each of the experiments discussed; both suggest that further physiological data are needed. These experiments are intended to be illustrative of those made possible by heuristic models. Any factor affecting any scaling parameter needs to be evaluated in the complex environment in which the suite of characters contributing to it has evolved. The model presented by Dunham et al. (1989) and Dunham (1993) provides the ability to do that, one part of which is illustrated here. Any experimental manipulations can produce data indicative of potential variation; the key is to have the biological content in which to evaluate the importance of resulting patterns.

Acknowledgments

This study required large amounts of resources, some of the most important of which were people. Among the more valued of the multitude of field assistants were Graham Watkins, Farah Bashey, Eva Byer, Dave Dunning, Greg Haenel, Ali Jalili, Kirstin Nicholson, and Paul Super. Field assistance, physical support (muscle), and advice was also provided by Ray Huey and Steve Beaupre. Ray Huey was invaluable in planning the variable temperature experiment. Warren Porter graciously loaned me one of the CR7X data loggers. Financial support was generously provided by Sigma Xi, the Theodore Roosevelt Memorial Fund of the American Museum of Natural History, a WARF Fellowship from the University of Wisconsin-Madison, three Davis Fellowships, an NSF grant (BSR 8805641) and three University of Pennsylvania Research Foundation Grants to Arthur E. Dunham, and an NSF Doctoral Dissertation Improvement Grant to the author. The Veterinary School of the University of Pennsylvania provided incredible amounts of logistic support. Art Dunham provided invaluable intellectual stimulation, only constructive criticism, non-stop moral support, and did a lion's share of the back-breaking physical work hauling data loggers to the field and maintaining them despite formidable challenges. He also provided, as he does for so many, invaluable statistical and graphical expertise. Two anonymous reviewers spent a large amount of time critically reading this manuscript; many of their comments served to clarify some sections and strengthen others; I am grateful.

CHAPTER 4
EXPERIMENTAL TESTS OF
REPRODUCTIVE ALLOCATION PARADIGMS

Barry Sinervo

Variation among individuals is the raw material for natural selection (Darwin 1859). Natural selection acts on variation in traits, but because of genetically based trade-offs, adaptive compromise among traits is an implicit assumption of most life-history modeling (Williams 1966a; Gadgil and Bossert 1970; Stearns 1977; Partridge and Harvey 1985; Reznick 1985, 1992; Pease and Bull 1988; Partridge 1992). For example, the trade-offs involving maternal allocation to reproduction are frequently expressed as two life-history paradigms: (1) the trade-off between offspring number and offspring quality (Lack 1954; Williams 1966a; Smith and Fretwell 1974) and (2) the trade-off between allocation to current versus future reproduction that is reflected in cost of reproduction (Williams 1966a; Pianka 1976; Reznick 1985).

Variation in maternal investment is readily generated by the techniques of "allometric engineering," an alteration of maternal investment that results in changes in hatchling size (Sinervo 1990a, 1993; Sinervo and Huey 1990; Sinervo and Licht 1991a,b; Sinervo et al. 1992) (Fig. 4.1). Allometric engineering places the study of variation in offspring size in an experimental context. Adding this experimental context is useful for three reasons.

First, such experimental manipulations supplement natural variation in offspring size (Sinervo 1993) and thus facilitate the detection of natural selection (Endler 1986; Mitchell-Olds and Shaw 1987; Schluter 1988; Wade and Kalisz 1990). Second, these manipulations verify causal links between offspring size and offspring quality as indexed by survival in natural populations (Endler 1986; Mitchell-Olds and Shaw 1987; Pease and Bull 1988; Schluter 1988; Wade and Kalisz 1990). Third, allometric engineering also makes the analysis of covariation among traits accessible to a causal analysis. Some of the changes in hatchling size are achieved by manipulating the physiological mechanisms that control clutch size and egg size (Sinervo and Licht 1991a,b). Manipulating the mechanistic coupling between clutch size, egg size, and total clutch mass permits an experimental analysis of the consequences of reproductive trade-offs for both the female (costs of reproduction) and her offspring (optimal size) (Sinervo and Licht 1991a,b). By carrying out such manipulative experiments in nature, the proximate physiological trade-offs that underlie covariation are placed in a more ultimate evolutionary context that involves natural selection on correlated traits (Sinervo 1993).

In this chapter, I synthesize results from laboratory and field experiments that provide an experimental test of the premises of the theory of optimal offspring size. David Lack is widely known for his formulation of the evolution of clutch size in birds. In contrast, his formulation of the problem of clutch size and offspring size for lizards is not well known (Lack 1954). For organisms without parental care (e.g., the example used by Lack was *Sceloporus* in California), Lack (1954) realized that natural selection should favor a compromise (optimizing selection) between the quantity and quality (size) of offspring. His prediction was based on two premises: (1) egg size is inversely related to clutch size (egg number), but (2) offspring quality, and hence offspring survival, is positively related to egg size (see Smith and Fretwell 1974 and Brockelman 1975 for elaborations of this theory). Optimizing selection results from the conflict between fecundity selection (presumably favoring many small eggs) and survival selection (presumably favoring a few large eggs). The joint action of fecundity selection and survival selection yields an optimal egg size that is, by definition, the reproductive strategy producing the most surviving offspring at maturity. At evolutionary equilibrium, the average egg size should match the optimum. This theory was refined by Williams (1966a) who included a consideration of the costs of reproduction that result from a trade-off between investment in current reproduction and future reproductive success.

The analysis focuses on life-history variation of the side-blotched lizard (*Uta stansburiana*), a small phrynosomatid (see Frost and Etheridge 1989) weighing 3-10 g and reaching sexual maturity in one year. I describe the mechanistic basis of the physiological trade-off between clutch size and offspring size in side-blotched lizards. The joint action of fecundity selection (arising from the clutch size and offspring size trade-off) and survival selection of offspring to maturity yields an optimal egg size that is, by definition, the reproductive strategy producing the most surviving offspring at maturity.

Optimal egg size is also constrained by functional limitations on maximum offspring size that a female can produce. These functional constraints may result in selective mortality of females that produce large offspring, and I present preliminary data illustrating that survival of the female parent is dependent on maternal allocation per offspring.

The traits making up a life-history tactic (e.g., clutch size, offspring size, total clutch mass) are not only coupled physiologically, but natural selection acts on the traits in both the female parent and her offspring resulting in a "multi-trait" and "multi-generational" selective trade-off. An analysis of optimal offspring size based on the trade-off between offspring number and offspring survival is incomplete without an analysis of the trade-off between current reproduction and future reproductive success of the female parent. In light of these observations, I outline future directions in lizard ecology that could be taken using this experimental approach and extensions of the approach to life-history variation expressed at higher taxonomic levels.

The Physiological Trade-off between Offspring Size and Number

Jones (1979) proposed the mechanism of "follicle selection" to explain patterns of clutch size regulation in the vertebrate ovary. The follicular selection model has since been refined (Sinervo and Licht 1991a,b) to explain both clutch and egg size regulation in the reptilian ovary. Presumably, follicles compete for circulating gonadotropin (FSH) to maintain growth rate during early stages of vitellogenesis. Those follicles with greater vascularization than neighboring follicles may also have higher uptake of FSH and yolk (Jones et al. 1976; Jones 1978) and this may lead to even greater vascularization at the expense of neighboring follicles. Follicular atresia provides an additional mechanism for eliminating follicles from the pool of potential follicles that could be recruited into the clutch. If a follicle does not receive sufficient FSH to maintain growth, it will undergo follicular atresia at a relatively advanced stage of development.

The model of "follicular selection" is consistent with experimental manipulations of clutch size conducted on the side-blotched lizard (Sinervo and Licht 1991a,b; Fig. 4.1). Female *U. stansburiana* that are administered exogenous gonadotropin during the earliest stages of vitellogenesis produce larger clutches (average clutch size is 6 eggs) relative to controls (4.3 eggs; Fig. 4.2). Note that FSH treatment is only effective if initiated very early during the reproductive cycle, i.e., when follicles are less than 3 mm in diameter. A typical cycle between clutches for *U. stansburiana* is 24 d and the 3 mm follicle size is reached within the first half a week. If treatment with exogenous FSH is initiated after this point, clutch size and egg size are unaffected (Sinervo and Licht 1991a). Thus, recruitment of additional follicles, the first phase of "follicular selection," takes place during early vitellogenesis (< 3 mm in diameter). Even though upward adjustments of clutch size are not observed after the 3 mm stage of follicle development, the follicle can still undergo follicular atresia after the 3 mm size if the follicle does not receive sufficient FSH to maintain growth. Follicular atresia can occur in follicles that are as large as 5 mm in diameter, which is reached between approximately 7–10 days in the reproductive cycle (Sinervo and Licht 1991a). Ovulation of follicles occurs at approximately 8 mm (about 14 days), and fertilized eggs remain in the oviduct for about 10 days during which they are shelled and development of the embryo is initiated.

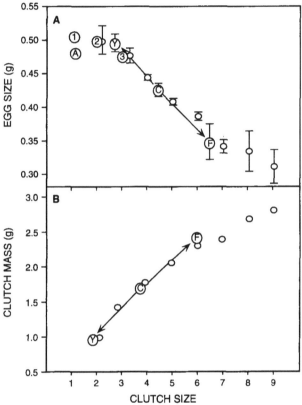

Figure 4.1. (A) Physiological trade-off between clutch size and egg size. The relationship between clutch size and mean (*S.E.*) egg size (O) among female side-blotched lizards from the Coast Range of California. Egg size decreases with increased clutch size among female side-blotched lizards. Experimental manipulations that increase average clutch size by administration of exogenous gonadotropin (ovine follicle stimulating hormone, FSH) decrease average egg size relative to controls (C→F). Conversely, experimental manipulations that decrease clutch size to ①, ②, or ③ eggs) increase average egg size relative to controls (C→Y, average egg size and clutch size for yolkectomy procedure are pooled from clutches of ①, ②, and ③ eggs). The trade-off between clutch size and egg size resulting from these complementary experimental manipulations of clutch size (arrows between groups C→F or C→Y) tracks the phenotypic trade-off between clutch size and egg size among unmanipulated female side-blotched lizards quite closely. Also shown is the average egg size for an *Anolis* lizard that naturally lays single-egged clutches (Ⓐ). Egg size is calculated for a female *Anolis* with a 4 g postpartum mass (Andrews and Rand 1974), which is comparable in size to the average postpartum mass of female side-blotched lizards in the Coast Range of California. (B) Effect of experimental manipulations on total clutch mass. The effect of experimental manipulations of clutch size on total clutch mass (arrows between groups C→F or C→Y) is very similar to the relationship between clutch size and total clutch mass found among female (O) side-blotched lizards [From Sinervo and Licht 1991a,b].

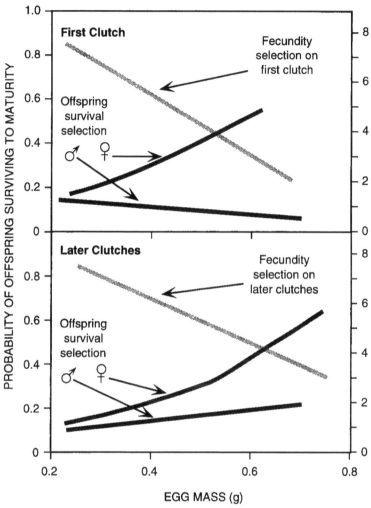

Figure 4.2. Natural selection on egg size in side-blotched lizards on the first clutch of the reproductive season and on later clutches. In both cases, fecundity selection reflects the higher fitness of a female that produces a large clutch of small eggs. Selection for offspring survival to maturity is partitioned into female and male offspring. Larger offspring generally have a higher probability of survival to maturity. However, on the first clutch, the male offspring have a higher survival probability to maturity (from Sinervo et al. 1992).

Manipulation of clutch size by exogenous gonadotropin also has a correlated effect on average egg size (Fig. 4.2). Females receiving FSH produce large clutches but smaller eggs (0.35 g) relative to controls (0.43 g).

Presumably, a female undergoing atresia would shunt energy that would otherwise have been invested in the atretic follicle into the remaining follicles. An experimental analog, follicle ablation, has been developed for the natural process of follicular atresia. Follicle ablation verifies the coupling between the control of clutch size and the control of egg size (Sinervo and Licht 1991a,b). If a subset of the follicles is ablated during midvitellogenesis (by "yolkectomy," i.e., removal of yolk from a follicle), females typically produce a smaller clutch (2.3 eggs) consisting of much larger eggs (0.50 g; Fig. 4.1).

The underlying physiological and hormonal coupling between clutch size and egg size forms the mechanistic basis of a proximate trade-off between these two traits (Fig. 4.1B) (Sinervo and Licht 1991a,b). It has been hypothesized (Sinervo and Licht 1991a,b) that the additional coupling between clutch size and total clutch mass (Fig. 4.1B) would arise if the levels of vitellogenin produced by the liver in response to estrogen (Yaron and Widzer 1978; Ho et al. 1982; Gavaud 1986) is also quantitatively related to the number of yolking follicles and their estrogen production. Increased number of follicles would increase the total surface area of follicles, which is the site of estrogen production. Presumably, elevated levels of estrogen associated with a large clutch would enhance total clutch mass through its effects on circulating levels of vitellogenin in the bloodstream and subsequent uptake by follicles.

The hormonal cascade that I have elaborated also provides a mechanistic framework for interpreting seasonal changes in clutch size and egg size that appear to be prevalent in lizards (Nussbaum 1981; Ferguson and Snell 1986; DeMarco 1989; Sinervo 1990a; Sinervo and Licht 1991a). For example, other hormones that are correlated with reproductive condition (e.g., corticosterone levels; Wilson and Wingfield 1992), might modulate circulating levels of gonadotropin and as such, would have proximate effects on clutch size and egg size through a process of follicular recruitment during the follicle selection phase or through follicular atresia.

Measuring Natural Selection on Offspring Survival

The allocation model presented above is the essence of the physiological trade-off which results in fecundity selection, the first episode of selection involved in determining optimum egg size in nature. However, the high fecundity of a female that lays a large number of small eggs must be discounted by the probability of survival of these offspring to maturity, the second episode of selection.

The survival component of natural selection on offspring size was experimentally assessed using the techniques of allometric engineering on a natural population of side-blotched lizards (Fig. 4.2). During 1989 and 1990 survival of 978 neonates was followed in a natural population of side-blotched lizards located near Los Baños Grandes, California (complete details can be found in Sinervo et al. 1992). Because side-blotched lizards increase clutch size throughout the reproductive season (Nussbaum 1981; Sinervo and Licht 1991a; Sinervo et al. 1992), the analysis was partitioned into offspring from the first clutch and offspring from later clutches. The seasonal shift in egg size has been interpreted in an adaptive context: selection on later clutch offspring is thought to be more intense because hatchlings from later clutches must compete not only with other later clutch hatchlings, but also with first clutch hatchlings, which were larger by virtue of age.

Females with large vitellogenic follicles or oviductal eggs were obtained from the study site and maintained in the laboratory until they laid their eggs. Females were returned to the study site after they had oviposited their eggs. Freshly laid eggs from these females were treated as follows. Miniaturized offspring (approximately 18% smaller than controls) were obtained by aspirating 18% of the yolk by mass using a sterile syringe. Control offspring were obtained by inserting a sterile syringe into eggs but not aspirating yolk (sham manipulated). These offspring served as natural-sized controls (see Fig. 4.4). A split clutch design was used, in which half the eggs from each clutch were miniaturized (total miniaturized $N = 313$) and half the eggs were sham manipulated ($N = 380$).

A second group of females was brought into the laboratory during the earliest stages of vitellogenesis. Gigantized eggs and offspring were obtained ($N = 87$) from some of these females (approximately 20% larger than controls) by ablating a subset of their follicles (approximately 1/2 of the follicles on both ovaries). Controls offspring ($N = 150$, of which 88 were unmanipulated and 62 were miniaturized) were obtained from females that were maintained in the laboratory for a comparable length of time (approximately 2–3 weeks). In addition, eggs ($N = 48$, of which 26 were unmanipulated and 22 were miniaturized) oviposited by females that underwent sham surgery (surgery with no follicle ablation) served as a second type of control offspring.

The patterns of natural selection detected using the restricted sample of natural variation (not shown) in all cases agreed with analyses based on natural and experimentally enhanced variation. Moreover, a comparison of control, sham-manipulated, and experimentally manipulated offspring indicates that no pathological artifacts arose from manipulations (Sinervo et al. 1992). Thus, patterns of survival described for experimentally manipulated and control offspring reflect the causal impact of egg size on offspring survival.

Significant phenotypic selection was detected for offspring survival to maturity by fitness regression (Arnold and Wade 1984; Sinervo et al. 1992). In general, large offspring tended to have higher rates of survival (Fig. 4.2). Survival selection on female offspring was very strong and consistently favored the largest female hatchlings. However, there were notable exceptions for male offspring. For example, the smallest male offspring that were produced on the first clutch had higher rates of survival compared to larger male hatchlings. This trend is reversed in later clutches where the largest male offspring had the highest survival rates.

Measuring Natural Selection on the Egg Size and Egg Number Trade-off

Because fitness effects act multiplicatively across different episodes of selection (e.g., fecundity and survival selection; Arnold and Wade 1984), the fecundity of females producing different-sized eggs (Fig. 4.2) is adjusted by the probability that these offspring survive to maturity (Fig. 4.3). This yields an "optimal offspring size" that describes the net effects of fecundity and survival selection (Fig. 4.4, Sinervo et al. 1992).

Unique optima for female offspring produced in the first clutch and for male offspring produced in later clutches were observed. In these two cases, female parents that laid intermediate-sized eggs had more offspring that survived to maturity than female parents with either smaller of larger offspring. However, for male offspring from the first clutch, selection favored the female parent that produced the smallest offspring (e.g., miniaturized or naturally small males). For female offspring from later clutches, selection favored the largest individuals (e.g., gigantized or naturally large females). For male offspring from the first clutch and female offspring in later clutches, selection was directional, and optimum egg size was defined as the reproductive strategy, e.g., phenotype in the population, that produced the most offspring that survived to maturity.

Optimum egg size in later clutches was larger than optimum egg size in the first clutch for female and male offspring. Thus, there was selection for the seasonal shift in egg size that is a typical reproductive pattern in side-blotched lizards (Nussbaum 1981; Sinervo and Licht 1991a; Sinervo et al. 1992). The first clutch optimum for male and female offspring size bracketed the observed egg size. In this case, observed egg size in the first clutch reflects a compromise between selection on male and female offspring size. However, later clutch optima for males and females were much higher than observed egg size, suggesting that these optima are perhaps beyond the functional limits of offspring size that female side-blotched lizards are capable of producing. Thus, seasonal shifts in observed egg size of side-blotched lizards do not track optimum egg size because of functional constraints.

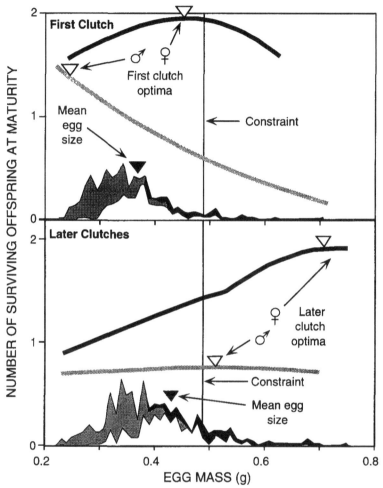

Figure 4.3. Optimal egg size in natural populations of side-blotched lizards for the first clutch of the reproductive season and for later clutches. The curves that describe the number of offspring that survive to maturity as a function of egg size were obtained by multiplying the functions that describe fecundity selection on offspring size and survival selection to maturity on offspring size (from Fig. 4.2). Distributions show relative numbers of control (open histogram), miniaturized (hatched), and gigantized hatchlings (solid) that were released on each clutch. Observed egg size (solid) and optimal egg size (open) are shown for each clutch. The observed egg size should be reflected in an adaptive compromise between selection on both sexes. This appears to be true for the first clutch in which the optima for the sexes bracket the observed egg size. However, the optima for male and female hatchlings on later clutches are far larger than the observed egg size.

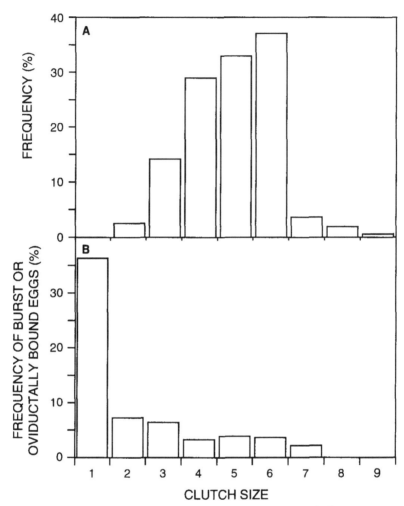

Figure 4.4. (A) Frequency histogram of clutch size for side-blotched lizards in the coast range of California. Note the low frequency of two-egged clutches and the absence of one-egged clutches in natural populations. (B) Frequency of reproductive difficulties as a function of clutch size for unmanipulated clutches and experimentally reduced clutches. Reproductive difficulties included eggs that burst at oviposition or eggs that became oviductally bound. Difficulties increased with decreased clutch size (from Sinervo and Licht 1991b).

Functional Constraints on Offspring Size and Costs of Reproduction

Classically, costs of reproduction have been thought to result from a large energetic investment in total maternal investment (e.g., total clutch mass) that decreases the probability of survival or future reproductive success (Reznick 1985, 1992). Trivers (1972) also considered the simple trade-off between egg size and clutch size to reflect a cost of reproduction (Trivers 1972; Coleman and Gross 1991). By producing large offspring to increase the offspring's probability of survival (future reproductive success), a parent diverts resources from the production of more offspring (current reproductive success; Trivers 1972). In addition to such broadly defined costs, are the costs of reproduction arising from the effects of maternal investment per offspring on survival of the female parent. These costs may be related to functional constraints on maximum offspring size.

Functional constraints may explain why females do not track the seasonal shifts in optimum egg size. Females with experimentally decreased clutch size or naturally small clutches produce large eggs that often become lodged in the reproductive tract or burst at oviposition (Sinervo and Licht 1991b). A pronounced increase in the probability of such reproductive difficulties occurs at egg sizes > 0.48 g (Figs. 4.1 and 4.4; Sinervo and Licht 1991b), well below many of the later-clutch (seasonal) optima detected in these studies. Females that become egg bound would ultimately die in natural populations, and thus the production of excessively large eggs required to match the optimum egg size on later clutches would result in a "cost" that decreases future reproductive success of females.

This presumed cost of reproduction is reflected in decreased survival of females that produce very large eggs. The size of the eggs that a female parent produces affects her own survival (Fig. 4.5). In later clutches, survival of the female parent was under stabilizing selection. Females producing very large eggs in later clutches (May) had lower probability of survival until the end of the breeding season (September) compared with females with intermediate-sized eggs.

A female producing large eggs relative to the population average at the end of the season has greater reproductive success because these offspring generally have superior survival to maturity (Fig. 4.4). However, this increase in reproductive success enhances the risk of the female becoming egg bound and thus lowers probability of future reproductive success either in other late season clutches or during the subsequent reproductive season (Fig. 4.5). Females that produce the largest eggs on the first clutch also tend to produce the largest eggs on later clutches.

If functional constraints on maximum offspring size were the only factor

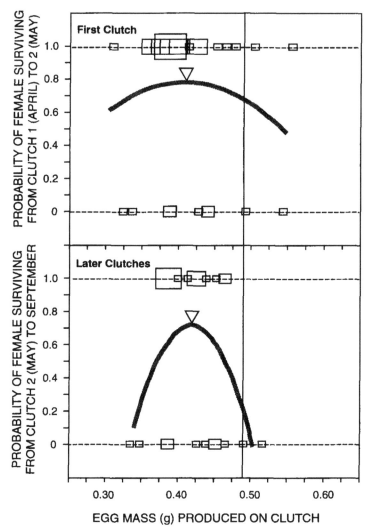

Figure 4.5. Natural selection on adult side-blotched lizard females as a function of maternal investment (egg size) on the first clutch and later clutches. Female survival was partitioned from survival from the first clutch (April) to the second clutch (May) and survival from the second clutch to the end of the reproductive season (September). The distribution of females that survived each period of selection are shown at the top of each panel, and the distribution of females that died during each period of selection are shown at the bottom of each panel (point size indicates multiple observations). Egg size did not significantly influence survival on the first clutch of the reproductive season. However, females that laid intermediate-sized eggs in later clutches had higher rates of survival than females that laid smaller or larger eggs. The significant optimizing selection ($P < 0.05$) that describes these survival differences among adult females reflects a cost of reproduction arising from maternal investment per offspring.

influencing survival of the female parent, females laying the smallest eggs would be expected to have the highest survival. In contrast, females that produced very small eggs had a low probability of survival to the end of the reproductive season. The lower survival of females laying small eggs may arise from correlated selection on other reproductive traits. Females that lay the smallest eggs tend to lay the largest clutches (Fig. 4.1A) and produce the largest total clutch mass (Fig. 4.1B). Females that produce the largest total clutch mass tend to have low survival compared to females that produce smaller total clutch mass (Sinervo unpubl.). This effect is presumably related to the cost of reproduction arising from total reproductive investment in a clutch.

These preliminary results show that costs of reproduction arising from maternal investment per offspring can also act directly on survival and future reproductive success of the female parent. Because selection on offspring survival (based on maternal investment per offspring) is correlated with traits that affect survival of the female parent (maternal investment per offspring and total investment per clutch), a comprehensive test of theories of optimal egg size is intimately related to a test of the theories of costs of reproduction.

Functional Constraints on Maximum Egg Size in a Comparative Framework

Implicit in considerations of sets of traits that covary consistently among phylogenetically related groups of organisms is that traits are constrained by an underlying mechanistic trade-off (Sinervo 1993). The trade-off between egg size and clutch size is a classic example of such a constraint, and the hormonal and physiological processes involved in maternal allocation per offspring provide the mechanistic bases for such a constraint. In addition to this fairly intuitive constraint on reproduction, other "functional" constraints probably impose further limits on adaptive evolution of reproductive tactics of reptiles. The diameter of the pelvic girdle has been suggested to limit adaptive evolution of egg size in turtles (Congdon and Gibbons 1987). Likewise, adaptive evolution of egg size has been suggested to be constrained by functional limitations on maximum egg size in anoles and geckos which are phylogenetically conservative in their production of one or two eggs (Andrews and Rand 1974; Vitt 1986; Sinervo and Licht 1991b).

Owing to the egg size and egg number trade-off, taxa that produce only one or two eggs might be expected to have much larger eggs than more typical multi-egged phrynosomatid lizards such as *Uta*. By comparing *Anolis* lizards (family Polychridae) that exclusively lay one-egged clutches with experimentally induced clutches of one egg in *U. stansburiana*, functional constraints are placed on maximum egg size in *U. stansburiana* in a comparative context (Sinervo and Licht 1991b). Presumably, *Anolis* lizards have few difficulties laying single-egged clutches compared to side-blotched lizards that never lay single-egged clutches. As mentioned earlier, the number of eggs that became oviductally bound or eggs that burst at oviposition increase with decreased clutch size in *U. stansburiana*, and this effect becomes most pronounced in clutches of three and fewer eggs (Sinervo and Licht 1991b). Noteworthy in light of these experimental observations is the average egg size of a typical *Anolis* lizard with a comparable body size of *U. stansburiana* (approximately 4 g). *Anolis* lizards that exclusively lay one-egged clutches lay eggs that are slightly smaller (0.48 g) than those of *U. stansburiana* from experimentally induced clutches of one, two, and three eggs, specifically those clutch sizes and egg sizes with which *U. stansburiana* experienced the greatest increase in burst or oviductally bound eggs.

Such structural constraints on the evolution of maximum offspring size also limit the evolution of total clutch mass or relative clutch mass (RCM) of lizards (Sinervo and Licht 1991b). In small clutches, particularly those with three and fewer eggs (Fig. 4.2B), the size of eggs in *U. stansburiana* is not as large as one would predict from a simple yolk allocation model (e.g., see Sinervo and Licht 1991b) and for very good reasons. Eggs that are too large either burst at oviposition, resulting in the death of the embryo, or become oviductally bound, inevitably resulting in the death of the female. Egg size, egg number, total clutch mass, and RCM are physiologically coupled life-history traits. It is interesting that the functional limits on egg size are ameliorated to a certain extent in lizards because as eggs become large, either by experimental manipulation within a species or through evolutionary change among species, egg length increases dramatically relative to the egg width (Vitt 1981; Werner 1989; Sinervo and Licht 1991b).

Phylogenetic Constraints and the Evolution of an Invariant Clutch Size

Given these constraints against large eggs, it is puzzling that *Anolis* lizards lay only one-egged clutches (Smith et al. 1973; Andrews and Rand 1974; Vitt and Congdon 1978; Shine and Greer 1991; Sinervo and Licht 1991). As *Anolis* lizards evolved a fixed, one-egged strategy from the life-history strategy of large clutches more typical of iguanian lizards, and presumably the primitive condition, given that related outgroups are multi-egged (Dunham and Miles 1985; Vitt and Seigel 1985; Etheridge and de Queiroz 1988), relatively small changes would have been required in the mechanisms controlling reproduction. For example, as clutch size regulation evolved, regulation of egg size and total clutch mass would also have evolved in a concordant fashion owing to the physiological and hormonal coupling between these traits.

Phylogenetic conservatism and the evolution of an invariant clutch size in both *Anolis* and Gekkonid lizards have led several authors (e.g., Andrews 1974; Shine and Greer 1991; Vitt and Congdon 1978; Vitt 1986) to suggest three nonmutually exclusive explanations for the origin and maintenance of invariant clutch size. Stringent natural selection maintains single-egged clutches because (1) reduced clutch mass ameliorates costs of reproduction or (2) intense offspring survival selection favors large eggs. Selection is also operating in extant populations to maintain invariant clutch size. (3) Alternatively, such selection occurred in a distant ancestor to these groups, which also subsequently lost the capacity to produce clutches greater than one or two eggs. Subsequent ancestors and extant representatives maintain the pattern of invariant clutch size owing to a phylogenetic constraint.

There is evidence providing partial support for the notion of a mechanistic, phylogenetically based constraint in *Anolis* that prohibits the evolution of multi-egged clutches. *Anolis carolinensis* apparently is not responsive to FSH in the same way as *Uta*. When *Uta* is administered exogenous FSH, the ovary recruits additional follicles. In contrast, a wide range of doses of exogenous FSH (10–100 μl) does not alter the normal ovarian pattern in *A. carolinesis*, and there is at most one large follicle in each ovary at any one point in time (Licht 1970; Jones et al. 1975, 1976). Moreover, the hierarchical pattern of follicle sizes typical of *Anolis* is not disrupted. Thus, the ovary of anoles may no longer be responsive to FSH in the same fashion as the ovary of *Uta*, and this may reflect a potential mechanistic constraint alluded to by the phylogenetic conservatism of anoles with regards to the evolution of an invariant clutch size (Smith et al. 1973; Andrews and Rand 1974; Vitt and Congdon 1978; Vitt 1986; Shine and Greer 1991; Sinervo and Licht 1991b). Anoles apparently do not have genetic variation for responding to FSH levels in the same way as a typical sceloporine lizard. The FSH regulatory pathway in anoles may have been co-opted for another purpose. For example, the rate of recruitment of follicles into the ovarian hierarchy of follicle size is apparently controlled by FSH in *Anolis* (Jones et al. 1975, 1976).

A complete answer to this question will ultimately entail measurements of natural selection and further analysis of the mechanisms of hormonal regulation. Without direct evidence concerning the selective pressures, the possibility that natural selection is operating on extant populations to (1) reduce clutch mass and thus ameliorate costs of reproduction or (2) favor large offspring size cannot be ruled out. For example, if such experiments were to show chronic and strong selection favoring large offspring, but no response to selection, this would constitute evidence of functional constraints that limit the adaptive evolution of egg size in anoles. The techniques I describe for manipulating egg size provide an experimental avenue to measure natural selection on these traits. By experimentally manipulating offspring size and releasing these neonates into nature to index their subsequent survival and reproduction, one can measure the fitness optima for egg size of a single-egged species like *Anolis* (Sinervo 1990a; Sinervo and Huey 1990; Sinervo and Licht 1991a,b).

Discussion

Life-history analysis in lizard ecology

Previous approaches to allocation strategies in lizard ecology have described life-history variation in terms of purely phenotypic traits (compiled in Dunham and Miles 1985; Vitt and Seigel 1985). Comparison of life-history traits within and among populations or species have demonstrated substantial variation within life-history tactics of lizards. More recently, focus has been applied to proximate sources of variation in life-history traits (e.g., growth, time and size to maturity, allocation to reproduction) arising from environmental differences among populations (e.g., food availability, Ballinger 1977; thermal environment, Sinervo and Adolph 1989; Sinervo 1990b; incubation environment, Packard and Packard 1988; Overall this volume). Such proximate effects on life-history variation can be measured in field (Ballinger 1979; Niewiarowski and Roosenburg 1993; Niewiarowski this volume) or laboratory experiments (Sinervo and Adolph 1989; Sinervo 1990a). In addition, pioneering studies of natural selection on life-history traits (Fox 1975, 1978; Ferguson and Fox 1984) have added an important ecological context to analysis of life-history variation.

The techniques of allometric engineering described in this paper are a relatively new approach in which proximate mechanisms are used in conjunction with studies of natural selection to understand the selective basis of life-history variation in nature (Sinervo 1993). Techniques of allometric engineering also provide insights into the mechanistic causes underlying life-history trade-offs. Although I have restricted my current discussion to manipulation of reproductive allocation, in principle it is possible to apply such mechanistic hormonal analyses to a wide variety of life-history phenomena (e.g., growth, maturation).

The techniques of allometric engineering that I have described fall into two categories. Manipulation of hatchling size achieved by removing yolk from freshly laid eggs is a "single-trait" manipulation that has cascading effects on fitness (survival) or performance traits related to fitness (e.g., stamina, sprint speed; Sinervo 1990a; Sinervo and Huey 1990; Sinervo et al. 1992). Alternatively, manipulating a single physiological mechanism that has simultaneous effects on two or more traits involved in a life-history trade-off is a "multi-trait" manipulation (Sinervo and Licht 1991a,b). For example, manipulation of clutch size also has direct effects on egg size, and this coupling forms the basis of the clutch size and egg size trade-off.

Single-trait manipulation, e.g., offspring size manipulation, can be used to assess certain consequences of the trade-off (in this case survival selection). However, in the absence of the multi-trait manipulations, single-trait manipulations cannot be used to comprehensively assess life-history trade-offs (Reznick 1992). Multi-trait manipulations directly assess the physiological basis of the trade-off. Multi-trait manipulations presumably maintain the essence of any genetically based correlations that might be expected to arise from pleiotropy. For example, genetic differences in clutch size and egg size among females in a population are presumably attributable to either the amount of FSH produced by the pituitary, or perhaps the tissue sensitivity of the ovary to FSH. Given the common hormonal pathway for clutch size regulation and egg size regulation, a fairly strong genetic correlation between the traits would be expected owing to pleiotropy. However, knowledge of the quantitative genetic basis of reproductive traits in lizards is limited to a handful of measurements. Future studies should focus on the relationship between quantitative genetic and mechanistic analyses of life-history trade-offs (Sinervo 1993).

Combining multi- and single-trait manipulations could provide a synthetic analysis of life-history trade-offs. Analyses to date have largely focused on the effects of offspring size on offspring survival. I have presented preliminary data that suggest selection on offspring size also acts through survival of the female parent. This reflects a cost to reproduction associated with maternal investment per offspring. Presumably, a large total investment (e.g., per clutch) by a female may result in additional ecologically mediated costs in natural populations (Schwarzkopf this volume). Such ecologically mediated costs of reproduction in lizards might arise from decreases in sprint performance due to the burdening effect of a clutch of eggs that may increase predation risk (Shine 1980; Bauwens and Thoen 1981; Sinervo et al. 1991), or physiological costs arising from increased foraging effort. The techniques of ovarian manipulation that I have described would permit an experimental analysis of costs of reproduction. Such an analysis should entail multi-trait manipulations involving the clutch size, egg size, and total clutch mass character complex that is presumably related to physiologically and ecologically mediated costs of reproduction.

Manipulations of the mechanisms of reproduction may also be useful in comparative analyses of life-history variation of higher taxonomic levels.

Recent comparative analyses of life-history variation among lizard taxa have documented considerable variation, or the lack of variation (reflected as phylogenetic conservatism), that is attributable to ordinal, familial, and generic taxonomic levels (Dunham and Miles 1985; Dunham et al. 1988; Vitt and Seigel 1985). For example, the invariant clutch size of anoles relative to sceloporine lizards may be due to stringent extant natural selection for such reproductive traits, or perhaps to an underlying pleisiomorphic mechanistic constraint that is shared by all extant representatives. Mechanistic comparisons of representative anoles and sceloporine lizards indicate that there are perhaps fundamental differences in ovarian regulation (responsiveness of the ovary to gonadotropin). While such comparisons lend some support for the notion that the phylogenetic conservatism of anoles may have an underlying mechanistic basis, further verification must also come from an analysis of the selective basis of such life-history variation in natural populations.

PART II
BEHAVIORAL ECOLOGY

The aim of this introductory essay is to search for connections among each of the four chapters on behavioral ecology that follow (you are encouraged to read them yourself) and with behavioral ecology in general.

Behavioral ecology is a relatively new name for a field of long-standing interest to students of lizards and one to which lizard studies have contributed both examples and ideas. "The overall theme [of behavioral ecology] is still that of how behaviour is influenced by natural selection in relation to ecological conditions" (in Krebs and Davies 1978, p. ix). Though there are students of lizards doing behavioral ecology, relatively fewer people think of themselves as behavioral ecologists working with lizards. For example, in the last International Behavioral Ecology Congress (1991) only 6 of the 554 oral and poster presentations dealt with lizards. (Snakes and crocodilians were totally unrepresented, and the only other reptile paper was on turtles.) Behavioral ecology is central to the study of lizards but lizard studies are not, at present, central to behavioral ecology.

The sample of four papers included here is too small to cover all of the currently fashionable topics in behavioral ecology, but many are represented. Among the themes common to two of the chapters in this section, and several others in this volume, is the use of phylogenetic analysis in investigating the evolution of traits. The two chapters here have very different emphases; Martins concentrates on the techniques and how they are best used, while Cooper emphasizes the adaptive significance of the phylogenetic concordance that he describes.

Lizard behavioral ecology has made a series of contributions to the study of territoriality such as Judy Stamps' paper in the 1983 Lizard Ecology volume. Martins continues this tradition with her illustration of the uses of phylogenetic comparative techniques with an example for lizard behavioral ecology. Some of her results on the distribution of kinds of territoriality within lizard families are particularly exciting in relation to those of Cooper.

Cooper describes the phylogenetic distribution of chemical prey detection and foraging mode among lizard families. He shows that both chemical prey detection and foraging mode are evolutionarily stable and strongly correlated. Comparison of the phylogenetic distribution of types of foraging that Cooper describes with the distribution of the amount of area defended that Martins presents shows a striking concordance: sit-and-wait foragers defend most of their generally small home ranges while active foragers defend only specific areas within larger home ranges or defend only their immediate vicinity.

Perhaps it is not too surprising that sit-and-wait foragers primarily use vision to detect larger mobile prey while active foragers and herbivores largely use chemical cues to detect immobile and highly clumped prey, but I was surprised by the evidence that when a primitively active foraging lineage, such as the geckos, evolved sit-and-wait foraging, chemical prey detection was lost. The implications of changes in prey-detection modality for social signaling would be worth exploring.

Another question that Cooper's paper raised for me is, "Why are lizards so different from birds, many of whom, particularly among the passerine birds, forage actively for immobile prey which they detect visually?" An ornithologist colleague suggests that foraging in birds involves a great deal of learning of where to look for food and what to recognize as prey and that lizards are too stupid to do this. Rising above my volcanic reaction to his ornithocentricity, it does seem possible that prey location and detection in lizards depend much more heavily on hard-wired cues than it does in birds.

Craig Guyer presents a very nice meld of observational and experimental techniques on what limits local populations of *Anolis* [*Norops**] *humilis*. When he supplemented food, populations of both sexes increased; females because they survived better, and males because they immigrated from elsewhere. Correlative data suggested to him that males were responding to numbers of females in addition to, or instead of, food. Direct experiments increasing numbers of females in an area increased the number of males present. This increase in density due to immigration was presumably local and on a slightly larger scale would represent a change in distribution rather than a change in population size.

One of the questions that Guyer's chapter raises is of the role of territory in his anoles. I would have expected aggressive interactions between males to be one of the important factors limiting local male density. Is *A. humilis*, because of its terrestrial habits, less effective at defending a territory than other anoles? Were the migrating males floaters who were drifting through the area anyway and who only stayed when more females were available?

An example of the value of long-term studies is provided by Bull's report on the population dynamics and social organization in the sleepy lizard, a giant Australian skink that is long lived and produces very small litters of large young. His study reveals an unexpected social organization based on nearly permanent monogamy.

These lizards clearly recognize their mates as individuals. Long known in birds and mammals, individual recognition has been repeatedly suggested for lizards on the basis of anecdotal observations. Sleepy lizards join the species where convincing evidence is accumulating showing that individual recognition is quite widespread in lizards.

*Editors' note: Nomenclature of anolines remains controversial; *Anolis* was split by Guyer and Savage (1986), reconsidered by Cannatella and de Queiroz (1989) and Williams (1989), and further reconsidered by Guyer and Savage (1992).

The sleepy lizards offer an opportunity to use a popular technique that, as yet, has been little used in lizard studies. DNA fingerprinting can identify kinship relationships—particularly paternity. Ideas about bird mating systems are being completely rethought as DNA fingerprinting studies reveal that, in many apparently monogamous pairs, both males and females are achieving fertilizations outside of the pair. If extra-pair fertilizations occur in Bull's monogamous sleepy lizards they might explain the observed close association of the pair during much, but not all, of the breeding season. The possibility that this is mutual mate guarding could more easily be evaluated if things like frequency of extra-pair fertilizations (EPFs), sperm storage, and sperm priority could be determined. Possibly the females that Bull "divorced" by removing their mates did not pair again because available males were already paired and could not divide their time between two females. Nevertheless males might visit and mate with the divorcees without establishing pair bonds. DNA fingerprinting might help sort out these possibilities. I suspect that there are currently a number of graduate students scattered around the world snipping bits out of lizards in order to use DNA fingerprinting to establish their relationships.

As we have seen, new techniques of phylogenetic analysis are encouraging studies comparing lizards at very different levels: populations, species, genera, and even families. These comparative studies are only possible using data from detailed life-history studies of specific species. And, as the chapters here stress, comparisons are often hampered by a paucity of data. One reason is that the data they need often can come only from long-term studies, particularly those that combine experimental and observational approaches.

Though natural history studies may be difficult and arduous, for many of us they are fun to do: one works with live animals in natural environments, students can usefully participate, ingenious experiments must be designed, visits to interesting places are involved, and, when modern laboratory analyses are required, one can often find somebody else to do the lab work. Further, even if detailed natural history studies may not be publishable easily in the high-profile journals like *Nature* and *Science*, they have a much longer useful life and may continue to be cited long after the theories that prompted their initiation have been forgotten.

I hope that the traditions of collecting basic life-history data on lizards and interpreting them in larger contexts illustrated by the four chapters in this section continue to thrive.

A. Stanley Rand

CHAPTER 5
PREY CHEMICAL DISCRIMINATION, FORAGING MODE, AND PHYLOGENY

William E. Cooper, Jr.

Historical and adaptive components of the current expression of behavioral and morphological traits are often difficult to separate, but recognition of their contributions, entailing consideration of the selective factors leading to adaptation, can provide insights into the distribution of characteristics in a clade and suggest reasons for the evolution of related traits. This is especially likely for interrelated suites of traits serving shared function. In lizards, detection and identification of prey chemicals on environmental substrates typically involves lingual sampling. The presence or absence of this chemosensory behavior is strongly affected by foraging mode. Evidence of an adaptive relationship between foraging mode and identification of prey mediated by tongue-flicking could thus contribute not only to our understanding of the taxonomic distribution of the chemosensory behavior and the selective factors that determine it, but also could suggest a partial explanation for the distribution of a suite of related morphological features of the tongue and vomeronasal system. Because lingual morphology has long been important to the definition of higher squamate taxa (e.g., Camp 1923), the relationships among foraging mode, importance of tongue-flicking to detection of prey, and morphological features of the tongue and vomeronasal system may well offer a key to understanding some major aspects of the evolutionary diversification of lizards.

Use of the tongue as a chemosensory sampling device for detection of prey and putative relationships between such tongue-flicking and the phylogeny and foraging mode of squamates have been the subject of many recent investigations (partially reviewed by Cooper 1990a; Halpern 1992). The purely phylogenetic aspect of this interest is based largely on the traditional and current importance of chemosensory structures, including the tongue, in defining major taxonomic groups of lizards (e.g., Camp 1923; Estes et al. 1988; Schwenk 1988). Chemosensory tongue-flicking is also interesting ecologically because there is a strong relationship between foraging mode and ability to identify prey chemicals by tongue-flicking in lizard families: species belonging to families of active foragers detect prey by tongue-flicking, whereas species belonging to families of ambush foragers do not (Cooper 1989a; 1990b).

In an important paper on ecological consequences of foraging mode, Huey and Pianka (1981), after Regal (1978) and Evans (1961), postulated a

correlation between use of chemoreception to locate prey and wide foraging. The strong association between type of foraging mode and presence or
absence of prey chemical discrimination is consistent with the hypothesis
that foraging ecology influences chemosensory behavior. It has not been
clear, however, to what extent this relationship might be the consequence of
phylogenetic inertia rather than adaptation or even current selective benefit.
Emphasizing this uncertainty, Schwenk (1993) has recently cautioned
against interpreting prey chemical discrimination by tongue-flicking as an
adaptation to the active foraging mode because variation in chemosensory
structures and behaviors has a strong historical component.

Here, I present preliminary evidence from ongoing investigations that
suggests phylogenetic influences on both foraging behavior and prey chemical discrimination and the adaptive adjustment of the chemosensory behavior to foraging mode. Evidence is adduced regarding the stability of
presence or absence of prey chemical discrimination and of the qualitative
type of foraging mode within families of lizards. Separate influences of both
phylogeny and foraging ecology on prey chemical discrimination behavior
are discussed. Evolutionary interrelationships among foraging mode, prey
chemical discrimination, more specialized chemosensory behaviors, and
structural specializations of the tongue and vomeronasal organ to enhance
lingual sampling of chemicals are considered speculatively.

Stability of Foraging Mode and Prey Chemical Discrimination

The type of foraging mode and presence or absence of prey chemical discrimination not only correspond in most lizard families (Cooper 1989a,
1990b), but appear to be stable over an impressive range of lizard taxa. Evidence regarding the degree of stability is discussed in this section using the
taxonomic categories of Estes et al. (1988), modified to incorporate the findings of Frost and Etheridge (1989) for iguanian families, Kluge (1987) for
geckos, and Lang (1991) for gerrhosaurids and cordylids.

Foraging mode: Categories, distribution, and stability

Lacertilian foraging modes are commonly characterized as belonging to one
of two types, ambush (or sit-and-wait) foraging versus active (or wide) foraging (e.g., Pianka 1966; Huey and Pianka 1981; Vitt and Price 1982). Lizard
ecologists have long recognized that variability occurs in each of the categories and have sometimes used the categories in an explicitly relative sense
(e.g., Huey and Pianka 1981). Variability within categories has been
approached in two ways. Some investigators have suggested that ambush
and wide foraging represent opposite extremes of a continuum of movement

patterns (e.g., Magnusson et al. 1985); others have erected intermediate categories such as cruise foraging (Regal 1978), in which search is conducted during relatively slow locomotion.

Although the division of constellations of foraging behaviors into named modes is often taken to imply qualitative differences, there is quantitative variation between modes in average speed of movement during an interval, average speed while actually moving, and percent of time moving while foraging (Magnusson et al. 1985; Pietruszka 1986). Beyond these quantitative differences are qualitative ones considered in community studies (e.g., Pianka 1986 and references therein) related to diet, means of searching for food (e.g., digging, burrowing), diel activity pattern, and microhabitat (e.g., arboreal, terrestrial). Despite this variation, the accumulated quantitative data on frequency of movement from several studies on lizards reveals a bimodal distribution of movement patterns corresponding to the two polar foraging modes (e.g., McLaughlin 1989).

Three foraging modes are recognized in this study: active, ambush, and herbivorous. Active foraging consists of search for prey during locomotion and/or requiring frequent locomotion and other active search such as digging. This differs from wide or intensive foraging (Regal 1978, 1983; Anderson 1993) in that the distance traveled and speed need not be great. Active foraging thus encompasses cruise foraging, wide foraging, and intensive foraging. Ambush, or sit-and-wait foraging, is defined in the traditional sense of low frequency of movement and high percentage of capture attempts initiated from a standstill.

The third category is herbivorous foraging. This necessarily differs from the ambush foraging of insectivores and other carnivores because the lizards must search for plant food rather than wait for it to approach. Although a dietary definition of foraging mode may be less than satisfactory, a shift to herbivory implies a major reorganization in foraging behaviors. I have chosen to place herbivores in a separate foraging category as a better alternative to categorizing them as active foragers in the absence of quantitative studies on their search patterns. Another difficulty with this category is that many lizards have mixed diets containing a variable percentage of plant material. In this paper chemosensory data are included for only one primarily herbivorous species, *Dipsosaurus dorsalis*.

An incomplete but fairly extensive review of the literature was conducted to determine the distribution of foraging mode within and between families of lizards. The results summarized in Table 5.1 are based on 257 species and 65 references. Within Iguania the vast majority of species and higher taxa are largely carnivorous ambush foragers. This is true of all species in five of the eight iguanian families and in 16 of 17 species from a sixth family. The only exceptions in Table 5.1 are the single large herbivorous

Table 5.1. The taxonomic distribution of foraging mode in lizard families. In families believed to have more than one foraging mode, genera having both modes are counted twice, but species are counted only once (See text for details). Foraging modes are A = active, H = herbivorous, M = mixed, SW = sit-and-wait (ambush), and ? = poorly known.

Taxa	Foraging mode	# Genera	# Species
Iguania			
Chamaeleonidae	SW	7	16
	H	1	1
Corytophanidae	SW	2	2
Crotaphytidae	SW	2	2
Iguanidae	H	6	9
Phrynosomatidae	SW	8	29
Polychridae	SW	2	9
Tropiduridae	SW	3	4
Scleroglossa			
Anguidae	A	3	4
Cordylidae	SW	1	1
	H	1	1
Eublepharidae	A	1	2
Gekkonidae	SW	18	52
	A–M	6	8
Gerrhosauridae	?	–	–
Helodermatidae	A	2	2
Lacertidae	A	13	63
	SW	3	7
Pygopodidae	SW	1	1
Scincidae	A	8	15
	M	1	1
Teiidae	A	3	16
Varanidae	A	1	12
Xantusiidae	?	2	2

chamaeleonid species *Uromastyx microlepis* (Arnold 1984) and the herbivorous family Iguanidae. Herbivory and the necessary adjustment of foraging mode are clearly autapomorphic in *Uromastyx* with respect to other agamine chamaeleonids and will not be considered further here. The predominant strategy in scleroglossan lizards is active foraging (Table 5.1). Ambush foraging is the sole foraging mode in the small gekkonoid family Pygopodidae and is the predominant mode in Gekkonidae and Cordylidae, but active foraging is the sole or primary mode in the other seven families for which foraging mode could be determined. Assignment of foraging modes was usually clear cut, but other cases require further discussion.

Active foraging appears to be the sole mode in the families Helodermatidae, Teiidae, and Varanidae. In addition, all species surveyed in the Scincidae forage actively, although one species, *Mabuya heathi*, also uses ambush (Vitt and Price 1982; Vitt 1990). Anguid lizards are also considered to be active foragers here. Vitt and Price (1982) characterized one species as a sit-and-wait forager due to its much slower and less extensive movements than lizards such as teiids. However, descriptions of the foraging behavior by the same species and a congener (Fitch 1935) and by members of two other genera (Fitch 1989; Karges and Wright 1987) indicate active search for food. Eublepharids are also designated active foragers. Eublepharids move slowly while foraging, which led Vitt and Price (1982) and Vitt and Congdon (1978) to describe *Coleonyx variegatus* as a sit-and-wait strategist with limited foraging. However, others have noted that this and one other species forage actively (Kingsbury 1989; Shenbrot et al. 1991) despite their slowness in comparison with other active foragers such as skinks and teiids.

The greatest intrafamilial variation in foraging behavior occurs within Gekkonidae and Lacertidae. Most gekkonid species are ambush foragers, but a small number (3) have been described as active foragers, and several others have been stated to show mixed foraging behavior with some active foraging added to a basic ambush strategy. Gekkonids are thus predominantly ambush foragers, but a few species may in certain circumstances forage actively as well. Quantitative data are needed to permit evaluation of the degree of movements of these species, especially those stated to forage actively (Henle 1990, 1991; Shenbrot et al. 1991), and other geckos. One of the three species stated to sometimes forage actively has also been called an ambush forager (Vitt and Price 1982). If the three species are truly active foragers, the condition is presumably derived within Gekkonidae and shows intrageneric variation (Henle 1990, 1991).

In Lacertidae the large majority of genera and species are active foragers. However, a few species have been described as sit-and-wait foragers. Quantitative data have been published for some of these. In the Kalahari desert the proportion of time spent moving was much lower for *Pedioplanis lineoocellata* and *Meroles suborbitalis* (0.143 and 0.135, respectively) than for four other lacertid species (0.502–0.574; Huey and Pianka 1981). However, the number of movements per minute for these species was an order of magnitude greater than that for eight species of ambush foragers belonging to other families (Moermond 1979b; Magnusson et al. 1985). The distance moved per minute was greater than for eleven of twelve species of ambush foragers (Moermond 1979b; Magnusson et al. 1985; Pietruszka 1986); the other species, *Gambelia wislizenii*, is a considerably larger lizard.

Therefore, although *Pedioplanis lineoocellata* and *M. suborbitalis* are relatively much less active foragers than the other lacertid species studied by Huey and Pianka (1981), I classify them here as active foragers for the purpose of interfamilial comparison.

In the only other paper providing quantitative data on lacertid movements, Perry et al. (1990) reported that four Israeli species spent 0.288–0.305 of the time in motion whereas in a fifth species, *Acanthodactylus scutellatus*, the proportion of the time moving was only 0.077. However, the number of movements per minute was six or more times greater than for the eight species of ambush foragers studied by Moermond (1979b) and Magnusson et al. (1985). Another lacertid species (*A. pardalis*) was classified as an ambush forager by Vernet et al. (1988) despite their statement that this species is very active.

In light of these considerations, it seems likely that although there is great variation in foraging strategy within Lacertidae, a great majority of species, perhaps even all, are active foragers in comparison with members of other families classified as ambush foragers. Quantitative study of foraging behavior in the genus *Pedioplanis*, which varies intragenerically, and comparison with that of sympatric ambush foragers would be illuminating. If some lacertids are true ambush foragers, ambush foraging would appear to have been derived from active foraging within the family. For the remainder of this paper, lacertids are considered active foragers.

Less information is available for the remaining families. Cordylid lizards would appear to be primarily ambush foragers (Branch 1988), but quantitative studies are lacking. Foraging modes could not be assigned for Gerrhosauridae or Xantusiidae due to absence of data for Gerrhosauridae and inconclusive descriptions for xantusiids. Two xantusiid species were called sit-and-wait foragers by Vitt and Price (1982), but one of them, *Klauberina riversiana* was described by Fellers and Drost (1991) as slowly walking, which would have excluded them from the widely foraging category used by Vitt and Price (1982). As noted by Vitt and Congdon (1978), these animals occupy very restricted habitats. Some occur primarily inside rotting Joshua trees or under rocks, places where it is very difficult to observe foraging behavior. Because no direct observations of foraging in such places have been published, I consider the foraging mode of xantusiids undetermined. In summary, the data on foraging mode in lizards indicate great stability of foraging mode within genera and families. Even in the higher taxonomic categories Iguania and Scleroglossa there have been few changes. Although foraging mode is often a fixed familial trait (e.g., Vitt and Price 1982; McLaughlin 1989), variation in degree of activity may be substantial. More examples of intrafamilial variation in foraging mode will undoubtedly be discovered, but the striking pattern of intrafamilial stability will remain as a major feature of the distribution of basic foraging modes.

Prey chemical discrimination: Taxonomic distribution and stability

Prey chemical discrimination mediated by tongue-flicking has been studied in only a tiny fraction of lizard species, but results have been remarkably consistent in indicating that presence or absence of this ability, or at least the ability to detect food by tongue-flicking is fixed within many higher lizard taxa. Among both iguanian and scleroglossan lizards this is true for all families studied (Table 5.2). Most data in Table 5.2 are based on experimental data, observational data being included only when there is direct reference to the absence of tongue-flicking to locate prey. References to high and low tongue-flicking rates per se, while not included, very likely correspond to presence and absence of the use of the tongue in foraging. If so, stability at the familial level noted in Table 5.2 would be based on far more species and genera. For example, my personal observations of phrynosomatids show that *Holbrookia propinqua, Uta stansburiana, Cophosaurus texanus*, and several species of *Sceloporus*, among others, have low tongue-flick rates and do not use the tongue to locate or identify food prior to capture. The following discussion is based largely on the species in Table 5.2.

Among iguanian families the typical condition is absence of lingual prey chemical discrimination. In Chamaeleonidae, true chameleons have tongues specialized for projectile prey prehension (Bellairs 1970) and do not use the tongue to chemically investigate prey, which are captured from a distance. In addition, representatives of two genera of agamines do not detect prey lingually. Among phrynosomatids, formal studies have been limited to several species of the large genus *Sceloporus*, but, as noted above, field observations reveal similar behavior in members of other genera. Polychrids and tropidurids in more than one genus each fit the broad iguanian pattern.

The only iguanian known to employ the tongue in detecting food chemically is the single iguanid species, *Dipsosaurus dorsalis*, in which vomerolfaction (Cooper and Burghardt 1990) is employed to discriminate food chemical cues from control stimuli (Cooper and Alberts 1991). *Iguana iguana* also appears to use the tongue more extensively than do iguanians studied in other families (data in Burghardt et al. 1986), suggesting that prey chemical discrimination may be widespread in Iguanidae. Testing of additional species of iguanids is needed. Scleroglossans typically detect prey by chemical cues obtained by tongue-flicking. This has been demonstrated in few species per family (Table 5.2), but in most cases the species represent diverse groups within families. This diversity applies in Eublepharidae, Lacertidae, Teiidae, and Anguidae. As Helodermatidae and Varanidae are monogeneric families and the former contains only two species, these families are well represented. Gekkonidae, represented by two genera, and Scincidae, represented by two genera and two subfamilies, are quite large families that have been less adequately sampled. Gerrhosauridae is

Table 5.2 The taxonomic distribution of lizard species shown to detect and/or identify prey chemicals. + = present; − = absent; E = experimental; O = observational.

Taxa	Chemical detection	Type of data	Sources
Iguania			
Chamaeleonidae			
Agama agama	−	E	Cooper in press a
Calotes mystaceus	−	E	Cooper 1989a
Chamaeleo sp.	−	O	Bellairs 1970
Iguanidae			
Dipsosaurus dorsalis	+	E	Krekorian 1989; Cooper & Alberts 1990
Phrynosomatidae			
Sceloporus clarkii	−	E	Cooper in press a
S. jarrovi	−	O	Simon et al. 1981
S. malachiticus	−	E	Cooper 1989a
S. poinsettii	−	E	Cooper in press a
S. undulatus	−	E	Noble & Kumpf 1936
Polychridae			
Anolis carolinensis	−	E	Cooper 1989a
A. lineatopus	−	E	Curio & Mobius 1978
Polychrus acutirostris	−	O	Vitt & Lacher 1981
Tropiduridae			
Leiocephalus inaguae	−	E	Noble & Kumpf 1936
Liolaemus zapallarensis	−	E	Cooper unpubl.
Tropidurus torquatus	−	O	Swain 1977
Scleroglossa			
Eublepharidae			
Coleonyx brevis	+	E	Dial 1978
C. variegatus	+	E	Dial et al. 1989
Eublepharis macularius	+	E	Cooper in press b
Gekkonidae			
Gekko gecko	−	E	Cooper in press b
Thecadactylus rapicauda	−	E	Cooper in press b
Scincidae			
Eumeces fasciatus	+	E	Burghardt 1973; Von Achen & Rakestraw 1984
E. inexpectatus	+	E	Loop & Scoville 1972
E. laticeps	+	E	Cooper & Vitt 1989
Scincella lateralis	+	E	Nicoletto 1985

Table 5.2 (cont.)

Taxa	Chemical detection	Type of data	Sources
Cordylidae	–	E	Cooper & Van Wyk unpubl.
Gerrhosauridae			
Gerrhosaurus nigrolineatus	+	E	Cooper 1992
Lacertidae			
Acanthodactylus scutellatus	+	E	Kahmann 1939
Podarcis hispanica	+	E	Cooper 1990b
P. muralis	+	E	Cooper 1991a
Teiidae			
Ameiva exsul	+	E	Noble & Kumpf 1936
A. saurimanensis	+	E	Kahmann 1939
A. undulata	+	E	Cooper 1990b
Tupinambis nigropunctatus	+	E	Noble & Kumpf 1936
T. rufescens	+	E	Cooper 1990b
T. teguixin	+	E	Yanoskey et al. 1993
Anguidae			
Elgaria coeruleus	+	E	Cooper 1990c
E. multicarinata	+	E	Cooper 1990c
Ophisaurus apodus	+	E	Kahmann 1939
O. attenuatus	+	E	Von Achen & Rakestraw 1984
Helodermatidae			
Heloderma suspectum	+	E	Cooper 1989b
Varanidae			
Varanus bengalensis	+	E	Auffenberg 1984
V. exanthematicus	+	E	Cooper 1989b
V. komodoensis	+	O	Auffenberg 1981
Amphisbaenidae			
Blanus cinereus	+	E	Lopez & Salvador 1992

represented by a single species that detects prey chemicals by tongue-flicking. Gerrhosauridae has very recently been separated from Cordylidae (Lang 1991). Using Lang's terminology, two cordylid species have much lower tongue-flicking rates than do two species of gerrhosaurids (Bissinger and Simon 1979). Tongue-flicking rates of the cordylids were even lower than those of a phrynosomatid species (Bissinger and Simon 1979), suggesting a likely difference between cordylids and gerrhosaurids in use of the tongue to detect prey.

The only Scleroglossa known to lack lingually mediated chemical identification of prey is a single cordylid species, *Cordylus cordylus* (Cooper unpubl.), and two species belonging to different genera of Gekkonidae, *Gekko gecko* and *Thecadactylus rapicauda*, neither of which tongue-flicked prey chemicals at all in an experimental study (Cooper in press b). Additional gekkonids representing phylogenetically and ecologically diverse taxa should be tested for prey chemical discrimination.

Phylogeny of Foraging Mode

The distribution of the major foraging modes among some higher taxa of lizards is mapped in Fig. 5.1 on a phylogeny of higher lizard taxa based on Estes et al. (1988), modified by the subsequent refinement of iguanian relationships by Frost and Etheridge (1989), the gekkonoid phylogeny of Kluge (1987), and the recognition of Gerrhosauridae by Lang (1991). The two major squamate taxa in a recent classification (Estes et al. 1988) are Iguania, which consists primarily of ambush foragers, and Scleroglossa, which includes primarily actively foraging lizards plus snakes. Unresolved relationships among iguanian families other than Chamaeleonidae are indicated by a polytomy.

Virtually all iguanians studied except iguanids (sensu Frost and Etheridge 1989) are ambush foragers. In conjunction with the insectivorous diets of the large majority of iguanians in families other than Iguanidae, this indicates that the absence of sit-and-wait foraging in iguanids is derived, presumably as a direct result of the adoption of herbivory. Because iguanids are not constrained by the need to remain immobile to avoid detection by prey, they are freer to search for plant food, which appears to be done visually, and to assess it chemically by tongue-flicking (Cooper and Alberts 1990).

In scleroglossans the large majority of families consist of active foragers, as noted above. Only gekkonids and cordylids are known to be ambush foragers. As the lack of digital specialization for arboreality indicates that Eublepharidae is the most primitive family of geckos (Kluge 1987; Grismer 1988), ambush foraging by gekkonids most likely was derived from actively foraging ancestral geckos. Among the numerous autarchoglossan families, active foraging is symplesiomorphic with the exception of cordylid lizards and possibly some lacertids.

Hypotheses regarding characteristics of foraging behavior by primitive squamates can best be formulated by outgroup comparison with the tuatara (Gauthier et al. 1988). No detailed quantitative information on foraging movements has been published for *Sphenodon punctatus*, but the description of foraging by Walls (1981) indicates that the tuatara is an ambush for-

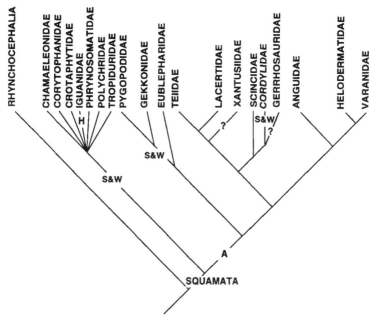

Figure 5.1. Cladogram of foraging modes in Lepidosauria. Some taxa for which no data are available are omitted. A = active, H = herbivorous, S&W = ambush (sit-and-wait).

ager. Although the interpretation is based on the single extant species of rhynchocephalian, the most likely foraging mode of basal squamates is ambush.

Foraging mode is evolutionarily quite stable among higher lizard taxa. Only a few shifts account for the observed distribution of foraging mode in all lizards. Under the hypothesis that primitive squamates were ambush foragers, foraging behavior was modified in Iguanidae, presumably as a response to the evolution of herbivory. Likewise, active search for prey has been retained in most major scleroglossan taxa, but has been abandoned for ambush foraging by cordylids and by most gekkonoid species other than eublepharids.

For the vast majority of species and the majority of families, then, the qualitatively defined broad foraging modes of lizards cannot be considered to be adaptive responses to current ecological conditions such as abundance, type, and size distributions of prey, or to current competition and predation. Such an interpretation is contradicted by their phylogenetic stability. Yet foraging mode may be a key factor in the organization of lizard life histories, one that may have had important historical effects in shaping them and that affects current fitness. However, the extreme conservatism of foraging mode greatly limits opportunities for testing the hypothesis that prey chemical discrimination is an adaptation to active foraging.

Phylogeny of Prey Chemical Discrimination

A major phylogenetic influence is evident in the apparent fixity of the presence or absence of chemosensory prey discrimination within all families of lizards for which data are available (Table 5.2). Detection of prey by chemosensory means would seem to be crucially important to successful foraging in many taxa. From a broad phylogenetic perspective, the ability to detect prey chemicals and to discriminate them from other substances is largely absent except in Scleroglossa, but occur in almost all major scleroglossan families and in most higher taxa of scleroglossan lizards (Fig. 5.1) and snakes (e.g., Burghardt 1970a; Chiszar and Scudder 1980; Chiszar et al. 1986a,b; Cooper 1990a, 1991b, 1992). Prey chemical discrimination thus appears to have arisen in the shared ancestor of scleroglossans, so that its presence is a synapomorphy among all scleroglossan groups studied other than Gekkonidae and Cordylidae, in which absence of prey chemical discrimination was derived secondarily (Fig. 5.1; Cooper in press b).

Because tongue-flicking is absent in *Sphenodon* (Schwenk 1986; Schwenk and Throckmorton 1989), the best outgroup for squamates, prey detection and prey chemical discrimination by tongue-flicking must be derived squamate characteristics. Although the phylogenetic relationships within the iguanian families formerly included in Iguanidae (Frost and Etheridge 1989) are uncertain, it is very likely that the primitive iguanian condition was absence of prey chemical discrimination. If so, its presence in Iguanidae (sensu Frost and Etheridge 1989) is autapomorphic and merely convergent with similar behavior in Scleroglossa (Fig. 5.1; Cooper in press b). Although iguanids are atypical iguanians in having large size and herbivorous diets, suggesting that they are likely to have evolved from smaller, insectivorous iguanians, it remains possible that prey chemical discrimination is plesiomorphic in iguanids and derived in other iguanians.

From the foregoing, it is apparent that the major determinant of the distribution among squamates of ability to discriminate prey by tongue-flicking is a single event early in squamate history. This corroborates Schwenk's (1993) caveat regarding ecological interpretations of squamate chemosensory behavior because much of its current distribution among species within families and even among families may be explained by phylogenetic inertia. This is not to say that adaptation has not occurred or that ecology has no current selective effects on chemosensory behavior; rather, the state of chemosensory behavior is plesiomorphic in many higher squamate taxa and therefore cannot be attributed to adaptation, which must be inferred from differences in chemosensory behavior and ecology among taxa.

Prey Chemical Discrimination and Adaptation

Foraging mode and prey chemical discrimination

What is the relationship between foraging mode and prey chemical discrimination? Is it adaptive or even functional? For the most part, scleroglossan lizards are active foragers that chemically discriminate prey, and iguanians are ambush foragers that do not. This correspondence between foraging mode and chemosensory behavior suggests that change in foraging mode induces selection leading to change in the character state of lingually mediated prey chemical discrimination. Viewed at a nominal level of measurement, the presence of prey chemical discrimination and active foraging appear to be fixed in most scleroglossan taxa at and above the familial level. However, ongoing selection on chemosensory detection of prey has produced a graded series of increasing specialization of lingual structure for chemical sampling in those lizard groups farthest from the iguanian-scleroglossan divergence, especially among the most active foragers, such as teiids and varanids (see McDowell 1972 regarding lingual structures). The importance of prey chemical discrimination and related chemosensory behaviors appears to vary accordingly (Auffenberg 1981; Cooper 1989b,c, 1990b, 1991a, 1993).

If the argument is restricted to the presence of the two behaviors (prey chemical detection and discrimination) and their phylogeny, the best evidence that use of chemosensory tongue-flicking to identify prey has arisen or been lost as an adaptation to foraging mode and is not merely a reflection of a single bifurcation early in the evolution of lizards would be demonstration of (1) shifts from the absence of chemosensory tongue-flicking in ancestral groups of ambush foragers to its presence in association with active foraging and (2) similar changes in the opposite direction. Even such cases of correlated change, however, are subject to undetected influences of other variables.

Support must be obtained, therefore, from cases of historical changes, especially those counter to the broad predictions from phylogenetic patterns. In this vein, prey chemical discrimination in Iguanidae appears to be one such adaptive shift in chemosensory behavior naturally selected by the adoption of a new type of foraging. The largely herbivorous iguanids are not active foragers in the sense that most scleroglossan lizards are, but they have clearly abandoned ambush foraging. It has been suggested that the iguanid *Dipsosaurus dorsalis* identifies plant food and may assess its nutritional qualities chemically (Cooper and Alberts 1990).

The initial shift in foraging mode at the branch point establishing iguanian and scleroglossan lizards involved another change supporting the
hypothesized adaptive relationship. At the origin of these groups, either
ambush foraging arose in iguanians with an associated loss of prey chemical
discrimination by tongue-flicking or active foraging arose in scleroglossans
with the concurrent advent of lingually mediated prey chemical discrimination (Cooper in press b). The absence of prey chemical discrimination by
tongue-flicking and the presence of ambush foraging in the tuatara strongly
suggest that prey chemical discrimination and active foraging both arose in
Scleroglossa.

Variation in prey chemical discrimination within Gekkonoidea provides a
third case supporting the hypothesis that the presence or absence of prey
chemical discrimination is adaptively coupled with foraging mode in lizards.
The presence of prey chemical discrimination involving tongue-flicking has
now been confirmed in both species of eublepharids studied (Dial 1990;
Cooper in press b) and prey chemical detection has been demonstrated in a
third species (Dial 1978). Thus, prey chemical discrimination and active foraging are presumably plesiomorphic in Gekkonoidea. The two gekkonid
species studied do not investigate potential prey lingually or tongue-flick at
high rates while foraging, suggesting that the absence of prey chemical discrimination may have been derived as an adaptive response to the adoption
of ambush foraging (Cooper in press a,b).

A final case of correlated change in foraging mode and lingually mediated prey chemical discrimination occurs in the scleroglossan family Cordylidae. Cordylid lizards have been described qualitatively as ambush foragers
(Branch 1988), which I have confirmed quantitatively for several species in
the Cape Province of South Africa (Cooper unpubl.). As their autarchoglossan ancestors were active foragers (Fig. 5.1), this represents a shift in foraging mode. The hypothesis that prey chemical discrimination is adaptively
adjusted to foraging mode predicts that prey chemical discrimination is
absent in Cordylidae. Recently collected data for a single cordylid species,
Cordylus cordylus, confirm this prediction (Cooper and Van Wyk unpubl.).

There are thus four known cases of linked changes in foraging mode and
prey chemical discrimination and no contradictory evidence, providing
strong preliminary support for the hypothesis that the chemosensory behavior evolves adaptively in tandem with foraging mode. I am undertaking an
explicitly phylogenetic analysis of these data to determine whether the evidence for adaptation is statistically significant. Additional data are being
gathered in an ongoing research program. Although a statistical demonstration of adaptation is thus far missing, the data at hand are quite convincing.
It is striking that no change in foraging mode has occurred without a shift in
prey chemical discrimination, and that the taxa showing changes of polarity

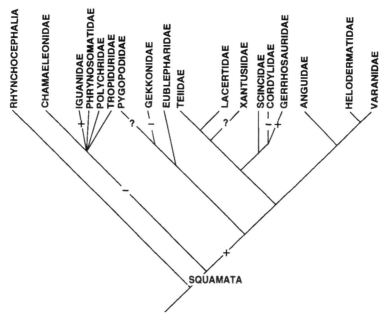

Figure 5.2. Cladogram of prey chemical discrimination in Lepidosauria. Some families of lizards for which no data are available are omitted

are dispersed widely through the cladograms (Figs. 5.1 and 5.2), emphasizing their independence of phylogeny.

Beyond confirming the broad phylogenetic relationship between foraging mode and prey chemical discrimination mediated by tongue-flicking, the present findings based on all exceptions from this prevailing phylogenetic association are novel in establishing an adaptive relationship between foraging mode and prey chemical discrimination. It is striking that the appearance of prey chemical discrimination followed or developed simultaneously with the advent of both active foraging and herbivory and that its loss ensued upon reversion to ambush foraging. The phylogenetic conservatism of foraging mode suggests that ambush foraging and active foraging may lie in rather widely separated adaptive zones that may be crossed only very rarely. Once a lizard taxon has passed from one adaptive zone into the other, however, natural selection would then favor not only evolution of a different foraging mode, but of morphological and behavioral features optimal for the new foraging mode.

Selection, foraging mode, and chemosensory tongue-flicking

Beyond its effects on prey chemical discrimination, foraging mode influences several important ecological and behavioral features of lizards, including defensive behavior and diet (Vitt and Congdon 1978; Huey and Pianka 1981; Huey and Bennett 1986). The relationships between foraging mode, defensive behavior, and diet, as well as the defenses of prey, would appear to strongly affect the options of lizards to detect prey by chemosensory tongue-flicking (Table 5.3). Active foragers, including most families of scleroglossan lizards, use rapid running as the primary means of escape in encounters with predators (Vitt and Price 1982) because their frequent movements render them readily detectable to predators. Tongue-flicking movements during foraging, being unobtrusive relative to locomotion, would be unlikely to increase the likelihood of detection by a predator. Even if they are occasion-ally the stimuli causing initial detection, the probability of capture by the predator is unlikely to increase by any appreciable amount because the fre-quent bodily movement of active foragers would itself attract the attention of predators shortly after any tongue-flicking. Avoidance of predation in active foragers depends primarily on escape, not avoidance of detection. In contrast, ambush foragers, including most iguanian families, rely on crypsis maintained by immobility to avoid being detected, running only when threatened by a predator's approach (Vitt and Price 1982). Tongue-flicking while waiting immobile for prey would disrupt crypticity, materially increas-ing the chances of detection and capture by a predator.

Foraging mode appears to affect diet in part because active foragers, searching over a wider area than otherwise comparable ambush foragers, are more likely to encounter patchily distributed sedentary prey (Huey and Pianka 1981). As such prey tend to be locally abundant small inverte-brates such as termites, active foragers are expected to consume prey that are of smaller average size than are those of ambush foragers of the same body size.

Both the patchy distribution of food and the diet of active foragers may place a premium on use of chemical cues for location of prey. In searching for food patches, actively foraging lizards could improve the likelihood of finding food by concentrating searches in areas where prey chemical cues are detected. Behavior consistent with such search has been reported in *Varanus bengalensis* (Auffenberg 1984). For ambush-foraging lizards, how-ever, immobility during foraging precludes use of chemical cues to continu-ously or frequently search widely for patches or to use area-concentrated search upon locating chemical prey cues. In contrast to ambush-foraging liz-ards, rattlesnakes do use chemical cues during wide searches for patches of sparsely distributed large prey, but then remain immobile for long periods while waiting in ambush (Duvall et al. 1990). The difference in behavior may relate to degree of patchiness of insects versus rodents and size of prey. Patchily distributed prey items such as termites are often invisible on the surface, but their presence may be detected chemically, allowing them to be

Table 5.3 Relationships of defense by lizards, defense by prey, and diet to foraging mode and hypothetical relationships to and effects on utility of prey chemical discrimination (PCD). TF = tongue-flicking

	Active foragers	Ambush foragers
Antipredatory defense		
Primary defense	rapid running	crypsis with immobility
Effect on PCD	TF does not increase detection by predators	TFing disrupts crypticity
Defense by prey		
Visual sensitivity to movement: effect on PCD	little effect because lizards move while foraging	TFing increases chances of being detected by prey, possibly of avoiding detection or escaping
Diet		
Prey distribution	patchy	dispersed
Effect on PCD	PCD helps locate patches, area-concentrated search	PCD not useful to find patches due to relative immobility
Prey size	smaller	larger
Effect on PCD	might favor PCD if small prey are harder to detect visually	can be detected visually, especially if moving in open
Prey location	hidden and exposed	exposed in open
Effect on PCD	PCD helps locate prey that cannot be detected visually	

dug out of their nests. Small prey size per se may also favor use of the chemical senses to some extent if small size decreases the likelihood of detection by vision alone, especially of prey species that tend to remain immobile. The chemical senses may be especially important for detecting any prey, whether patchily distributed or not, that is hidden from view. Actively searching lizards may locate by chemosensory means prey that usually remains hidden under objects in sand or soil (Vitt and Cooper 1986; Cooper and Vitt 1989).

Ambushers remain in one location for some time while searching for prey visually. They therefore tend to consume a higher proportion of food items that are widely dispersed rather than clumped, are active on the surface and therefore visually detectable, and may be of somewhat greater

average body size than those consumed by active foragers. As ambush foragers tongue-flick primarily immediately at the termination of locomotion
(Cooper et al. 1994), it is possible that ambush foragers might chemically
assess a foraging post by tongue-flicking immediately upon arrival, having
longer giving-up times at sites where the substrate bears prey chemical cues.
This possibility has not been studied.

The sensory capacities of prey and their ability to escape after being
detected are additional factors possibly having differential effects on active
and ambush foragers. For prey sensitive to motion, tongue-flicking might
reveal the presence of an ambush predator that would otherwise go undetected, whereas the presence of an active forager would be revealed by its
locomotion. This would affect the predator-prey relationship only to the
extent that the prey were able to avoid being detected or to enhance its ability to escape by being aware of the predator prior to being attacked.

Chemosensory structures, behavior, and higher squamate taxa

Despite its relative stability in Iguania and Scleroglossa, lingually mediated
prey chemical discrimination, along with related structures and behaviors,
would appear to have strongly influenced the evolution of higher squamate
taxa. Both foraging mode and attendant changes in chemoreception may
have contributed to the divergence of these basal groups from a common
ancestor. Their effects are more apparent within Scleroglossa, which is a
simple consequence of the absence of active foraging in Iguania and the
selection against tongue-flicking by ambush foragers while occupying an
ambush post. In actively foraging scleroglossans, natural selection favoring
prey chemical discrimination may be conceived as favoring both morphological and behavioral enhancements of detection of prey by vomerolfaction.
The distribution of these improvements appears to be closely related to
squamate macroevolution, particularly within Autarchoglossa.

Within Iguania, lingual structure is relatively uniform and generalized
(McDowell 1972), and the percentage of chemoreceptive cells in the vomeronasal system is lower than in scleroglossans. However, among the species
for which data are available (Gabe and Saint Girons 1976), it is highest in
Iguanidae, the family showing lingually mediated prey chemical discrimination and altered foraging mode. Much greater variation in structure and
behavior is found in Scleroglossa. Lingual structure remains generalized in
Gekkonoidea (McDowell 1972).

Although prey chemical discrimination appears in Eublepharidae, and
the percentage of vomeronasal chemoreceptor cells is somewhat greater
than in Iguania (Gabe and Saint Girons 1976), specialized lingual morphology and more advanced chemosensory behaviors are lacking. The greatest

diversification has taken place in Autarchoglossa. Lingual structure varies from the generalized condition in Scincidae, to the slight elongation with short tines in Lacertidae and Anguidae, to the more greatly elongated form with deep forking in Teiidae, Varanidae, and Serpentes (McDowell 1972). The percentage of vomeronasal chemoreceptors increases in rough correspondence with the degree of specialization in lingual morphology (Gabe and Saint Girons 1976). In addition to lingually mediated prey chemical discrimination, which has been the focus of this paper, a number of other chemosensory behaviors related to feeding have evolved in squamates, including scent-trailing of prey (Auffenberg 1981), strike-induced chemosensory searching (i.e., use of tongue-flicking to relocate prey that has been bitten, but dropped, escaped, or released), and a strike-release-trail strategy, the latter used only by highly venomous snakes for handling dangerous prey (Chiszar and Scudder 1980). The more advanced chemosensory behaviors thus far appear to be limited to those taxa showing the greatest lingual specialization and the highest percentages of vomeronasal chemoreceptors. I am collecting quantitative data on lingual morphology and chemosensory behaviors to establish the reality and tightness of these perceived relationships.

These considerations suggest that evolution of the lingual-vomeronasal system for chemosensory sampling and analysis and evolution of associated chemosensory behaviors are tightly linked to the evolutionary diversification of lizards. Examination of numerous morphological and behavioral traits pertaining to vomerolfaction, including data for numerous families on prey chemical discrimination, allows construction of a phylogenetic tree for lizard families that is quite similar to one based on those plus numerous other anatomical characters (Schwenk 1993). Despite the undoubtedly substantial correlation among the various traits related to vomeronasal chemoreception, Schwenk's (1993) findings demonstrate well the close relationship between chemosensory features and squamate evolution.

The adaptive relationship between lingually mediated prey chemical discrimination and active foraging suggests that active foraging provided the crucial selective impetus for the initial evolution of prey chemical discrimination. I have argued elsewhere that pheromonal communication was probably the first lingually mediated chemosensory function to evolve after the opening of the vomeronasal ducts into the mouth in basal squamates (Cooper in press c). Discriminative capacities of vomerolfaction combined with repeated exposure to prey chemicals may have preadapted lizards for prey chemical discrimination in a selective environment favorable to active foraging. It seems likely that the sensory capacity for at least rudimentary prey chemical discrimination enhanced the efficiency of early attempts at active foraging, which in turn induced selection for greater ability to identify prey chemicals. The close correspondence between higher squamate taxa on the one hand and lingual, vomeronasal, and behavioral aspects of chemoreception on the other strongly suggests that after prey chemical discrimination appeared, subsequent morphological and behavioral evolution of the lingual-vomeronasal system among active foragers became a major factor in evolution of new higher lizard taxa.

Although there is still some controversy, snakes appear to have evolved from varanoid, probably varanid lizards (McDowell 1972; Schwenk 1988), which have highly developed chemoreceptive abilities (Auffenberg 1981; Cooper 1989b,c, 1993), vomerolfactory systems (Gabe and Saint Girons 1976), and snakelike tongues (McDowell 1972). The most primitive snakes, then, presumably were equipped with highly sophisticated lingual-vomeronasal structures and behaviors useful for foraging. Further evolution of chemosensory behavior in snakes after divergence from lizards may have been at least partially decoupled from foraging mode because both active and ambush foragers employ the strike-release-trail strategy (Chiszar and Scudder 1980; O'Connell et al. 1985), because rattlesnakes, which in most circumstances are ambush foragers, use chemical cues to locate ambush sites (Duvall et al. 1990), and because the ambush-foraging ball python, *Python regius*, exhibits lingually mediated prey chemical discrimination (Cooper 1991b). One possible explanation for this decoupling might be that snakes or their predecessors had attained a performance level high enough to have crossed a threshold for utility in a wide variety of foraging circumstances. Another reason might be that the foraging behavior of snakes, which has not been studied as thoroughly as that of lizards, differs markedly enough from that of lizards to bear a different selective relationship to lingually mediated prey chemical discrimination.

Directions for Future Work

Further verification of the predicted relationship between foraging mode and prey chemical discrimination is desirable to give the hypothesis even firmer footing. For example, although both foraging mode and prey chemical discrimination appear to be for the most part fixed within families, data on foraging mode are primarily qualitative. Furthermore, prey chemical discrimination has been studied in only a few species of any one family. Further study of prey chemical discrimination in families having unknown foraging mode and of foraging mode in families for which the presence or absence of prey chemical discrimination is known are especially desirable. Studies of variation in foraging mode and its relationship to prey chemical discrimination within taxa such as Lacertidae, Gekkonidae, and Serpentes, for which foraging mode is not fixed, could lead to valuable insight into the speed of evolutionary adjustments of prey chemical discrimination to shifts in foraging mode.

The full range of complexity of foraging behaviors in lizards can be represented only incompletely by three categorical foraging modes. Beyond the quantitative variation mentioned above, some lizards are omnivorous and others may exhibit mixed foraging strategies. Although these exceptions have not been treated explicitly, they may be readily incorporated into the evolutionary considerations. For example, lizards having a mixed strategy including active foraging and ambush foraging at different times are presumably subjected alternately to the selective regimes characteristic of each mode. Their tongue-flicking behavior in the field may be predicted to conform to that selectively favored by the current foraging behavior. In the laboratory, lizards that have this mixed strategy and belong to families of predominantly active foragers may be expected to retain lingually mediated prey chemical discrimination. Study of taxa having various combinations of mixed strategy should prove illuminating.

For natural selection to effect adaptive changes in prey chemical discrimination as a consequence of changes in foraging behavior, there must be underlying genetic variation. Studies of garter snakes of the genus *Thamnophis* clearly indicate that such genetic variation exists and that microevolutionary changes corresponding to dietary shifts occur in chemosensory response to food chemicals. In *T. elegans* a geographic change in food preference is accompanied by a similar change in the magnitude of tongue-flicking response (Arnold 1981). Chemosensory behavior showed heritable variation in both inland and coastal populations. Intraspecific variation in tongue-flicking and attack responses to chemical stimuli from several prey species similarly showed geographic variation corresponding to prey availability in *T. sirtalis* (Burghardt 1970b).

Such studies show that current garter snake populations contain sufficient heritable variation in chemosensory responses. They further suggest that garter snakes have had sufficient genetic variation to allow microevolutionary shifts adapting populations to local prey availability. Although the changes between foraging modes and between presence and absence of prey chemical discrimination are likely to involve greater genetic changes, the genetic control of chemosensory and prey attack responses strongly suggests that they can be affected by natural selection.

My research on lizards has been focused thus far on broad characterization of both foraging mode and presence or absence of lingually mediated prey chemical discrimination. As even these broad features have not been studied in the vast majority of lizard species and in some entire families, exceptions to the simple pattern that I have detected will doubtless be discovered. Nevertheless, enough information is available to establish the importance of both phylogeny and foraging mode to prey chemical discrimination. Another important avenue for future study will be comparison of variation in finer aspects of chemosensory roles in foraging with intrafamilial variation in quantitative and qualitative aspects of foraging ecology. Such studies have not been attempted in lizards, but will be important in determining the degree to which prey chemical discrimination is adjusted to foraging mode.

The appearance of alternate foraging modes may have created sufficiently great differences in selective environments to bring about major evolutionary changes in lizards, especially in chemosensory behaviors and structures, but also in features such as defensive behavior and reproductive characteristics (Vitt and Price 1982). There is thus a great need for quantitative studies of foraging behavior in a variety of lizard taxa. Qualitative information is available for many species, but quantitative estimates of percentage of time spent moving, number of movements per minute, distances moved, rates of capture and attempted capture, and the like will be important for determining possible influences of foraging behavior on degree of several aspects of chemosensory behavior and morphology.

To augment our sketchy knowledge of relationships between chemosensory structures and behaviors and their association with foraging mode, I have undertaken quantitative comparative studies of variation in several aspects of lingual morphology. The findings will be important for determining any correlation with quantitative estimates of development of the vomeronasal system, with various aspects of chemosensory behavior, and with foraging behavior. The ultimate goals of these and related studies are to trace the evolution of the suite of morphological and behavioral traits important to use of vomeronasal chemoreception in acquisition of food, to understand the selective roles of foraging mode and the associated evolutionary responses, and to ascertain probable influences of foraging mode and chemosensory behaviors on the evolution of higher squamate taxa.

Acknowledgments

This work was partially supported by the Department of Biology of Indiana University-Purdue University Fort Wayne. Thanks are due to Laurie Vitt, who reviewed an early draft of a portion of the manuscript, to Ray Huey and Wayne Maddison for information and advice about phylogenetic inference, and to anonymous reviewers. I am grateful to the many biologists who have collectively amassed an enormous amount of data on the foraging, vomerolfaction, and phylogeny of squamates, especially lizards. In writing this paper, I have tried to synthesize some of that information and have not hesitated to generalize from incomplete, sometimes very small samples, and to speculate regarding evolutionary relationships. My hope is that the paper will serve as a stimulus for future research bearing on these matters.

CHAPTER 6
PHYLOGENETIC PERSPECTIVES ON THE EVOLUTION OF LIZARD TERRITORIALITY

Emília P. Martins

The evolution of ecological and behavioral traits is difficult to study because of the absence of such traits from the fossil record. For example, variation in lizard territoriality both within and among species may be due to differences in life-history patterns, availability of ecological resources, access to mates, or ontogenetic and phylogenetic constraints. Many studies at the intraspecific level have attempted to uncover reasons for differences among individuals, between the sexes, and across populations in territorial behavior using correlational analysis, experimental manipulations, and theoretical considerations (see reviews in Carpenter 1967; Rand 1967; Stamps 1977, 1983). However, patterns of natural and sexual selection found within extant species may not be the same as those that have been operating through evolutionary time, and the hypotheses resulting from intraspecific ecological studies cannot easily be confirmed through reference to the fossil record. At the interspecific level, studies of territoriality have been limited primarily to considerations of the importance of factors such as sexual dimorphism, foraging mode, and taxonomic status in the evolution of territorial behavior without any statistical means of taking phylogenetic information into account (e.g., see references above). In recent years, however, the development of new comparative analysis techniques and modern phylogenies make it possible to infer the patterns and processes of ecology and behavior in ways that were previously impossible. Here, I illustrate the use of such techniques to study the (1) evolutionary origins, (2) adaptive function, and (3) evolutionary processes underlying lizard territorial behavior. This paper is not intended to serve as a review of the available literature on lizard territorial behavior (several important papers have not been included), and the results of the analyses included herein are decidedly preliminary due to several limitations of the data and techniques that are described in greater detail below. Rather, the main goal of this paper is to illustrate the use of phylogenetic comparative techniques to a specific question in the evolution of lizard behavioral ecology, and to generate new hypotheses about the evolution of territoriality that lend insight into fruitful areas for future research at both intra- and interspecific levels.

Definitions of Territoriality

Noble (1939) defined a territory as "any defended area." Brown and Orians (1970) extended this definition by requiring that a territory satisfy three conditions. A territory must be (1) a fixed area that is (2) defended with behavioral acts that evoke escape or avoidance so that (3) the area becomes one of exclusive use with respect to rivals. Stamps (1977) suggested that lizards defend only three types of areas: (1) all or a large portion of the home range, (2) small areas within the range such as basking or shelter sites, and (3) no specific geographic area, but a "personal space" surrounding the individual animal. Although only the first two of these have been traditionally considered to be "territories," several species of lizards (e.g., most teiids and lacertids) fall into the third category by displaying considerable aggression towards other individuals without consistently defending any particular geographic area.

The second part of Brown and Orians' (1970) definition can also be extended. Several authors (e.g., Pitelka 1959; Schoener 1968a; Rose 1982; Smith 1985) have suggested that a territory is an area of exclusive use whether or not behavioral acts that evoke escape or avoidance are observed. Using this idea, I have developed a second categorization scheme to consider the different types of defense behavior or levels of aggression that are generally observed in lizards. First, lizards often engage in combat behavior –aggressive acts involving physical contact that may result in injury. Wrestling, biting, and other forms of direct physical aggression would fall into this category. Lizards also engage in a number of threat behavior patterns–aggressive communicative displays produced in response to another animal but without physical contact or risk of injury. Agonistic or "challenge" push-up displays, "full shows," and other visual displays that are directed towards another animal but which are produced from a distance are examples of this category. Finally, lizards may defend particular areas using a combination of indirect displays such as chemical signals or broadcast displays (e.g., "assertion" push-up displays) that evoke escape or avoidance. These indirect displays occur in the absence of direct contact with other animals and when the defender is not at risk of injury.

Putting the two categorization schemes together, nine types of spatially relevant aggressive behavior are possible, ranging from combative defense of all or a large part of the home range to nonrandom distributions of animals maintained with indirect displays but with no fixed geographic areas of exclusive use (Table 6.1). This categorization scheme differs from earlier schemes in a number of ways. Only the three types that refer to defense of all or a large part of the home range are what has traditionally been referred to as a "territory." Only the first six types that refer to defense of a particular

Table 6.1. Types of aggressive defense behavior exhibited by lizards. Categories of defense "area" were taken from Stamps (1977). Types of defense style are defined in the text. "Combat" refers to aggressive acts involving physical contact that may result in injury (e.g., wrestling, biting), "Threat" refers to aggressive communicative displays produced in response to another animal but without physical contact or risk of injury (e.g., "agonistic," "challenge," or "full show" push-up displays). "Avoidance" refers to the use of indirect displays such as chemical signals or broadcast displays (e.g., "Assertion" push-up displays). Although only Types I, II, and III have traditionally been considered to be "territorial" behavior, all 10 types may be involved in determining the particular spatial distribution of individual lizards. Each animal may exhibit several of these types in different parts of their temporal and spatial range.

	Defense Area		
Defense Style	All or part of Home Range	Specific Site (basking, shelter)	No Area (self)
Combat	Type I	Type IV	Type VII
Threat	Type II	Type V	Type VIII
Avoidance	Type III	Type VI	Type IX

Type X = Affiliative aggregations or random distribution of animals

geographic area (i.e., home range or specific site defense) are consistent with Brown and Orians' (1970) requirement that the animals be site-tenacious. If we add a tenth category to encompass the less frequently mentioned random spatial distributions or aggregations formed through affiliative behavior, this categorization scheme allows us to describe the full complement of spatial patterns occurring in lizards. Most importantly, this categorization scheme allows us to distinguish between the observation of a particular type of behavior (spatially relevant aggression) and the proposed functions of that behavior (traditionally, the defense of a resource), and emphasizes the need to demonstrate empirically any relationship between form and function before assuming that it is there.

An individual animal may exhibit several of these 10 types of behavior in different parts of its range, during different seasons or even at different times of the day. For example, a lizard might defend its entire home range using indirect displays and avoidance, while defending a smaller part of its range using combat or threat behavior (e.g., Fig. 6.1). Other animals exhibit aggressive defense behavior during the breeding season, but form aggregations during the non-breeding season (e.g., *Sceloporus jarrovi*; Ruby 1978). Finally, individuals of some species partition their home ranges temporally, and maintain areas of exclusive use at particular times of the day or on particular days simply by varying their levels of activity (e.g., *Sceloporus jarrovi*; Simon and Middendorf 1976). For example, a lizard may defend a shelter

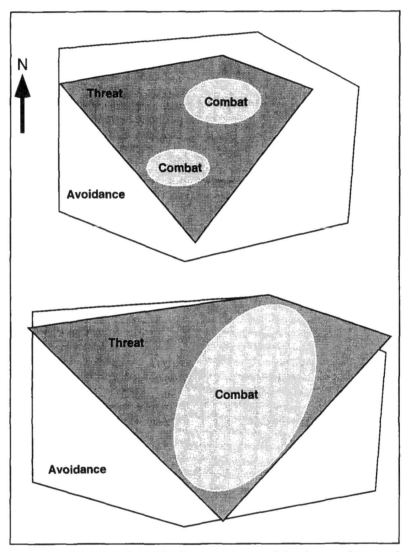

Figure 6.1. Illustration of possible changes in the spatial distribution of territorial defense style in an individual lizard with time.

site using combat behavior during most of its "inactive" periods, exhibit threat displays toward intruders in a larger subset of its home range during hours of the day when it is most active, and continuously defend its entire home range using indirect displays to encourage avoidance. Defense area, behavioral defense style, and spatio-temporal variation are all important components of a lizard's territorial behavior, and will contribute substantially to its ability to maintain a stable position in the overall spatial distribution of that species. Unfortunately, such detailed information on the territorial behavior of even a single species of lizard is rare.

I. Inferring the Origins of Lizard Territoriality

One common use of comparative studies in ecology is to infer the evolutionary origins of particular behavioral or ecological patterns. The evolutionary origins of territorial behavior can be inferred from comparative data using standard systematics techniques such as maximum parsimony and outgroup analysis (e.g., Fitch 1971; see Maddison and Maddison 1992 for a summary of available techniques and friendly computer programs to conduct the analysis; and Brooks and McLennan 1991 for a general description of the use of parsimony reconstruction to infer the evolution of phenotypes). As with most comparative studies, all of these techniques require averaging over broad forms of within-species variation (e.g., individual, age, sex, seasonal, and population differences) and the assumption that differences among species, genera, families, or other taxonomic levels are substantially greater than differences within that level. I will discuss this problem at greater length below.

Territory area

Data and methods

In her extensive review of the literature, Stamps (1977) found that despite some variation within families, most species of phrynosomatids and chamaeleonids (including former agamids) exhibit defense of large parts of their home ranges. In contrast, although anguids, teiids, lacertids, and varanids often exhibit aggressive behavior or defense of their immediate surroundings, they do not generally defend specific geographic areas within their ranges. Scincids, xantusiids, and cordylids seem to be intermediate, and defend small parts of their ranges containing sheltering sites (this last categorization is somewhat questionable given the very small amount of information available about the territorial behavior of these lizard families). Gekkonids are somewhat ambiguous as they include some species that clearly exhibit home range defense and others that seem to have only specific site defense. Although I have scored them as primarily having home range defense to conduct the analysis, this characterization is questionable. (See Stamps 1977 and references therein for detailed descriptions. More recent information on the territorial behavior of many species of lizards is available in several modern studies, but as the broad classification of lizard families is still roughly correct in the light of this new information, I do not review that information herein.)

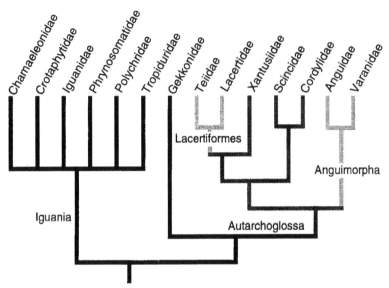

Figure 6.2. Phylogeny for 14 lizard families for which there is information regarding the amount of area defended. Cladogram and metataxa names were developed as a combination of information from Estes et al. (1988) and Frost and Etheridge (1989), both of whom used morphological characters and parsimony algorithms to infer evolutionary relationships. Information regarding spacing patterns is summarized by Stamps (1977), who considers three types of territorial behavior. Black refers to those families that exhibit defense of much or all of their home ranges. Dark gray refers to families that seem to exhibit defense of specific areas within their ranges (e.g., shelter sites, burrows, or basking sites). Light gray refers to those families that do not show defense of any particular geographic area. Standard parsimony reconstruction techniques were used to infer the ancestral states of this character (see text for further explanation).

Although Stamps (1977) considered differences in territorial area defended by different families of lizards in the context of Camp's (1923) classification of those families, the same data can be used to infer the evolutionary origins of territory area if standard systematics techniques are used to overlay territorial behavior on a phylogeny. Essentially, parsimony reconstruction argues that evolutionary changes are rare, and that the ancestral state of a character in a group of organisms is likely to be the most common character exhibited by that group. We can recreate the ancestral state of a clade by beginning with the most recent set of sister taxa (e.g., teiids and lacertids) and estimating the state of their most recent common ancestor (at Lacertiformes) as being equal to the most common state observed in the set of sister taxa (no defense of a specific geographic area; see Fig. 6.2). This information is then used to infer the state of the next ancestor and so on.

In some cases, this simple procedure may not be sufficient to determine all ancestral states of a character in a phylogeny. For example, the ancestor of teiids, lacertids, and xantusiids might exhibit either defense of no area (as do the teiids and lacertids) or the defense of specific sites (as do the xantusiids). More generally, if equal numbers of the sister taxa exhibit different character states, more information is necessary to determine which of the states was shared by the hypothetical ancestor of the clade. Outgroup analysis can help to resolve these problems. For example, given that the next closest relatives of this clade (scincids and cordylids) exhibit specific-site defense, fewer total evolutionary changes would be required if the ancestor of teiids, lacertids, and xantusiids also exhibited specific-site defense (see Brooks and McLennan 1991 for a general description of the uses of outgroup analysis).

Ranking of the types of possible evolutionary changes or character "ordering" can also help resolve such problems. For example, using simple parsimony reconstruction and outgroup analysis, the ancestor of all Autarchoglossa might have exhibited home range defense (as do the gekkonids), defense of no particular geographic area (as do the Anguimorpha), or specific-site defense (as do the other groups within the Autarchoglossa). Without further information, it is difficult to infer the state of this ancestral group. However, if we also assume that shifts from the defense of no geographic area to defense of specific sites are easier to produce and therefore more common than shifts from the defense of no area to full home range defense, it seems most likely that the ancestor of the Autarchoglossa will have exhibited specific-site defense, thereby providing a transition between the home range defense of the primitive ancestors and the defense of no particular area exhibited by the anguids and varanids. Various other options are possible in the parsimonious reconstruction of character evolution and numerous algorithms have been developed to estimate the states of all hypothetical ancestors on a phylogenetic tree from comparative data (see Maddison and Maddison 1993 for detailed descriptions of available algorithms).

Results

Using data reviewed by Stamps (1977), a phylogeny created combining work of Estes et al. (1988) and Frost and Etheridge (1989), and parsimony techniques described above, it is possible to infer evolutionary origins of the amount of area defended by lizards in general (Fig. 6.2). Given the preponderance of home range defense within the Iguania it seems highly likely that the ancestor of this group also had home range defense. In fact, if the three types of territorial area are also assumed to lie on a continuum from defense of a large area to defense of no area at all it seems that home range defense was probably the primitive condition for all lizards. Furthermore, most evolutionary changes that have occurred among lizards have involved decreases

in the amount of area defended. For example, at least two independent evolutionary changes from specific-site defense to defense of no fixed area are
likely to have occurred–first in the divergence of the Lacertiformes (teiids
and lacertids) and second in the divergence of Anguimorpha (anguids and
varanids–although lack of information regarding the latter group makes this
conclusion rather tentative). A third evolutionary change probably led from
home range defense to specific-site defense at the formation of the Autarchoglossa. More evolutionary changes are likely to have occurred within the
Gekkonidae and Chamaeleonidae to account for substantial variation within
these families in the amount of area defended, but further studies of these
lizards would be needed to determine the details of such changes.

Assumptions and possible problems

The results of the analysis described above should be considered cautiously as they depend on only a few main points, and changes in the available information regarding the behavior of certain lizard families could
result in drastically different conclusions. First, although the conclusion that
home range defense is the primitive condition for all Iguania is relatively
robust to changes in the available data, the conclusion that home range
defense is primitive to all lizards depends substantially on the home range
defense presumed to be characteristic of the Gekkonidae. For example, if
the gekkonids as a family were categorized as having site-specific rather than
home range defense, it would be impossible to infer the ancestral state of all
lizards without reference to an outgroup (e.g., tuataras). In this situation, we
would conclude that the ancestor of all lizards probably exhibited the same
type of behavior as the outgroup. Unfortunately, categorization of the
Gekkonidae is difficult given the limited information available, and as stated
earlier, many species of geckos seem to exhibit site-specific while others
exhibit home range defense. As the inferred primitive condition for all lizards depends on the type of territoriality exhibited by the Gekkonidae (and
possibly tuataras), it is likely to change as more data regarding these groups
become available.

Similarly, conclusions regarding the placement of evolutionary changes
on the phylogeny depend in large part on the assumption that site-specific
defense is intermediate between defense of no geographic area and defense
of entire home ranges, and on the observation of site-specific defense in xantusiids, scincids, and cordylids. Unfortunately, information about the specific
mechanisms of territorial behavior including the costs of specific-site
defense and aggression without defense of specific areas is also unknown,
and it is difficult to justify the assumption behind the ordering of this character with real data. Furthermore, information on the areas defended by these
three lizard families is sparse, and it is quite possible that further study will
demonstrate that these groups do not actually exhibit true defense of spe-

cific geographic areas. Changes in either the assumption of character order-ing or the categorization of these families could lead to dramatically different hypotheses regarding the amount of area defended by hypothetical ancestral species and in where specific evolutionary changes occurred.

Suggestions for future research

Despite possible problems, results of this analysis lead to several sugges-tions of interesting areas for future research. First, since the parsimony reconstruction strongly suggests that defense of all or a large part of a home range is the primitive condition for all Iguania (and less strongly that home range defense is primitive for all lizards), one simple explanation for the existence of home range defense in modern Iguania is that the ancestors of this group also exhibited home range defense and that either there is little genetic variation in the trait or selection has not acted sufficiently strongly to eliminate the behavior. An alternative hypothesis is that stabilizing selection may have acted to maintain the defense of home ranges. However, there is no direct evidence in the comparative data to support this hypothesis, and the same patterns can easily be explained solely on the basis of phylogenetic history without need for further adaptive explanations. Selection is more likely to have acted in those species that do not exhibit home range defense, where there may have been a cost to the defense of home ranges leading to the adaptive loss of traditional territorial behavior. Thus, future studies of the mechanisms underlying the defense of home ranges including the potential physiological or energetic costs to territorial defense may be par-ticularly useful, as would be studies of the spatial behavior of geckos and chameleons that exhibit variation in the amount of area defended.

Comparisons of the ecologies of teiids and/or anguids to those of scincids and/or xantusiids may also be helpful in determining why defense of no area has evolved from site-specific defense. Further determination of the origins of home range defense in lizards might be better obtained through compar-ison of the social behavior of lizards to the spacing patterns of tuataras, tur-tles, birds, or even mammals rather than solely through comparing the territorial behavior of different species of lizards. Although detailed studies of the territorial behavior of any single species may provide insight into the evolutionary forces that maintain territorial behavior, they are unlikely to say much about the origins of home range defense in this group because of the prolonged temporal distance from the evolutionary adoption of home range defense to extant species. Similarly, elucidation of the primary forces and any general patterns constraining and maintaining "true" territorial behavior in lizards is probably better obtained through examining species within those clades in which this type of territorial behavior is thought to be the pri-mary pattern (e.g., Phrynosomatidae and Agaminae).

Territory defense style

The same sort of analysis can be repeated looking at different types of defense behavior (i.e., combat, threat, or indirect displays) rather than the amount of area that is defended. Unfortunately, there does not seem to be sufficient quantitative evidence in the literature to suggest that different families of lizards use different types of defense behavior preferentially. It may be that variation within families is comparable to, or even exceeds, variation among families and that a phylogenetic comparative study would not be useful at this level or simply that the available information is insufficient. Nevertheless, it is possible to look at this question at the genus or species level within a smaller clade. For example, all 13 species of *Sceloporus* for which data are available exhibit some sort of home range defense (see references in Table 6.2). Although *S. orcutti* and *S. magister* show very little direct agonistic behavior, they are site-tenacious, and individual males and females seem to maintain nonoverlapping home ranges through broadcast or other types of indirect displays (Type III; e.g., Mayhew 1963; Tanner and Krogh 1973; Tinkle 1976; Tinkle and Dunham 1986). Although very little has been published about the aggressive behavior of *S. woodi*, *S. variabilis*, and *S. clarkii*, they seem to follow a similar pattern (Type III; Fitch 1973; Lee 1974; Tinkle and Dunham 1986; and pers. obs.). In contrast, both male and female *S. jarrovi*, *S. poinsettii*, and *S. olivaceus* exhibit combat and threat behavior in defense of large parts of their home ranges during the breeding season (Types I and II), but form nonagonistic aggregations in the winter months (Type X; e.g., Blair 1960; Ballinger 1973a; Ruby 1977). The remaining species (*S. merriami*, *S. graciosus*, *S. virgatus*, *S. undulatus*, and *S. occidentalis*) seem to be more toward the other extreme in which territories are defended aggressively throughout most of their active seasons using combat and/or threat behavior patterns (though the intensity of aggression may decrease somewhat after the breeding season; Types I and II; e.g., Milstead 1970; Vinegar 1975a; Rose 1982; Davis and Ford 1983; Martins 1991, 1993).

Although *Sceloporus* are likely to be monophyletic (Wiens 1993), specifics of the phylogenetic relationships among *Sceloporus* are still under considerable debate (see Sites et al. 1992 for a review), and it is difficult to have much certainty in any one phylogenetic hypothesis. In one recent study, Mindell et al. (1989) provided a hypothesis of the phylogenetic relationships among species of 19 *Sceloporus* based on allozyme data that include 9 of the species considered in this study. In earlier work, Larsen and Tanner (1974, 1975) provided an analysis of morphological data that include the four remaining species (*S. magister, S. woodi, S. orcutti,* and *S. graciosus*). A combination of these two hypotheses is presented in Fig. 6.3. This

Table 6.2. Means of the measurements available from the literature of 13 species of *Sceloporus* lizards, used for interspecific comparisons in this study. In studies containing data from several populations or subspecies, each population was counted as a separate measurement. Estimates of body size from Fitch (1978) were included in all cases. HR is home range.

Species	HR Size (m²)		Density (lizards / ha)			Body Size (mm)		References
	M	F	M	F	All	MSVL	FSVL	
jarrovi	551	258	15	22	73	83	74	Ballinger 1973a; Ruby 1978, 1981; Beuchat 1982; Ruby and Dunham 1984; Ruby and Baird 1994
graciosus	47	–	–	–	80	55	55	Stebbins 1944; Tinkle 1973; Burkholder and Tanner 1974; Deslippe and M'Closkey 1991
virgatus	970	349	35	38	79	54	60	Smith 1981, 1985; Rose 1981; Vinegar 1975a,b
undulatus	683	271	–	–	25	59	63	Crenshaw 1955; Kennedy (unpubl. in Turner 1969); Ferguson and Bohlen 1972; Tinkle 1972; Tinkle and Ballinger 1972; Ferner 1973; Vinegar 1975a,c; Jones and Droge 1980; Tinkle and Dunham 1986; Jones and Ballinger 1987; Jones et al. 1987b
olivaceus	684	293	–	–	51	83	93	Blair 1960
variabilis	580	–	–	–	–	66	53	Fitch 1978
occidentalis	72	56	–	–	32	72	75	Tanner and Hopkin 1972; Davis and Ford 1983
merriami	201	84	56	56	112	52	50	Dunham 1980; Ruby and Dunham 1987b
poinsettii	–	–	–	–	96	116	97	Ballinger 1973a
orcutti	5385	3563	–	–	57	102	92	Mayhew 1963; Weintraub (unpubl. in Turner 1969)
woodi	–	–	–	–	35	47	50	Jackson and Telford 1974; Lee 1974
magister	–	–	–	–	7	104	93	Parker and Pianka 1973; Tanner and Krogh 1973; Tinkle 1976
clarkii	–	–	–	–	24	103	89	Tinkle and Dunham 1986

phylogeny is almost certainly imperfect and will change as systematists continue to work with this group of lizards. As the categories of defense style proposed above are also rather rough, available information for the different species is poor, and there is considerable within-species variability in the types of aggressive defense used (some of which is reviewed below), all results from this analysis should be considered with caution. Again, my purpose herein is not so much to reach firm conclusions regarding the evolution of territorial behavior, but to illustrate the use of certain techniques and generate new hypotheses for future research.

Nevertheless, if we overlay types of defense behavior on this phylogeny using parsimony reconstruction (including character ordering and outgroup analysis) as described in the previous section, we find that combat or threat defense of the home range throughout the active season seems to be the primitive condition in *Sceloporus* (Fig. 6.3). Evolutionary decreases in aggression probably occurred at least twice. There seems to have been a single change from aggression throughout the active season to aggression only during the breeding season in the clade of live-bearing *Sceloporus* (i.e., the group including *S. jarrovi*). A second change resulting in home range defense through indirect displays rather than combat probably occurred along the branch leading to *the S. orcutti* group. As in the phylogenetic reconstruction of the origins of territory area, to find the origins of combat or threat defense of the home range, we would need to look outside of the clade, and compare the defense style of *Sceloporus* with that of other genera of lizards. Similarly, under the null hypothesis of no selection, there seems to be a clearer need to posit adaptive reasons for the loss of aggressive defense style in some *Sceloporus* than for the continuing defense through combat exhibited by most of the species examined.

II. Inferring the Function of *Sceloporus* Territorial Behavior

Phylogenetic analyses can also be used to elucidate some aspects of the function of territorial behavior while taking phylogenetic information into account. For example, although combat or threat defense of home ranges seems to be the primitive condition in *Sceloporus* as shown above, the presumably high cost of such behavior due to risk of injury suggests that there may be additional adaptive explanations for the maintenance of these behavior patterns. For example, stabilizing selection may have been acting throughout the clade to maintain territorial behavior. Three hypotheses have been proposed for the function of territorial behavior in lizards: (1) to defend food and/or water resources, (2) to defend mates, and (3) to defend basking or other sites with important thermal properties. Which of these is

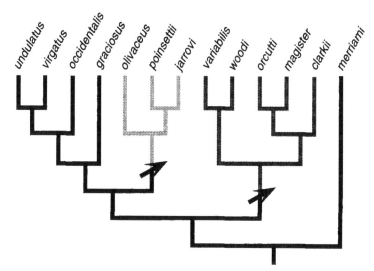

Figure 6.3. Phylogeny for 13 species of *Sceloporus* for which there is information regarding their territorial behavior. Phylogeny is a composite of that developed by Mindell et al. (1989) from allozyme information and that developed by Larsen and Tanner (1975) from morphological data. Branch lengths were created arbitrarily as described in the text. Black refers to those species that defend their home ranges using combat or threat behavior throughout their active seasons. Light gray refers to species that use combat or threat behavior during the summer and breeding season, but form large, nonagonistic aggregations during the winter months. Dark gray refers to those species that defend home ranges using chemical signals, broadcast displays, or other indirect displays as defined in the text. Again, parsimony was used to infer the ancestral states of this character (see text for further explanation).

most important in maintaining aggressive territorial defense is unclear, and may depend to some degree on the particular species being considered. Furthermore, the hypotheses are not mutually exclusive, and it is likely that more than one is acting simultaneously in the clade.

The general statistical relationship reported between home range size and body size in lizards has been used as evidence that one of the major factors determining spatial use by lizards is the availability of food resources (e.g., Turner et al. 1969; Christian and Waldschmidt 1984). Again among *Sceloporus*, experimental studies have provided similar conclusions. For example, in a classic experimental manipulation of food abundance, Simon (1975) found that food abundance was an important factor in determining the size of short-term aggressively defended territories of male and female *S. jarrovi*. The suggestion that territories are defended to maintain adequate access to food is further supported by the finding that juveniles of this species defend territories from each other at as early as 13 days of age (Simon and Middendorf 1980) and exhibit substantial aggression at a young age towards adults as well as to other juveniles (Ruby and Baird 1994). Agonistic behavior between different age classes may be held at a minimum despite the substantial overlap between the home ranges of juveniles and adults by temporal and microhabitat partitioning of the home range (Simon and Middendorf 1976).

In contrast, the results of Ballinger (1973a) and Ruby (1986) suggest that on a time scale longer than a few days there may be little or no relationship between availability of food and territory size for adult male *S. jarrovi*. These studies and the more detailed work of Ruby (1978, 1981) and Ruby and Baird (1994) suggest that the size of aggressively defended home ranges of adult male *S. jarrovi* may depend more on the number of females that they enclose than on any difference in food availability. Endocrinological studies seem to support this suggestion by showing that the aggressive territorial behavior of adult male *S. jarrovi* is mediated by testosterone and tightly linked to the reproductive behavior of these animals. Cycles of testosterone levels of male *S. jarrovi* parallel the yearly cycle of aggressive territorial defense with testosterone levels reaching an all-time low when the lizards are clumped in winter aggregations (Moore 1986; Ruby 1978). Males with testosterone implants have larger territories and exhibit more aggressive behavior (Moore 1988; Moore and Marler 1987), while castrated males show substantially decreased levels of aggression and have smaller territories (Moore 1987). Males with testosterone implants also suffer greater mortality (Marler and Moore 1988) probably due to the higher energetic cost of increased aggressive territorial behavior (Marler and Moore 1991). Thus, although males, females, and juveniles may all hold some minimal territory to defend food resources, the aggressive territorial behavior of adult males is also linked to reproduction and is probably due at least in part to a need to defend a large number of mates.

Variation in territoriality between the sexes and among different age classes has also been used to compare the importance of ecological to reproductive requirements. For example, *S. virgatus* females have clearly defined territories without fighting, while males have larger home ranges with considerable overlap (Smith 1985). Male home range size was positively related to the frequency of courtship and mating behavior in this species, indicating the importance of mate defense in determining the need for male territoriality. Similarly, Rose (1981) found a substantial decrease in activity levels of adult males after the breeding season that would make it impossible for males to defend areas of exclusive use after this period (although there was no corresponding decrease in aggression towards territorial intruders when the animals were actually active). This again suggests that territorial defense is for mates rather than for food resources. Variation in female territories was not explained as easily. Smith (1985) found little relationship between body size and home range size in *S. virgatus* in either males or females (though this result may be due to a small sample size, it is almost statistically significant), suggesting that territories are not held to defend food resources. Furthermore, females are usually larger in body size than are males, suggesting that they would probably need larger rather than smaller territories than males if their territories were used primarily to defend food. Rose (1981) suggested that females may still be holding territories to defend food resources, but can maintain smaller territories by tolerating less home range overlap.

Studies of *S. graciosus*, *S. occidentalis*, and *S. merriami* have illustrated a third factor by examining the importance of thermal requirements on lizard spatial distributions and territoriality. Adolph (1990a,b) showed that microhabitat use in *S. graciosus* and *S. occidentalis* depends both on the availability of thermally suitable microhabitats and on species-typical preferences for particular microhabitat structures. Grant and Dunham (1988, 1990) and Grant (1990) further showed that thermal requirements place a major constraint on the activity levels of *S. merriami* resulting in significant population differences in growth rates, adult body size, and age of first reproduction. The specific importance of thermal requirements to territorial behavior has not been studied in these species.

Other studies have not found conclusive evidence that either of the other two proposed hypotheses (i.e., territories for food or territories for mates) explain all of the variation in territorial behavior or have had contradictory results. Several have been unwilling to attribute a single function to territoriality. In a classic series of competition experiments, Dunham (1980) showed that population density of *S. merriami* is unrelated to the density of the sympatric *Urosaurus*, which are similar in size and ecological requirements and would almost certainly compete for food. Ruby and Dunham (1987b) confirmed this suggestion by examining home ranges of *S. merriami*, concluding that variation in home range size is not explained by any single factor such as food availability, density of either sex, or thermal requirements, but may be determined by a complex interaction of all three. Similarly, Rose (1976) found no relationship between prey size and body size within either *S. occidentalis* or *S. graciosus* and no indication that home ranges varied with either body size or food availability. Davis and Ford (1983) also found no relationship between home range and body size in *S. occidentalis* when comparing animals of different age and sex classes. They concluded that male territories are probably used in part to defend females as a decrease in aggressive behavior was observed after the breeding season had ended.

Similarly, Deslippe and M'Closkey (1991) found no differences among

S. graciosus home ranges in terms of food availability. Although they also found that experimental removal of female *S. graciosus* from male ranges did not lead to detectable changes in male home range size, they concluded that males may hold territories to guarantee long-term reproductive success. Ferguson et al. (1983) found that supplemental feeding of *S. undulatus* juveniles resulted in less overt aggression (i.e., territoriality), less dispersion, and no exclusive home ranges, suggesting that food is a critical reason for maintaining territories. In contrast, Jones et al. (1987b) found no effect of supplemental food and/or water on hatchling home range size in the same species. At the population level, Tinkle (1972) found a density of *S. undulatus* on a Utah site at the edge of their range with several other species of lizards that were likely to be competitors (e.g., *Uta*, *Cnemidophorus*) that was about twice that found by Jones and Ballinger (1987) for a population in Nebraska where mortality due to predation is low and food is probably not limiting. This result is contrary to what would be predicted if territories are being maintained to defend food resources.

Although there is little direct information on the territorial behavior of other species of *Sceloporus*, Ballinger (1973a) reports a pattern of territoriality in *S. poinsettii* similar to that found in *S. jarrovi* with aggregations of one adult male and several females and juveniles during the early summer. Pairs formed in the fall during the mating season and at least some individuals were often found repeatedly in certain predictable geographic centers of activity, suggesting that they may have distinct home ranges or territories. Mayhew (1963) reports that although individual *S. orcutti* were not observed to engage in aggressive territorial interactions and several males may share a single rock, these animals occupy certain preferred sites repeatedly, and may have territories through avoidance rather than defense. Tanner and Krogh (1973) report territorial behavior in *S. magister* but present few details as to the size and quality of those territories or to differences between the sexes or among age classes in territorial behavior. Some information on the territorial behavior or spatial distributions of *S. olivaceus*, *S. variabilis*, *S. woodi*, and *S. clarkii* is also available (and summarized in Table 6.2).

Overall, results of studies on various species of *Sceloporus* suggest that home range defense in *Sceloporus* is due to a complex interaction of factors including defense of food resources, mates, and sites with particular thermal properties, as well as phylogenetic constraints. However, defense of mates seems to be a primary factor for adult males, while adult females and juveniles may also defend territories for access to ecological resources such as food. More work is clearly needed to determine whether thermal requirements play a major role in the spatial distributions of lizards. These results are not entirely convincing because of variation among species and disagreements in the results of many studies. A comparative analysis at the interspecific level can be particularly useful in this sort of situation as a means of determining whether any of the observed patterns can be generalized. For example, a multiple regression of territory size on food resources, number of mates, and thermal characteristics could lend insight into which of these factors are more important than others at the interspecific level.

Methods

The data

Many complications arise when doing this sort of comparative study. Data collected by different investigators on different projects are likely to present a broader, more objective view of lizard behavioral ecology, swamping many of the biases that may be caused by the predispositions of any particular researcher. On the other hand, measurements made by different researchers may not reflect the same biological phenomenon, and are subject to interobserver error. In terms of the territorial behavior or spatial distributions of lizards, some researchers have estimated territory size by setting up artificial encounters at presumed boundaries. Others have determined home range size from sightings of undisturbed animals or by distinguishing areas of exclusive use and using any one of several different estimation techniques (e.g., see Rose 1982). Still others have established population density from direct counts, mark-recaptures or other density estimation techniques (e.g., see Turner 1977 for general discussion). Although all of these are relevant to the definition of spatially relevant agonistic behavior proposed above, they are unlikely to be directly comparable as required for interspecific analyses.

To my knowledge, published estimates of territory sizes from staged encounters are available for only one species of *Sceloporus* (S. jarrovi; Simon 1975; Simon and Middendorf 1976, 1980). Estimates of male home range sizes of nine species of *Sceloporus* are available, if all studies that have made some attempt to estimate home range sizes from sightings of lizards are considered, regardless of the estimation technique applied. Estimates of female home range sizes are only available for seven species, and as the results of analyses obtained using these were not different from those obtained when considering male home range sizes alone, results for females are not reported herein. Population density estimates are the most common sort of measurement of lizard spatial distribution, and are available for a total of 12 species of *Sceloporus*. Separate estimates of male and female densities are available for only three species. As these and all other species for which there is some qualitative information have been reported to have roughly equal sex ratios, only total adult population density is considered in the following analyses (Table 6.2). Needless to say, although these data represent a huge quantity of time and careful observation, they form a rather scanty, mixed set of data that will result in many problems with interpretation. Again, results should be judged with caution, and considered as a means of generating new ideas and testable hypotheses rather than as firm conclusions in and of themselves.

Phylogenetic analyses

Most statistical analyses of comparative data either implicitly or explicitly involve the use of phylogenetic information. As most statistical techniques (e.g., t-tests, regression, ANOVA, nonparametric statistics) require that the data be statistically independent of one another, not taking phylogenetic information into account is equivalent to assuming that the species diverged essentially instantaneously from a single ancestor in a "star" radiation. We can improve on this estimate by incorporating the phylogeny of Figure 6.3 in the analysis, using a technique such as those proposed by Felsenstein (1985), Cheverud and Dow (1985), Huey and Bennett (1987), Grafen (1989), Lynch (1991a) for continuously varying traits, or Maddison (1990), Janson (1992), or Sanderson (1993) for categorical or state variables (see Harvey and Pagel 1991; Losos and Miles 1994 for reviews). Any of the techniques for use with continuous variables (e.g., population density, home range size) can be used to transform species data into phylogenetically relevant and statistically independent variables that can then be analyzed using standard statistical approaches.

As an illustration, I will apply only Felsenstein's (1985) method of independent contrasts. Given a known phylogeny and modeling character evolution as a standard Brownian Motion process, this technique transforms raw species data into a set of "contrasts" or differences between pairs of species that are statistically independent of one another and that have been standardized to have a mean of zero and variance of one. This technique (and all of the others as well) requires information as to the branch lengths of the phylogeny in units of expected variance of phenotypic evolution (i.e., the amount of phenotypic change expected to occur along each branch). This quantity can be either known or estimated, but is generally difficult to obtain. For this study, branch lengths were created rather arbitrarily simply by setting the distance between sister species equal to one, and creating progressively longer branches as relationships became more distant (Fig. 6.3; for alternative ways of obtaining branch lengths, see Felsenstein 1985; Grafen 1989; Gittleman and Kot 1990; Martins and Garland 1991; Martins 1993). Of course, results of a phylogenetic comparative study can only be as reliable as the phylogeny on which they are based. The arbitrary nature with which these branch lengths were obtained provides one more reason for considering the final conclusions of this study with caution.

Using phylogenetic correlations to infer territorial function

A phylogenetic approach to the question of adaptive function involves looking for relationships among evolutionary changes in various traits on a macroevolutionary level. In terms of the evolution of territoriality, for example, we might expect that under the first hypothesis (defense of territories for food), evolutionary changes in spatial distributions or territorial behavior of lizards would be closely related to evolutionary changes in foraging patterns, including preferred diets, food availability, or foraging mode. Evolutionary relationships between the two types of traits could have existed in the past or may currently exist among species whether or not a relationship between food abundance and territory size exists within a single species. To test this hypothesis using comparative data, quantitative measures of the traits in several different species are needed. Although quantitative estimates of food abundance are not available for many of the 13 species of *Sceloporus* for which population densities and/or home ranges have been estimated, body size can be used as a rough indicator of food availability as larger animals will require more food to maintain the same basic activity levels as smaller animals. If territories are being maintained to defend food resources, then larger-bodied animals should also have larger territories and lower population densities. As mentioned earlier, this sort of argument has been made in comparing the food uses of male and female lizards, but might also be used to compare different species. Body sizes for most species of lizards are easily available from the literature (e.g., Table 6.2).

For the second hypothesis (defense of territories for mates), we might expect relationships between territorial quality and the number of mates. Although direct measures of number of mates for particular lizards are not usually available in the literature, degree of sexual dimorphism (e.g., in body size) can also be used as an estimate of mating system or the strength of potential sexual selection acting on different lizard species (e.g., Stamps 1983). Under this second hypothesis, we would predict a significant relationship between population density or home range size with sexual dimorphism. In the current study, residuals from a regression of male body size (mean snout-vent length [SVL]) on female body size were used as estimates of the degree of sexual dimorphism in different species. As the data used in this regression were species mean phenotypes, I applied Felsenstein's (1985) technique and the phylogeny of Fig. 6.3 to correct the data for statistical nonindependence due to phylogenetic relationships before conducting the regression and calculating residuals (forcing the regression through the origin as required by the method).

Unfortunately, there were not sufficient data to consider the importance of the third proposed function (defense of thermal resources) on territorial

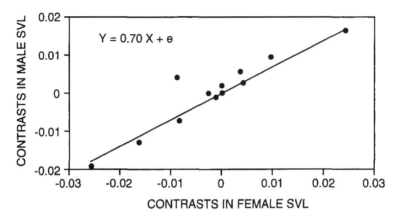

Figure 6.4. Relationship between Felsenstein contrasts in male and female SVL for 12 species of *Sceloporus* lizards ($r^2 = 0.865$; $F_{1,10} = 64.22$; $P << 0.01$).

behavior. To determine the relative impacts of the first two hypotheses, I calculated a multiple regression of home range size on body size and sexual dimorphism for the 6 species of *Sceloporus* for which data were available, using the home range sizes of adult male lizards only as there were not sufficient data to repeat the analysis for females or juveniles. Because this is an exceedingly small sample size of species, a second set of multiple regressions of population density on body size and sexual dimorphism was also conducted for the 12 species for which data were available. As in most cases, information on population density was available only for the two sexes together, analyses were run separately to consider the predictive value of male body size, female body size, and an average of the two in determining overall population density. Results of these did not differ from one another, so only the regression involving male body size is reported. In all cases, Felsenstein's (1985) method and the phylogeny of Fig. 6.3 were used to conduct phylogenetically relevant analyses.

Results

Not unexpectedly, there was a very strong relationship between Felsenstein (1985) contrasts in male and female snout-vent length (SVL: $r^2 = 0.86$; $F_{1,10} = 64.22$; $P << 0.01$; Fig. 6.4). Residuals from this line were used as the index of size sexual dimorphism. Multiple regressions of Felsenstein contrasts in population density on contrasts in both body size (either males or females) and sexual dimorphism found that neither variable had significant predictive abilities ($r^2 = 0.12$; $F_{2,9} = 0.63$; $P > 0.55$; Fig. 6.5). Although the lack of significant relationship between sexual dimorphism and population density may be due in part to small sample size ($r = -0.35$; $df = 9$; $P \approx 0.32$; Fig. 6.5b), there was not even a slight indication of relationship between body size and

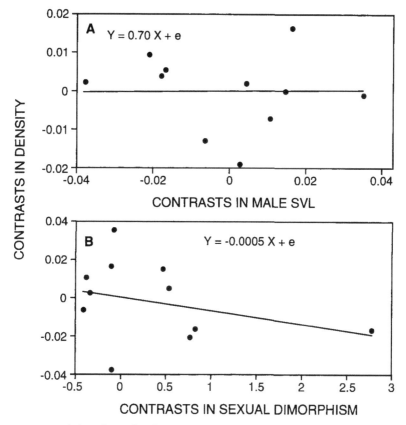

Figure 6.5. (A). Relationship between Felsenstein contrasts in male SVL and adult population density for 12 species of *Sceloporus* lizards (r = –0.13; P > 0.9). (B). Relationship between Felsenstein (1985) contrasts in size sexual dimorphism and density for 12 species of *Sceloporus* (r = –0.35; 0.3 < P < 0.33).

population density (r = –0.13 for male SVL; r = –0.11 for female SVL; df = 9; P ≈ 0.99; Fig. 6.5a). The even smaller sample of data available on home range size also made it difficult to obtain conclusive results. Phylogenetic multiple regressions showed that variation in body size (P < 0.01) but not sexual dimorphism (P > 0.05) could be used to predict home range size (r^2 = 0.585; $F_{2,6}$ = 6.65; P < 0.03 for the full model). However, this relationship was due to the extremely large home ranges reported for *S. orcutti* and disappeared when this species was not included in the analysis (Fig. 6.6). Moreover, when *S. orcutti* was removed from the analysis, the relationship between contrasts in sexual dimorphism and home range size became statistically significant (P < 0.05).

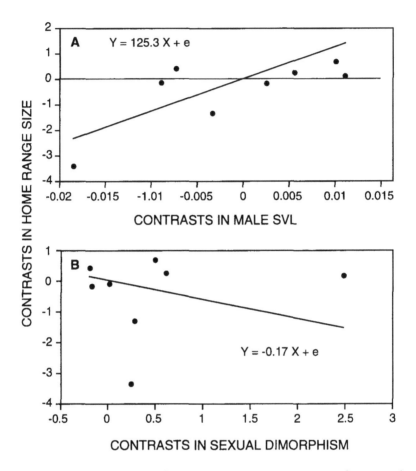

Figure 6.6. (A). Relationship between Felsenstein contrasts in male SVL and male home range size for 9 species of *Sceloporus* lizards (*P* < 0.1). Contrast between *S. variabilis* and *S. orcutti* is a highly influential point due to very large home range reported for *S. orcutti*. (B). Relationship between Felsenstein (1985) contrasts in size sexual dimorphism and male home range size for 9 species of *Sceloporus* (0.1 < *P* < 0.2).

Discussion

Overall, there seems to be little or no evidence to suggest that evolutionary changes in territorial behavior are related to evolutionary changes in either body size or sexual dimorphism in *Sceloporus* once phylogenetic relationships have been taken into account. These results are somewhat surprising given the positive relationship between body size and home range size found in earlier studies (e.g., Turner et al. 1969; Christian and Waldschmidt 1984), and emphasizes the importance of doing this sort of analysis in a phylogenetic context. As the above analyses are all phylogenetically based, they can also be used to infer patterns in the evolutionary changes that have occurred. If anything, the analysis suggests that further studies might find a relationship between population density and degree of sexual dimorphism if data from more species can be considered. It seems unlikely that more data will strengthen the predicted relationship between population density and body size. Analyses involving home range size suggest that studies of *S. orcutti* may be particularly useful to determine why it seems to exhibit such an unusually large home range.

III. Inferring Rates of Phenotypic Evolution

Model and method

A final potentially useful perspective is obtained by trying to infer the rate or "tempo" of the process underlying the evolution of territorial behavior from comparative data. Differences among species in their spatial distributions are the result of the particular evolutionary process underlying that trait. For example, imagine that territorial behavior is a purely neutral character, with all evolutionary changes being the result of random genetic fluctuations. In this case, differences in the territorial behavior of two related species should be directly related to the amount of time since they diverged from one another, the rate of mutation, and population size (Lynch and Hill 1986). On the other hand, if natural or sexual selection has been acting on territorial behavior, differences between species should also be a function of the nature, strength, and direction of the selective forces acting on the trait. Many ways of estimating rates of phenotypic evolution from comparative data have been proposed (e.g., Haldane 1949; Lande 1976, 1977; Gingerich 1983; Templeton 1986; Raup 1987; Baverstock and Adams 1987; Turelli et al. 1988; Lynch 1990; Garland 1992; Martins 1994). In this final section of the chapter, I will apply the technique described in Martins (1994) as an illustration of the types of insight that can be gained through such an approach.

Using comparative data and estimates of time since divergence of various pairs of species (i.e., a phylogeny with branch lengths in units of time), it is possible to obtain a visual depiction of the phenotypic divergence among species with time. For example, envision all *Sceloporus* as beginning at some initial time (i.e., at the base of the phylogeny of Fig. 6.3) with only a single ancestral species and consequently a between-species divergence or variance in territorial behavior of zero. As evolution and speciation proceed, territorial behavior begins to differ among species, increasing the level of among-species variance in territoriality. For example, under a neutral model of phenotypic evolution, species that have been separated for long periods of time are expected to have diverged to a greater extent than recently separated species. In this case, the relationship between among-species phenotypic variance and time could be described by a straight line with a positive or increasing slope. The slope of this line is the rate of phenotypic evolution for the particular trait being considered. Under a model of stabilizing selection, species are expected to diverge up to some asymptotic point at which they begin to evolve in parallel. A plot of the amount of divergence among species with time can thus be used to yield insight into the particular evolutionary processes underlying phenotypic traits, or to the differences in evolutionary processes underlying those traits.

Lynch (1991a) suggested a way to estimate the variance between pairs of species that can be used in creating such a scatterplot. Martins (1994) provides an extension of this technique that allows for estimation of the rate of phenotypic evolution (the slope of the best-fit line) while taking phylogenetic relationships into account. A line can be fit to the data using either a neutral model of phenotypic evolution or one incorporating stabilizing selection, and the technique includes a test to determine whether consideration of stabilizing selection improves the fit of the model. The technique can also be used in conjunction with methods to estimate evolutionary correlations (e.g., Felsenstein 1985; mentioned above under Section II) as a means of estimating branch lengths in units of expected variance of change for use with such methods. The technique described in Martins (1994) differs from most other phylogenetic methods described in this chapter in that it allows for incorporation of data on the level of within-species variability inherent in the traits of interest.

Results

Applying this technique to the various traits measured from *Sceloporus* (Table 6.2), I found that a simple model of neutral phenotypic evolution fit all of the variables reasonably well and that there is no need to posit the added forces of natural or sexual stabilizing selection (chi-squared tests; $P > 0.05$ in all cases). This does not necessarily mean that selection has not been acting on these traits, but rather that the available data and the particular technique chosen do not provide sufficient evidence to prove that it has. Assuming a neutral model of phenotypic evolution (i.e., Brownian motion), the estimated rates of phenotypic evolution in male body size, female body size, and adult population density are significantly different from zero (rate $\pm$ SE = 156.70 $\pm$ 75.0 for male SVL; 74.70 $\pm$ 36.2 for female SVL; and 587.02 $\pm$ 180.3 for population density; Figs. 6.7 and 6.8), while estimated rates of phenotypic evolution of sexual dimorphism and home range size

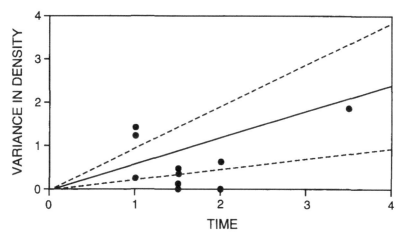

Figure 6.7. Pattern of evolutionary divergence in adult population density for 12 species of *Sceloporus* lizards. Final estimate of beta = 587.02 (*SE* = 180.328). Dashed lines depict upper and lower 95 percentiles.

were not significantly different from zero (*P* > 0.05 in all cases). In the latter case, again, the results do not necessarily mean that sexual dimorphism and home range size have not evolved. The results simply suggest that the available data and the technique do not have sufficient statistical power to demonstrate conclusively that differences among species cannot be explained using measurement error and within-species variability alone.

Differences among the relative rates of phenotypic evolution in *Sceloporus* show that male body size seems to have evolved about twice as quickly as female body size. Sexual dimorphism in body size has evolved so slowly in comparison as to be indistinguishable from a zero rate of evolutionary change. These comparisons assume that phenotypic variances of these characters have been measured on similar scales. The few estimates of standard errors that are available for male and female body size suggest that these two variables were generally measured with similar amounts of accuracy (Table 6.2). If we further assume that other forms of within-species variability (e.g., population variation) are similar for the two sexes, and that there is no overall bias towards one sex being larger than the other (in fact, 6 of the 13 species in Table 6.2 have larger females than males while 7 species exhibit the reverse), then the above comparisons of male body size, female body size, and degree of size sexual dimorphism would be reasonable. If estimates of the within-species variation or measurement error in these and the other variables were available, it would also be possible to compare the rate of phenotypic evolution of population density to rate of evolution in body size. As it stands, however, these rates have been estimated in different units of measurement, and direct comparisons are not biologically

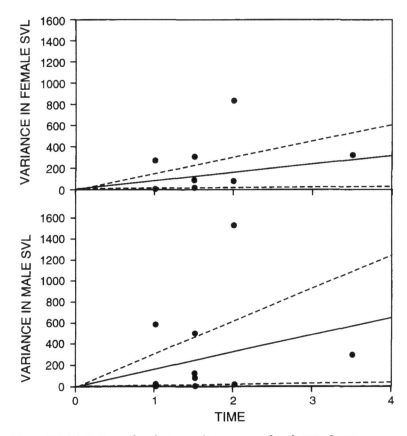

Figure 6.8. (A). Pattern of evolutionary divergence in female SVL for 12 species of *Sceloporus* lizards. Final estimate of beta = 74.40 (*SE* = 36.211). Dashed lines depict upper and lower 95 percentiles. (B). Pattern of evolutionary divergence in male SVL for 12 species of *Sceloporus* lizards. Final estimate of beta = 156.70 (*SE* = 74.983). Dashed lines depict upper and lower 95 percentiles.

meaningful. For example, although it is tempting to conclude that body size has evolved more quickly than population density or home range size, it may be simply that the latter two variables have been measured with considerably more error.

IV. Conclusions and Suggested Areas of Future Research

One of the most successful uses of comparative analyses is to generate hypotheses for future research (e.g., Brooks and McLennan 1991). The analyses presented here suggest a number of general patterns in the evolution of lizard territorial behavior and a large number of areas in which further information could be useful. Possibly the most striking quality of an interspecific comparison of the territorial behavior of lizards is the huge quantity of within-species variation in this character and the great need for more studies in this area. Lizards can vary in type of area defended and type of aggressive behavior used in defense and also exhibit spatial and temporal variation in these two factors. In addition, differences between the sexes and among different age classes are substantial in every species that has been considered, and population variation may also be important. In many species, male and female territories may not serve the same function, and territories of juveniles may also be maintained for different purposes. Unfortunately, it is still not possible to consider these factors at the interspecific level due to the paucity of information regarding territorial behavior of female and juvenile lizards, and the lack of quantitative information about both the spatio-temporal variation and the style of aggressive behavior patterns used by different species of lizards. To compare across species or to consider different types of traits, accurate quantitative estimates of the amount of within-species variation is crucial.

Using currently available information, however, results of this study also suggest a number of specific areas that may be of particular interest for future research. First, home range defense seems to be the ancestral condition for lizards, with evolutionary decreases in the amount of area defended occurring in several lizard families. A similar pattern was observed when considering levels of aggressive behavior within *Sceloporus*. Year-round aggressive defense using combat or threat displays seems to be the primitive condition, with evolutionary decreases in aggression leading to some species that are nonaggressive at particular times of the year and others that defend areas of exclusive use with indirect displays such as broadcast displays or chemical signals. More information is needed to know whether evolutionary decreases in territorial behavior are observed in other groups of lizards and to determine which potential costs of territorial behavior may have resulted in such losses.

Analyses undertaken here also emphasize the importance of phylogenetic context in determining the function of territorial behavior in lizards by suggesting that territorial behavior in modern lizard species may simply be the result of phylogenetic constraints. Empirical evidence in different species of *Sceloporus* suggests that all ages and sexes may defend territories to maintain access to food resources, while males may have an added need to maintain access to a large number of mates. At an interspecific level within *Sceloporus*, the current study found that neither proposed function seems to be of significant importance in determining variation in spatial distribution patterns. Once phylogenetic information was taken into account, there was really no evidence to suggest that variation in spatial distribution could be explained by variation in food resources. A very slight indication of a relationship between spatial distribution and sexual dimorphism, however, suggested that with a larger sample of species we might find that reproductive requirements are of some importance in maintaining the spatial distributions of adult males. Overall, however, it seems likely that phylogenetic constraints have played a primary role.

Finally, the consideration of processes underlying phenotypic evolution gave very little evidence that a response to selective forces in determining territorial behavior could be detected at the interspecific level. It was also difficult to compare patterns of phenotypic evolution across species without accurate estimates of within-species variation in all of the traits to be used for scaling. Considering only variables involving body size, it appears that male body size has evolved about twice as rapidly as female body size, and that both have evolved more quickly than size sexual dimorphism. Contrary to the results of Fitch (1978), this suggests that selection may be acting more strongly on male body size than on female body size. Intraspecific studies of the selective forces acting on males versus females might thus be particularly useful.

Overall, a phylogenetic perspective can be an important tool in describing general patterns and suggesting important areas for future research. The nonrandom spatial distributions of lizards are the result of a complex set of behavior patterns that have evolved due to complex interactions among various selective forces and phylogenetic history. Within-species variation in this character is also striking, with differences among individuals, sexes, age classes and populations only augmenting the spatial and temporal variation in the type of territorial behavior exhibited. All of these components are important, and all should be considered in any thorough discussion of lizard territorial behavior.

Acknowledgments

Thanks are due to David Cox, Lorraine Heisler, Mike Pfrender, Don Price, and Judy Stamps for many useful discussions, to Joe Felsenstein and Michael Lynch for general advice, and to Jonathan Losos, Judy Stamps, and Laurie Vitt for comments on the manuscript. This work was supported by an NSF postdoctoral fellowship.

CHAPTER 7
MATE LIMITATION IN MALE
Norops humilis

Craig Guyer

Long-term demographic data provide the foundation for much research in lizard ecology. Available data indicate a complicated pattern of environmental effects on life-history parameters constrained by phylogenetic history (e.g., Miles and Dunham 1992). One avenue of this research has been a comparative examination of the population ecology of anoline lizards on Caribbean islands with those on the mainland. The impetus of this research can be traced to Andrews' (1979) hypothesis that, for mainland forms, predators are more important than food in limiting lizard abundance and that the reverse is true for island forms. Long-term studies of Caribbean anoles have demonstrated unusual constancy of population density (Schoener 1985). For Caribbean species, food has demonstrable effects on fat storage (Licht 1974; Rose 1982), growth, and spacing patterns (Stamps and Tanaka 1981), but no effect on reproductive effort of females (Rose 1982).

Long-term studies of population dynamics of mainland anoles have demonstrated that lizard abundance changes dramatically among years (Andrews and Rand 1982; Andrews et al. 1983) and among seasons within a year (Andrews et al. 1983; Guyer 1988a). The factors responsible for these changes are complex. Reduced anole abundance is associated with high levels of egg predation, at least at one mainland site (Andrews 1982b). Additionally, population density of mainland anoles is positively associated with food abundance, although this pattern may change from year to year (Andrews and Rand 1982; Lieberman 1986).

Other factors, known or suspected to regulate population density have been largely ignored in studies of anoles. In a manipulative experiment (Guyer 1988a,b), I reported that artificially increased availability of food (arthropods) resulted in an approximate doubling of population density. For adults this resulted from increased survival of females and increased survival and immigration of males. Thus, food can limit density in at least one species of mainland anole. However, two observations indicated that males, rather than (or in addition to) responding to extra food, might have reacted to increased mating opportunities. One observation was that the sexes differed in foraging behavior. While experimenting with methods for altering food availability, I noticed that females seemed to approach and consume supplemental food more rapidly and over greater distances than did males, who were reluctant to accept prey items even when offered at short distances. The second observation was that sex ratios on plots with supplemental food differed from control plots. This difference was in the direction of relatively more adult females, indicating a stronger response to food in that sex than in males (Guyer 1988a). These observations indicated that mate limitation might be an overlooked factor in understanding regulation of abundance in anoles and in assessing Andrews' hypothesis.

In this chapter I summarize information that describes the relationship between food abundance and mate availability in regulating the abundance of male *Norops humilis*. Data discussed here were collected at various times over the past 15 years as part of my attempt to document cycles of abundance and to test factors regulating population density in this species. Data were collected at the La Selva Biological field station in Costa Rica run by the Organization for Tropical Studies. Descriptions of forest structure, climate, and other aspects of this site's natural history can be found in McDade et al. (in press) and references cited therein.

Correlative Evidence of Mate Limitation

Three lines of evidence argue for mate limitation in male *Norops humilis*. First, adult males outnumbered adult females two to one, a sex ratio that suggested limited mating opportunities for males. This pattern was first documented in my food manipulation study (Guyer 1988a). Additionally, repeated censuses made over the past six years on three plots (control plots of Guyer 1988a) indicate that a disparity in sex ratio is consistent from year to year, but differs between the wet and dry seasons (Table 7.1). This finding indicates that if competition for mates exists among males, then it should be more intense during the dry season when population density of *N. humilis* peaks and when females produce eggs at fast rates (Guyer 1988a,b). Such competition may be exacerbated by the observation that females may store sperm (Fox 1963). If these sperm remain viable so that females need not be inseminated each time an egg is ovulated, the operational sex ratio might be further biased towards males.

A second line of evidence involves patterns of overlap between sexes. If males responded to increased numbers of potential mates while females responded to increased food abundance on food-enhanced plots, then differences would be expected in the dispersion pattern of males relative to females and in feeding rates of individuals in the field. To examine potential mating opportunities of individual lizards, I selected target adults and counted the number of members of the opposite sex that overlapped the target during the time period in which the target was observed. Target individuals were restricted to animals observed 10 or more times. For these data,

°Editors' note: Nomenclature of anolines remains controversial; *Anolis* was split by Guyer and Savage (1986), reconsidered by Cannatella and de Queiroz (1989) and Williams (1989), and further reconsidered by Guyer and Savage (1992).

Table 7.1. Numbers of adult males and females captured during three separate censuses. Data are pooled for three 15 x 15 m sites (control sites of Guyer 1988a). Dry season samples were taken during January through March, wet season samples during July through August.

		Year					
		1988		1990		1992	
		Dry	Wet	Dry	Wet	Dry	Wet
Sex	Male	53	32	27	22	65	15
	Female	23	26	20	21	21	12

Year * Season $G = 6.08$; $P = 0.01$
Year * Sex $G = 2.82$; $P = 0.24$
Season * Sex $G = 10.63$; $P = 0.005$
Year * Season * Sex $G = 1.11$; $P = 0.57$

each plot was divided into an 11×11 m grid, creating 121 grid cells. Each cell was approximately 1 m². Each lizard captured during a census was recorded as being present in one cell (site). These data are from Guyer (1988a) with replicate sites pooled to reduce problems associated with small sample size of adult females on some plots.

Patterns of overlap with members of the opposite sex differed for males versus females (Fig. 7.1). On control plots, all females overlapped at least one male, with some individuals overlapping as many as five potential mates; the median value was three. A plurality of males on control plots overlapped no females, and the frequency of males decreased through higher categories of overlap with females (maximum overlap of seven females).

A plurality of females on plots with supplemental food overlapped no males (Fig. 7.1). This pattern differed from that described for control plots (two-sample Kolmogorov-Smirnov test, $X^2 = 6.5$, $P < 0.05$), indicating that females responded to factors other than the presence of more mates. The behavioral response of females to supplemental food observed by Parmelee and Guyer (unpubl.) suggests that females were attracted to, and restricted movements to, sites with extra food. For males, the proportion of individuals overlapping at least one female was unchanged relative to control plots (Fig. 7.1; two-sample Kolmogorov-Smirnov test, $X^2 = 0.6$, $P > 0.7$).

To examine the degree of overlap for pairs of adult males and females, I counted the number of sites occupied by target individuals and compared this index of home range size to the number of sites overlapped by individual members of the opposite sex. I used data from food supplemented plots because sample sizes of females were inadequate on control sites. However,

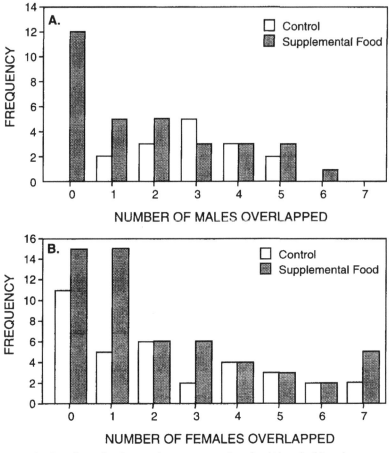

Figure 7.1. Number of males overlapping target females (A) and of females overlapping target males (B). Data are pooled from three control and three experimental plots of Guyer (1988a).

the pattern described below did not appear to differ on control plots. For females, target individuals occupied 2 to 12 sites (median = 6). The majority of males that overlapped target females did so on only one or two sites within the female's home range (Fig. 7.2). For target males, individuals occupied 2 to 21 sites (median = 7). The majority of females overlapping these males did so on only one site (Fig. 7.2). These data indicate that home ranges were small and equivalent between sexes, features noted with alternative analyses (Guyer 1988b). Also, pairs of potential mates overlap on only a small portion of each member's home range.

A final line of correlative evidence suggesting mate limitation involves differences in response of males and females to food. To examine behavioral

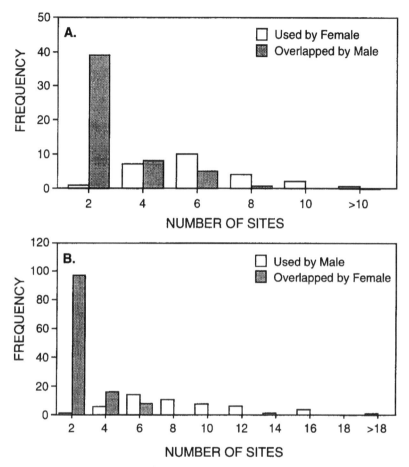

Figure 7.2. Home range size (number of sites used) and overlap (number of sites shared with individual members of opposite sex) of target females (A) and males (B). Data are pooled from three experimental plots of Guyer (1988a).

responses to food, Parmelee and Guyer (unpubl.) encased potential prey items in a clear plastic vial and then placed the vial at various distances (0.5–5 m) within the field of view of free-ranging male, female, and juvenile *N. humilis*. Data from this study demonstrated that females attack prey more frequently and at greater distances than do males. To determine whether this pattern reflects that of free-ranging lizards I counted the number of times that individuals were recorded as having food in their mouth during my food manipulation experiment. I expected to have encountered juveniles and females with food more often than males because of energetic demands for growth and egg production (Andrews and Asato 1977). How-

Table 7.2. Numbers of *Norops humilis* observed with and without food items. Data pooled for six plots (from Guyer 1988b).

	With Food	Without food
Juveniles	36	6012
Males	11	2402
Females	11	1728

$G = 0.74$
$P = 0.69$

ever, this expectation was not supported; instead, all three groups were equally likely to have recently captured food (Table 7.2). In examining these data I assumed that prey selection was roughly equivalent between sexes.

With the exception of data regarding feeding frequencies, all correlative evidence examined to date are consistent with the hypothesis that females are unlimited by mate availability but are food limited, and that males can be mate limited and may be unlimited by food availability. Current information suggests that males have restricted mating opportunities and that, in my manipulative experiment (Guyer 1988a), females responded behaviorally to increased foraging opportunities, and males migrated to and remained longer on plots with more females.

Experimental Evidence of Mate Limitation

To test the response of males to altered mate availability, I manipulated female abundance by adding females placed in clear plastic containers to experimental plots. In the presence of extra females, I predicted an increase in male abundance and/or an alteration of the dispersion pattern of males.

Lizard densities were monitored on six gridded plots (15×15 m) located near trail marker 950 on the Sendero Occidental (La Selva trail system). These plots were arranged in a fashion that was nearly identical to my food manipulation (Fig. 7.3). Mark-recapture data were gathered on all six sites. Four premanipulation samples were taken on each site to establish baseline information (31 Jan.–6 Feb. 1993). The experimental manipulation involved retaining individual females in clear plastic cups (Solo brand, 16 oz.) fitted with a stick (perch) and covered with clear plastic. Each cup had air holes and was inverted when placed on a plot. Water was always available as condensation on the plastic bottom, and food (small grasshoppers) was provisioned on alternate days. The side of each cup was wiped clear each morning to remove condensation that accumulated overnight, thereby maintaining visibility of the females by free-ranging diurnal males.

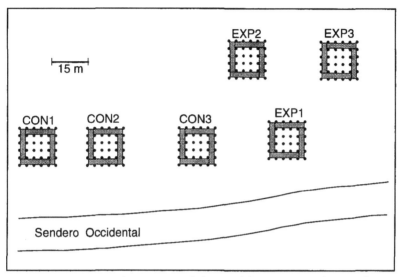

Figure 7.3. Location and orientation of plots used for experimental manipulation of female density. The shaded portion of each plot represents perimeter, and the unshaded portion interior areas for dispersion analysis. See text for explanation.

On each of three experimental plots, one female was added to each of 10 sites. This manipulation approximately doubled female abundance (pre-manipulation mean minimum number alive = 7.7 females; range = 6–11; N = 6). In my food manipulation I demonstrated that, when density increased females were added to sites previously unoccupied by other females (Guyer 1988a). Therefore, I placed caged females at sites randomly selected from the pool of sites not known to harbor females (based on pre-manipulation censuses; Fig. 7.4). The manipulation was maintained for a period of 15 days, the final four of which were used to gather postmanipulation data (13-17 Feb. 1993).

The other three plots served as controls. Empty cups were placed on these sites in a fashion similar to that on the experimental sites. Cups on the control sites were visited as frequently as those on experimental sites to control for the effect of my presence while maintaining the supplemented females. Control and experimental plots were paired into blocks so that both plots within a block could be censused during the same morning or afternoon period. This controlled for altered lizard activity associated with rain showers.

Population density was estimated with the Schumacher-Eschmeyer method for reasons outlined by DeLury (1958), and differences in density were evaluated by examining pattern of overlap of confidence limits for a minimum of three plots. Confidence limits were set at 90% because, if three

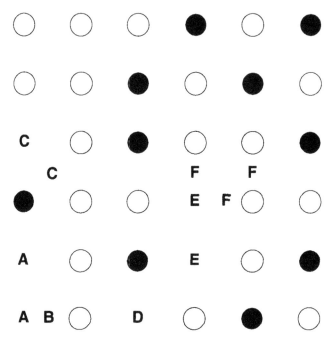

Figure 7.4. Representative example of manipulation of female abundance. Letters represent locations of individual females during four premanipulation sample periods. Darkened circles are sites chosen for placement of caged females. See text for explanation.

independent observations were made of nonoverlapping 90% confidence limits, then this would indicate a significant overall trend at alpha = 0.05 (Sokal and Rohlf 1981). Significant heterogeneity in estimated male abundance occurred during the premanipulation period (Fig. 7.5). Density of all other lizards (juveniles + adult females) displayed similar heterogeneity (Fig. 7.5). Therefore, data collected during the postmanipulation period were corrected for premanipulation densities in testing for a treatment effect. This was effected by subtracting premanipulation estimates from postmanipulation levels.

After 15 days of altered female density, males on all three experimental plots demonstrated increases in abundance relative to their premanipulation levels and their paired control plots (one-tailed 90% confidence limits about adjusted postmanipulation means did not overlap; Fig. 7.6). This pattern differed from that observed for juveniles and adult females combined; for these lizards, no consistent difference in adjusted abundance was observed on experimental plots relative to premanipulation densities or to levels on control plots (Fig. 7.6).

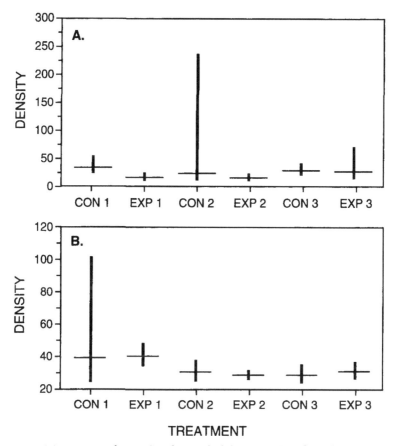

Figure 7.5. Premanipulation abundance of adult *Norops humilis* males (A) and females + juveniles (B). Data are mean (Schumacher-Eschmeyer estimate) and two-tailed 90% limits for three control and three experimental plots.

These data indicate a rapid response by males to female abundance. This treatment effect resulted in average increases of from 5-10 males per experimental plot. Because the experiment was conducted over such a short time, immigration must explain altered male abundance. This result conforms to the rapid response of males seen during food manipulation, where population trends were evident even after the first two-week experimental period (Guyer 1988a).

Because immigrants must enter from the perimeter of the grid, dispersion of males on experimental plots during the postmanipulation period should indicate increased use of the perimeter sites relative to their use during the premanipulation period. To examine the effect of extra females on male dispersion, I classified sites within each plot into two categories. One

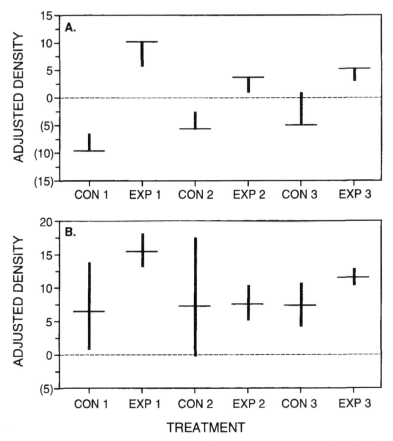

Figure 7.6. Postmanipulation abundance of adult *Norops humilis* males (A) and females + juveniles (B). Data are mean (Schumacher-Eschmeyer estimate) and 90% limits for three control and three experimental plots.

represented perimeter sites defined as those associated with the outermost grid locations (Fig. 7.3). The second category represented all other sites; (hereafter referred to as interior sites). To test the effect of female abundance on male dispersion, I counted the number of times perimeter versus interior sites were occupied by male lizards. For animals observed multiple times, only the first observation was used. These data were tallied separately for pre- and postmanipulation periods and compared with a log-linear test. The full model examined four levels: treatment (experimental vs. control), location (perimeter vs. interior), time (before versus after), and block (1, 2, or 3). The test of interest is the three-way interaction of treatment * location * time. Because there was no significant four-way interaction in the full

Table 7.3. Dispersion of male *Norops humilis*. Data are counts of number of males within four grouping variables: Time (before and after manipulation), Treatment (experimental and control plots), Location (perimeter and interior sites), and Block (three replicate experiments).

		Experimental		Control	
		Before	After	Before	After
Block 1	Perimeter	3	11	10	10
	Interior	7	6	3	3
Block 2	Perimeter	5	4	8	5
	Interior	7	9	6	8
Block 3	Perimeter	4	11	7	10
	Interior	14	5	8	7

Block *Location * Time * Treatment $G = 0.76$; $P = 0.68$
Location * Time * Treatment (blocks pooled) $G = 3.39$; $P = 0.06$

model (Table 7.3), I pooled all blocks in assessing the treatment effect on dispersion. There was a marginally significant three-way interaction for the pooled data, indicating that the relative number of perimeter sites used by lizards on experimental vs. control sites changed from the premanipulation period to the postmanipulation period (Table 7.3). This was in the direction of increased use of perimeter sites on plots with extra females. The above analysis was repeated for all other lizards (juveniles + females) to determine whether altered dispersion was a general phenomenon or was restricted to males alone. Once again, the four-way interaction was not significant (Table 7.4), justifying pooling across blocks. No significant change in the relative use of perimeter versus interior sites occurred after manipulation of female abundance (Table 7.4).

Implications for Anole Life History

By careful comparison of matched island and mainland species, Andrews (1979) first described a pattern of increased population density of anoles on Caribbean islands relative to mainland species and hypothesized that this resulted from ecological release; island species, with relatively few predators, increase in density until food becomes limiting, whereas mainland

Table 7.4. Dispersion of nonmale *Norops humilis*. Data are counts of number of juveniles and females within four grouping variables: Time (before and after manipulation), Treatment (experimental and control plots), Location (perimeter and interior sites), and Block (three replicate experiments).

		Experimental		Control	
		Before	After	Before	After
Block 1	Perimeter	8	11	13	10
	Interior	15	12	15	12
Block 2	Perimeter	8	10	10	9
	Interior	11	18	10	17
Block 3	Perimeter	10	12	14	10
	Interior	16	17	15	17

Block * Location * Time * Treatment $G = 0.04$; $P = 0.98$
Location * Time * Treatment (blocks pooled) $G = 1.17$; $P = 0.28$

species, with relatively many predators, maintain population densities below levels set by food availability. To address this hypothesis adequately, both factors (food and predation) must be examined simultaneously in a two-way design. Because accumulating evidence indicates that mainland and island species belong to largely divergent phylogenetic groups (Guyer and Savage 1992), comparisons within the species groups originally selected by Andrews (1979) to examine the hypothesis will be the most likely source of subsequent tests. My results indicate that interaction effects are expected to confound simple explanations. Additionally, factors other than predators and food need to be explored. Clearly, mate limitation for males should be entertained in subsequent hypotheses designed to explain life-history patterns of mainland anoles. Techniques for manipulating food and mate availability currently exist; further experimental evaluation of Andrews' hypothesis awaits an appropriate manipulation of predator pressure.

One intriguing implication of patterns of demography and behavior in *Norops humilis* is the difference in foraging effort between the sexes. The fact that females appear to be food limited whereas males are mate limited indicates that energetic needs must differ dramatically between the sexes. No energetic data have been examined in *N. humilis*, but data generated by Andrews and Asato (1977) for *N. limifrons* conform to patterns expected if male anoles are relatively unlimited by food. Thus, demands of growth in juveniles, and egg production in females appear to be significantly greater than demands resulting from territory maintenance and other social functions characteristic of male anoles. Adult *N. humilis* of both sexes maintain home ranges of roughly equivalent size (Guyer 1988b; this study) and so the only energy needs of males that are comparable to that of egg production in females are social interactions associated with intruding males and courtship. Thus, time available for social interaction is a key feature in the life history of male *N. humilis*, as it is for other mainland anoles (e.g., Andrews 1971). A trade-off may exist for males in environments of increased population density because time spent defending increasingly more precious mating opportunities should result in decreased time available for feeding to power such social behaviors.

The response of females to food resources appears to be consistent with energetic needs for reproduction. Because males outnumber females in the *N. humilis* populations which I have been monitoring, opportunities for mate choice by females exist. When population density increases in response to altered food resources, females experience reduced overlap with males and altered sex ratios, features that may limit mate choice.

Implications for Population Theory

Kiester and Slatkin (1974) argued that high levels of social interaction in anoles might function to allow individuals to track resources by assessing the behavior of neighboring lizards. The experimental data indicate that males can respond to the visual impression of altered female abundance. Males were observed to approach and display to females in the cups. However, this might be a general response to the addition of any new lizard. Additional manipulations will be necessary to distinguish whether male *Norops humilis* responded to female presence or to the presence of any new adult lizard. However, if food were the resource being evaluated by anoles (as discussed by Kiester and Slatkin 1974), then all ages and sexes should have responded to my manipulation. In fact, for energetic reasons associated with growth and reproduction (Andrews and Asato 1977), responses by juveniles and females should have been stronger than for males. The observation that only adult males changed in numbers and dispersion suggests that (1) males alone assessed resources via neighbor monitoring, (2) males recognized the supplemental individuals as being correlated with reproductive opportunities, and (3) food resources were not evaluated via neighbor monitoring. Additionally, current data indicate that, in *Norops humilis*, males conform to the time-minimizing, and females to the energy-maximizing strategies described by Kiester and Slatkin (1974).

In a modification of earlier work, Kiester (1985) used marriage dominance functions and two-sex demographic dynamics to predict intersexual differences in movement patterns of vertebrates in response to increased food resources. Three situations were modeled: (1) food increased at a point source with linear decay to background levels, (2) food increased at a point source with exponential decay to background levels, and (3) food uniformly increased over a given area (step function). Simulations demonstrated that females increase at areas of increased food resources in a form similar to the distribution of additional food. Males increase more rapidly and more strongly than females because both food and mating resources provide benefits (assuming females are the limiting sex).

Of the situations modeled by Kiester (1985) the third (step function) was most similar to my food and mate manipulations. Field observations conformed to his predictions in two important respects. First, males responded rapidly to altered resource levels via immigration (Guyer 1988b; this study). Second, females responded proportionally more strongly than did males (Guyer 1988a), a feature predicted by Kiester (1985) only for the case of even food resource alteration. My manipulation of female availability should have resulted in a more even distribution of this resource. Kiester's modeling results indicated that males should respond by increasing uniformly across these plots. My manipulation was insufficiently long to test this prediction. Additionally, alternative manipulative distributions of resources remain to be tested. However, the general features of the two-sex demography model may help explain results of my field manipulations of *N. humilis*, as well as altered female and male movement patterns and related behaviors observed in other lizard species (e.g., Hews 1993; M'Closkey et al. 1987b; Tokarz 1992).

Acknowledgments

I thank Laurie Vitt and Eric Pianka for the invitation to present this material at the third lizard ecology symposium and for numerous editorial suggestions and duties. Sharon Hermann, Ross Kiester, Ted Fleming, and an anonymous reviewer also read and commented on the manuscript. David and Debra Clark are thanked for logistical support of my research at La Selva.

CHAPTER 8
POPULATION DYNAMICS AND PAIR FIDELITY IN SLEEPY LIZARDS

C. Michael Bull

Like most other taxa, lizards have a wide range of life histories, reflecting diversity of phylogeny, habitat, and ecological niche (Ballinger 1983; James and Shine 1988; Shine and Greer 1991; Vitt and Breitenbach 1993). Consequently, new studies of lizard ecology are likely to uncover diverse population patterns rather than populations conforming to previously described patterns. In this chapter I present data from a long-term study of the population dynamics and social interactions of a long-lived Australian skink, the sleepy lizard, *Tiliqua rugosa* (previously *Trachydosaurus rugosus*).

I do not give any new generalizations about lizard ecology, but attempt to address a bias in many previous generalizations brought about by the under-representation of long-lived species. Specifically, I will suggest that in *T. rugosa* there is a complex social organization, based on long-term individual recognition, and that this behavioral complexity has evolved in response to population stability. I do not expect that this population structure will be unique among lizards, but extended studies of other long-lived species will be needed to determine general patterns.

Schoener (1985) suggested that lizard populations are unusually constant in size, compared with other animal taxa. Nevertheless, lizard populations still show annual variation in size, and we might expect different species to vary by different amounts. By analogy with birds and mammals, short-lived lizard species may respond strongly to annual variations in resources, climate, or other hazards, and show large year-to-year fluctuations in population size.

For instance, Turner et al. (1970) found that a population of *Uta stansburiana* increased and decreased as annual fecundity changed, presumably in response to resources or climate; Ferguson et al. (1980) described a population of *Sceloporus undulatus* with high variance in population size between successive springs, probably resulting from density dependent predation on egg-laying females; and Andrews (1991) showed that a population of *Anolis limifrons* changed density by as much as 5- to 8-fold from year to year, as a consequence, she suggested, of annual variation in rainfall and levels of food resources.

Longer-lived species of lizards should have smaller annual changes in population size. Strijbosch and Creemers (1988) found that populations of two *Lacerta* species, with longevities of 8–10 years, were relatively stable over time. Population dynamics of long-lived species will still be influenced by extreme environmental factors, but interactions within populations will play an increasing role in population dynamics. For instance, populations of the Galapagos marine iguana, *Amblyrhynchus cristatus*, suffered major declines during the 1982/83 El Niño event. Populations subsequently returned to previous levels, apparently set by individual spacing and access to feeding sites (Laurie 1989; Laurie and Brown 1990).

There are few studies of the population dynamics of longer-lived lizards, and we know relatively little about the complexities of social organization that may contribute to lizard population regulation. This paper describes a population study, from 1982–1992, of a long-lived Australian skink, the sleepy lizard, *T. rugosa*, which should at least increase the data base for long-lived lizard species.

Sleepy Lizards

My assistants and I have been studying a population of sleepy lizards for 11 years (1982–1992). Our study site, near Mt. Mary in the midnorth of South Australia, has a semiarid climate with an average annual rainfall of 250 mm. The vegetation is a mosaic of mallee scrub and open chenopod bushland (Petney and Bull 1984).

More than 20,000 random encounter captures of *T. rugosa* have been made during the 11 years, mostly by driving along more than 47 km of road transects in the study site during spring and early summer, when this species is most active. Upon capture, each new lizard is individually marked by toe clip. Sex is determined by cloacal examination (male hemipenes can be everted with ventral pressure), or by the relatively broader heads of the males. For all captured lizards, weight and snout-vent length (SVL) are measured before release at the point of capture.

Sleepy lizards are among the largest Australian skinks (Greer 1989). Mature adults, indicated by their participation in reproductive activity, weigh 600–900 g, and have SVLs of 28–33 cm. Individual lizards take at least 3 years to reach sexual maturity (Bull 1987), but then appear to stop growing. Longevity has not yet been estimated. Adult sleepy lizards caught and marked in 1982 were still present in the 1992 survey. The 1992 field data are not analyzed yet, but 28.4% of the 120 adult lizards caught in 1982 were recaptured in the 1990 or 1991 surveys and must have been at least 12 years old. Thus, sleepy lizards are long lived.

Tiliqua rugosa are most active in the spring (Bull 1987), when they forage for their omnivorous diet, mostly of flowers and berries, but also insects and carrion (Dubas and Bull 1991). Individual lizards occupy overlapping home ranges (Bull 1978; Satrawaha and Bull 1981; Dubas 1987), which are retained over several years (Bull 1987; Dubas 1987). Home range size does not differ between the sexes (Dubas 1987). Experimental manipulations of either food level (Dubas and Bull 1992) or density of neighbors (Dubas 1987) showed no influence on home range size. Despite intense observations we have seen no behavior related to home range defense.

Sleepy lizards are viviparous (Greer 1989; Bull, Pamula, and Schulze unpubl.). We captured 34 gravid females in December of 1984, 1986, and 1990–1992 and kept them in outside pens and laboratory cages. They produced litters of 1–4 live young (Fig. 8.1) in late March or early April of each year, after approximately five months of gestation. Mean litter size was 2.15 (SE 0.13; N = 34 litters) (Fig. 8.1). Young lizards weighed 60–140 g at birth. Relative clutch mass (RCM) (clutch [litter in this case] mass as a percentage of female gravid mass) averaged 28.3% (SE 1.1; N = 34 litters). This is high for lizards (Vitt and Price 1982) and amongst the highest values of RCM cited for Australian skinks by Greer (1989), but close to the average for eight skink species cited by Vitt and Price (1982). Vitt and Congdon (1978) and Shine (1992) have argued that RCM is related to female body shape of a species. Sleepy lizards have the short limbs and large body cavity morphology which are usually associated with high RCM. They are slow-moving lizards, a character also often associated with high RCM (Vitt and Congdon 1978).

The small litter of large offspring in *T. rugosa* is unusual. Other species in the genus *Tiliqua* produce from an average 5.2 offspring in *Tiliqua occipitalis* to 10.9 in *Tiliqua scincoides* (Greer 1989; Shea pers. comm.). *Tiliqua gerrardi*, which has now been moved to the related genus *Cyclodomorphus* (Shea 1990), can produce litters of over 50 live young (Greer 1989).

In *T. rugosa*, large size at birth may reduce the risk of predation on juveniles. It may also assist them to last through the winter. Young *T. rugosa* are born in March or April, just before the onset of the Australian winter. Juveniles, captured in the field early in the following spring (September), have the same body-size range as newborn young. They do not grow during the five cool months of autumn and winter and probably rely on fat reserves provided to them during gestation, or on food gathered in the short time before winter temperatures inhibit activity.

Population Dynamics

To illustrate the trends in population size of *T. rugosa* over the study period, I have analyzed data from one 6.6 km transect (transect 3). This transect was searched by car on at least 60 days over the four months of maximum lizard activity (Sept.–Dec.) in each year 1983–1991. An average of 218 different individual sleepy lizards were captured, by random encounter, each year.

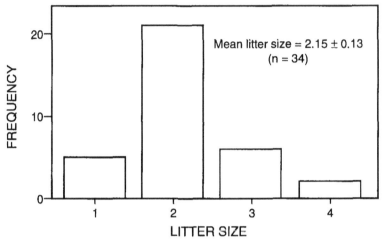

Figure 8.1. The frequency distribution of litter sizes from 34 gravid female *Tili-qua rugosa* caught in the field and held at the Flinders University campus until parturition.

I used Jolly-Seber analyses from recaptures of marked lizards to estimate population size along the transect in the years 1984–1990 (Fig. 8.2). Population size ranged from 250–380 individuals. The population grew over the three years 1984–1986, and then stabilized at an average 364 individuals, or about 5.5 lizards per 100 m of transect, over the four years 1987–1990.

Home ranges are not homogeneous in shape, but on average have diameters of about 200 m (Bull 1988). Thus a linear transect samples lizards in a 200 m wide belt, and population density (from 5.5 lizards per 100 linear m) can be estimated at about 2.7 lizards per ha. At its lowest density in this study (1984), transect 3 had 3.8 lizards per 100 m, or 1.9 per ha. These densities are similar to the 2–3 lizards per ha for the same species at a site near Tickera, South Australia (Bull 1987), but lower than the 8.9 lizards per ha estimated for *T. rugosa* at Cape Jaffa, South Australia by Yeatman (1988).

An intense drought in the area in 1982 and early 1983 may have caused the low population density at the start of the study (Fig. 8.3). Many lizards at that time were in very poor body condition. The deflated appearance of their tails indicated fat reserves close to exhaustion. Mortality from starvation was probably high.

For the remainder of the study period, rainfall values were average or above average (Bull 1992; Fig. 8.3). Since 1984, spring food has been abundant, and lizard body condition good. It appears that the population was depleted in 1982/1983, but since then has grown to a stable equilibrium. It is unclear what limits further population growth. Sleepy lizards do not appear to be limited by food resources (Yeatman 1988; Dubas and Bull 1991, 1992),

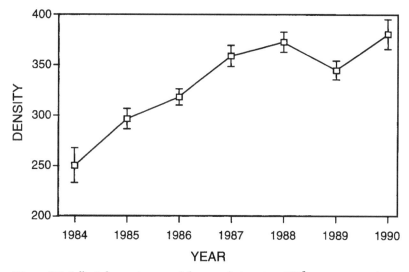

Figure 8.2. Jolly-Seber estimates of the population size of *Tiliqua rugosa* along transect 3 of the study area in the years 1984–1990. Vertical bars represent one standard error on each side of the estimate.

but some social interactions may impose an upper limit on population size. The major feature of the population dynamics of *T. rugosa*, however, appears to be year-to-year stability in population size.

Survival and Recruitment

Population stability in sleepy lizards is supported by other parameters derived from the Jolly-Seber analysis (Table 8.1). There was high (average 88.1%) annual survival of spring-captured individuals, a value that predicts about 32% survival after 10 years. This corresponds closely with the observed 28% survival until 1991 of the 1982 captures.

Adults usually have abundant spring food and are too large to be threatened by most predators. Sleepy lizards have only one further hazard. Their habit of foraging on weedy roadside vegetation, and slowly crossing roads, makes them tempting targets for irresponsible drivers on country roads. Over the course of the study 318 sleepy lizards were found dead on roads. This represents 6.6% of the total 4797 sleepy lizards encountered in the study, and a substantial proportion of average annual 11.9% population loss. This contrasts with the estimate by Ehmann and Cogger (1985) that 0.014% of all mortality in Australian frogs and reptiles is due to road kills, and suggests that *T. rugosa* is particularly vulnerable to this form of mortality.

The Jolly-Seber analysis also showed low recruitment to the population

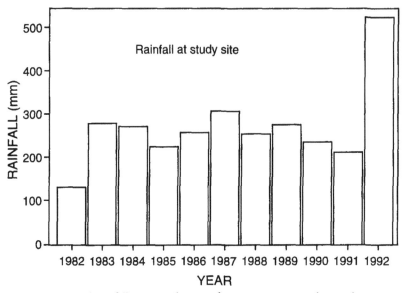

Figure 8.3. Annual rainfall measured at Bundey Bore Station in the northeast section of the study area from 1982–1992.

each year (Table 8.1). Low recruitment influences the population age structure (Fig. 8.4). Age is inferred from size; at birth SVL is 14–19 cm, young do not grow over winter, and recaptured juveniles are usually < 20 cm SVL in their first spring. Thus, all lizards caught in spring with SVL < 20 cm are classified as juveniles under 12 months old.

Sleepy lizards take at least three years to reach adulthood (Bull 1987). Spring-caught lizards, with SVL 20–28 cm, are considered subadults, aged 1–2 years. Some marked lizards took longer to reach mature size, so the subadult category probably also contains older lizards.

Adults were lizards with SVL > 28 cm. Most marked adults showed no growth over several years, so no attempt was made to estimate adult age from size. Adults make up the overwhelming majority, 69%–81%, of the sleepy lizard population, while juveniles make up only 0%–7% of individuals captured each year (Fig. 8.4). One interpretation of these data is that juveniles, although abundant, are more secretive and less often encountered than adults. However, this seems unlikely because it appears that juveniles forage actively over a wider range of climatic conditions than adults, perhaps because of a priority for rapid growth, and so should be more often encountered.

Alternatively, the low proportion of juveniles may reflect low levels of reproductive investment by females, or high rates of mortality of juveniles.

Table 8.1. Parameters derived from the Jolly-Seber analysis of the population of *Tiliqua rugosa* along transect 3, using data from captures over the years 1983–1991. Means ± 1*SE* are shown.

Year	Live Lizards	Population Size	% Survival	Recruits
1983	51			
1984	199	250 ± 17.0	88.4 ± 2.3	
1985	204	296 ± 10.3	90.5 ± 1.8	74.9 ± 17.5
1986	231	317 ± 7.70	88.4 ± 1.8	56.5 ± 10.2
1987	242	358 ± 10.5	91.5 ± 2.1	76.7 ± 9.8
1988	270	372 ± 10.0	85.0 ± 2.1	72.3 ± 10.0
1989	225	344 ± 9.60	83.9 ± 2.3	32.2 ± 8.6
1990	253	380 ± 14.6	89.2 ± 3.4	75.2 ± 9.6
1991	242			

The proportion of female *T. rugosa* that reproduces each year is unknown. Four out of seven radio-tagged females produced litters in the field following the 1991 and 1992 springs. If 50% of females produce an average two young per female per year, there should be one young for every two adults each year. In that case, spring populations with a ratio of 5 juveniles to 75 adults (Fig. 8.4) suggest about 85% juvenile mortality each year.

Small juvenile *T. rugosa* are vulnerable to predators like snakes and hawks. Many young probably also starve over winter. They are born at the end of the dry summer when food is scarce and they must survive the winter on the fat reserves provided by their mother, plus food found before cold weather inhibits activity. After the 1982 drought, no juveniles were found in the 1983 spring.

The population stability of sleepy lizards, the small litters of large offspring, and the low recruitment of juveniles into the population, all suggest a selective advantage for reproductive strategies producing high-quality offspring, rather than large numbers of offspring. This suggests a nonpolygamous mating system. Thus, population dynamics may strongly influence mating system and reproductive behavior of this species. Observations relevant to this hypothesis are presented next.

Reproductive Behavior

In road surveys, sleepy lizards are usually encountered alone in early spring (Aug.–early Sept.), but an increasing proportion are found as male-female pairs as spring progresses (Bull 1987, 1988). Pairs are found coiled in a refuge together, basking side by side, or foraging, with the male following the female and separated by only a few cm (Bull 1988; Bull et al. 1993).

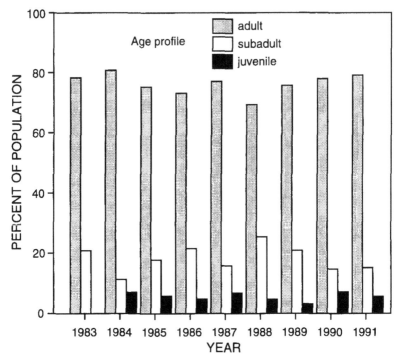

Figure 8.4. Proportion of adult, subadult, and juvenile *Tiliqua rugosa* among all individuals captured on transect 3 each year from 1983–1991.

Individuals that are randomly encountered in pairs more than once in a season are usually with the same partner (Fig. 8.5) over periods of up to eight weeks. This trend toward monogamy in *T. rugosa*, previously reported for smaller samples (Bull 1988, 1990), is now confirmed with this larger data set.

Monogamy is also confirmed by radio-tracking studies, conducted in the springs of 1990–1992, using 23 unmanipulated pairs of lizards. Each pair was observed on 10–54 days over a spring season. Three females changed partners during the spring. The rest were either attended only by a single male, or briefly contacted a second male before the first male reestablished the partnership.

Copulations have been observed in late October and early November of most years (Bull 1987, 1988). After copulation, pairs separate for ten months (Bull 1987, 1988; Bull et al. 1991). The following spring, sleepy lizards tend to pair again with the same partner (Fig. 8.6). Figure 8.6 extends and confirms results previously reported from a shorter time frame (Bull 1988, 1990). Pair fidelity, often over several years, is common. One pair has now been found together for 10 consecutive years (1982–1991).

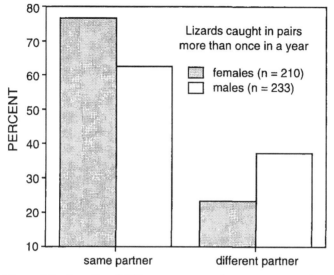

Figure 8.5. Data for all *Tiliqua rugosa* caught in pairs more than once in a season over the whole study area, showing the percentage found with the same partner and the percentage found with a different partner.

The proportion of individuals that continue with the same partner decreases with increasing number of years (Fig. 8.6). There are cases of one partner dying, or of lizards breaking one stable partnership to form another. However, in sleepy lizards the predominant pattern of mate choice is to select the same partner in successive years.

This pattern of mate choice cannot be explained by chance. Radio tracked lizards at Mt. Mary in spring 1992 had home ranges which overlapped or abutted with at least four other lizards of the opposite sex (Fig. 8.7). In these conditions, with random choice, the chance of selecting the same partner after one year is 25%, after two years is 6%, and after more than two years becomes vanishingly small. Although the number of potential partners will change with location and year, the observed incidence of pairing again with the same partner (Fig. 8.6) vastly exceeds any predictions based on chance. Each year, the majority of sleepy lizards appear to prefer, and actively select, previous partners. I know of no other reports of pair fidelity across years in any other lizard species.

Mechanism of Mate Recognition

Paired male lizards do not spend all of the spring period with their female partner. Pairs will separate for several hours up to several days at a time,

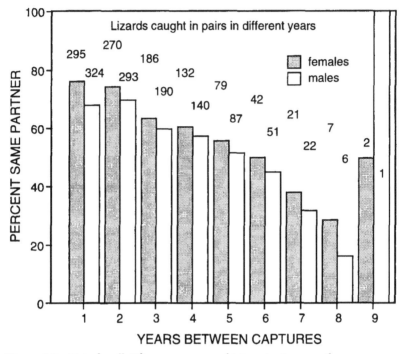

Figure 8.6. Data for all *Tiliqua rugosa* caught in pairs in more than one year, showing the percentage found with the same partner after 1–9 years since the first capture. The number above each bar shows the number of lizards of that sex caught in pairs that many years apart.

foraging apart and refuging in different sites, before relocating each other (Bull et al. 1991). In pairs that separate in this way, either the male or the female can initiate the reunion (Bull et al. 1993). For instance, females move to and locate the refuge site of their inactive male partner, after several days apart. Both sexes relocate partners after short periods of experimental separation (Bull et al. 1993).

We have observed sleepy lizards using frequent tongue-flicks. They tongue-flick at the substrate while following the exact path of their partner and they tongue-flick with their head in the air when they are moving directly towards their partner (Bull et al. 1993). These results suggest that sleepy lizards use both airborne odors and trail-following of substrate-deposited chemicals to locate their partners. Sleepy lizards show no obvious external secretory glands, but I assume there are either skin secretions or secretions from an undescribed gland, perhaps associated with the cloaca.

Observations also suggest that sleepy lizards search for their separated partners in familiar sites (Bull et al. 1993). This implies that they know, and can remember, behavioral patterns of other sleepy lizards.

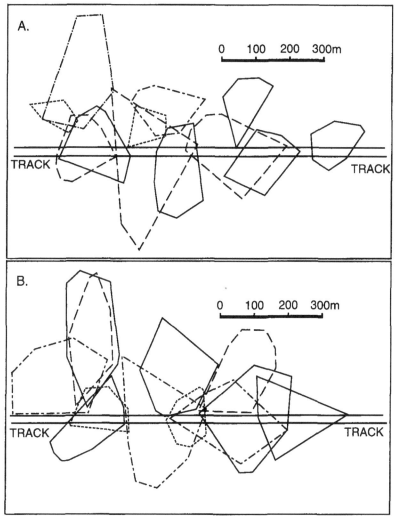

Figure 8.7. Maximum polygon home ranges for (A) 12 radio-tracked female *Tiliqua rugosa*; and (B) 12 radio-tracked male *Tiliqua rugosa* in the 1992 spring at a location on transect 3. Each home range results from over 100 observations of locations per individual on 28–54 days. Note there were other adult lizards in the area that did not have radio transmitters attached.

Pair fidelity implies individual recognition in *T. rugosa*. The cues that sleepy lizards use for this remain unknown. In other lizard species, individuals differ in the chemical composition of their secretions (Alberts 1991). Some lizard species use chemical cues to discriminate between conspecific individuals (Glinski and Krekorian 1985; Graves and Halpern 1991; Alberts

1992). A similar system probably operates in *T. rugosa*. Pair fidelity also implies that sleepy lizards retain individual recognition over several years. Long-term memory about other population members has seldom been suggested in other lizard species.

Functions of Monogamy: Mate Guarding

Monogamy, within a breeding season, has been recorded in many lizard species (Stamps 1983). Stamps (1983) proposed that monogamy is related to body size. In small lizard species, with small home ranges, a single male can defend a territory holding several females. In larger lizard species, females occupy larger home ranges, and the defense of extended territories becomes difficult and costly. Instead males guard single females. Thus, this hypothesis suggests that as body size increases, lizard species tend to change from a polygynous to a monogamous mating system (Stamps 1983).

Sleepy lizards fit the model in that they have large body size, large home ranges, and a monogamous mating system. Some evidence supports the mate-guarding hypothesis. Subjectively, captured males are more aggressive than females, particularly when they are in pairs. In the field, males are occasionally (two or three times in a spring) encountered fighting, but females never fight.

However, the mate-guarding hypothesis assumes females to be passive. This contradicts observations that separated females, as well as males, actively initiate pair reunion. Also, male sleepy lizards voluntarily leave females for several days at a time during the pairing season (Bull et al. 1991), an unexpected behavior if they need to defend females from rival males. These temporarily separated males rarely interact with other females in their home range, nor do other males commonly associate with the separated females, at least as far as can be determined. This suggests little need for mate guarding. Further evidence against the mate-guarding hypothesis emerges from experimental divorces of paired lizards.

Divorce Experiments

Females were experimentally removed (N = 7 cases) or males removed (N = 4 cases) from established pairs in late September or early October of 1990 and 1991. Separation was at least four weeks before copulation occurs in unmanipulated pairs. The remaining 11 divorced lizards were radio tagged, and observed on over 40 days over the following two months. Twenty three control pairs were also radio tagged and observed over 10–54 days in the same springs.

In the experimental design, it was anticipated that the divorced lizards would find new partners. This would allow comparison of the reproductive success of lizards with unfamiliar partners and control lizards pairing with familiar partners from previous years. However, divorced lizards generally did not form new partnerships in the year of their divorce (Fig. 8.8). None of the seven divorced females were found associated with another male, and only one of the four divorced males established a new partnership with a female. Pairs may have formed that we did not observe, but we detected pair associations in most of the control lizards that were monitored with equal or lower intensity (Fig. 8.8).

All divorced lizards were located within the home ranges of other radio-tagged lizards of the opposite sex. In 1991, male and female divorces were deliberately conducted in overlapping home ranges to provide adjacent unpaired lizards of each sex, but no new pairs were formed. Divorced males seldom interacted with other females. Similarly, paired males showed no attention to divorced females in their vicinity, either while paired (even on days of temporary separation), or after copulation, when the original pairs had split up. Since divorced males generally ignore both paired and unpaired females, and paired males ignore opportunities to court additional, unpaired females, the requirement for mate guarding seems questionable. In addition mate guarding does not explain why lizards should choose the same partner each year. Alternative functions for monogamy in sleepy lizards must be examined.

Functions of Monogamy: Mate Priming

Another interpretation of the function of monogamy is that females require male attention to bring them to reproductive condition. A histological study of *T. rugosa* suggested that some females do not reproduce in a year (Egan 1984). In our study, detection of pregnancy in live lizards has been difficult because females retreat into deep refuges, like rabbit burrows, in summer, prior to parturition. Seven paired females radio-tracked in the 1991 or 1992 spring were captured and caged in late summer. Only four of these produced young.

Male *T. rugosa* may prime females for reproduction. In other lizard species, female ovarian functions are stimulated by male courtship (Crews 1975). In sleepy lizards, male courtship may induce ovulation (usually in November; Egan 1984), or receptivity to copulation. There is certainly variation in the proportion of spring days that unmanipulated pairs spend together (Fig. 8.8). Of the seven caged females from the 1991–1992 springs, the four that produced young were those that ranked highest in male attendance time in the previous spring.

Monogamy in a season could be explained if a male has to continue to

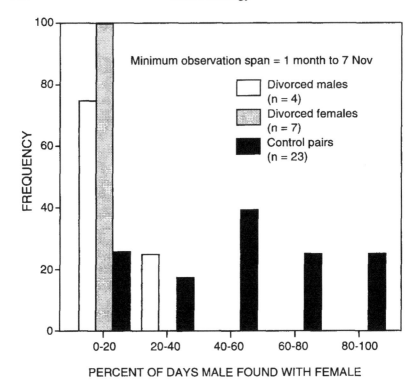

Figure 8.8. Frequency distribution of proportion of time radio-tracked *Tiliqua rugosa* were found in pairs. Data are given only for lizards observed over at least one month up to 7 Nov., a date by when most lizards had copulated and pairs were splitting up. Percent time together is divided into 20% blocks because of the potential errors from observing at intermittent intervals. Data are shown for 23 unmanipulated pairs, seven divorced females, and four divorced males.

attend the female to induce reproduction. He may temporarily leave her, for other activities, if he assesses his presence has not yet made her receptive, but will then return to continue to attempt to induce receptivity. Fidelity over different years may result if familiar males more readily induce female receptivity than unfamiliar males. No mechanism for this is yet known. Attempts to test this hypothesis, by inducing unfamiliar pairs to form after experimental divorces, have so far been unsuccessful, as explained above.

Functions of Monogamy: Attractive Females

Alternatively, females may control the pairing. During spring, a female may terminate reproductive activity if her energy reserves are too low to produce young that will survive. This may occur if current food resources are low or if body reserves are low, for instance as a result of producing a large litter in the previous year. On the basis of their reproductive condition, females may then encourage or discourage male partners. Encouragement may be through production of chemical signals, through actively seeking the male, or through accepting male courtship and copulation. Divorced females may cease reproductive activity because their social environment has been disrupted, and so they will not attract male attention.

This hypothesis could explain monogamy within a season because the reduced number of receptive females would require males to defend reproducing females from other males. The hypothesis fails to explain pair fidelity over consecutive years, because dominant males should theoretically switch from nonreceptive to receptive females.

Perhaps a combination of the two mechanisms, mate priming and female attractiveness, applies. If part of the requirement for a female to reproduce is regular male attendance, then male attention and female reproductive status (and hence attractiveness to males) will reinforce each other. However observations and field experiments are at too early a stage for further differentiation of the various hypotheses to explain the function of monogamy in sleepy lizards.

Conclusions

Two conclusions emerge from this long-term study: (1) Explanations of the organization and dynamics of the population of *T. rugosa* cannot be derived from short-term studies. Long-term investigations comparing individual reproductive success, survival of offspring, and social position of individuals in the population are needed to complement studies of population dynamics over varying environmental conditions. Similar long-term studies of other long-lived lizard species are needed to provide a richer perspective of lizard ecology. (2) Work to date has shown that reproduction in *T. rugosa* is not polygamous. A complex social organization features monogamy in a season and pair fidelity across years and implies long-term individual recognition. Presumably, adult longevity, coupled with stable population dynamics, allows development of this complex organization, while the low recruitment rate of juveniles to the population selects for behaviors leading to high-quality young, rather than large numbers of young.

Any generalizations about lizard ecology must include data from long-lived species like *T. rugosa*, yet that group is generally underrepresented in reviews. This continuing study of sleepy lizards contributes to lizard ecology by increasing the representation of long-lived species.

Acknowledgments

This research has been funded by various grants from what is now the Australian Research Council and by the Flinders University Board of Research. A large body of research students and assistants have helped in the project over the years, but I would particularly like to thank Dale Burzacott for his contribution to the transect survey and Jason Ferris for his help in radio tracking paired lizards. The landowners of the area, particularly Mr. D. Eberhardt and Mr. C. Jaensch, have been generous in their hospitality and in allowing access to their properties. Yvonne Pamula and Greg Johnston made valuable suggestions about a first draft of this chapter.

PART III
EVOLUTIONARY ECOLOGY

The world of evolutionary ecology has changed dramatically since Eric Pianka, Tom Schoener, and I edited the last edition of "Lizard Ecology." So in my introduction here, I want to summarize some of those changes, to evaluate some strengths and weaknesses in the field, and to suggest some new directions. My comments are intended as strictly personal, and not encyclopedic.

I have organized my commentary around the primary theme of "temporal" scale in evolutionary ecology. As ecological patterns can be analyzed on differing spatial scales, evolutionary patterns can be analyzed on differing temporal scales. For example, one can study the evolution of a trait in the past, the present, or the future.

A second common theme is that of integration. For me, a major focus of the excitement in evolutionary ecology has been the incorporation of new perspectives into traditional questions. For example, comparative biology–one of the oldest evolutionary approaches–has recently been enriched by the integration of a phylogenetic perspective.

The Comparative Method: Evolution in the Past

The comparative method–specifically examining patterns of trait variation among a diversity of extant species–was for many years a primary way that evolutionary ecologists investigated historical patterns of trait evolution. The comparative method can also be used to derive insight into historical trajectories leading to contemporary patterns. Accordingly, the comparative method is a keystone of evolutionary ecology.

Lizard ecologists have long utilized the comparative method. For example, Don Tinkle and his students pioneered the use of quantitative, interspecific comparisons of lizards to uncover suites of life-history traits. Similarly, Bill Dawson, Paul Licht, and their colleagues looked at interspecific patterns of covariation of thermal preferences and measures of physiological adaptation to the environment. Further, Tom Schoener quantified patterns of size differences among island populations of *Anolis*, thereby elucidating historical patterns of "Size Rules" for competitive interactions.

All of these comparative studies–and a host of similar ones–were enormously successful, provocative, and influential. Ironically, all would have trouble getting published today! The primary reason traces to insights of British evolutionary ecologists (notably Tim Clutton-Brock, Paul Harvey, and Mark Ridley) who argued that traditional comparative analyses were flawed statistically because species data are not statistically independent.

Mark Ridley, Joe Felsenstein, and Jim Cheverud noted that solutions to this statistical loggerhead could be achieved by incorporating an explicitly phylogenetic perspective. Their insights, and those of many others (e.g., George Lauder, Dave Wake), have recently led to an explosion of phylogenetically based comparative methods and to a revitalization of the entire field. Moreover, phylogenetic considerations are now appreciated to be crucial in selecting species for comparative study, not just for analyzing existing data.

Lizard ecologists have been active in developing and applying these new methods. Emília Martins and Ted Garland published (together and independently) important papers on "comparing comparative methods" and on appropriate statistical approaches; they have also developed software that facilitates efforts of many comparative biologists. Don Miles and Art Dunham have sustained the "Tinkleian" tradition of comparative studies of life-history evolution, but they have now done so in a phylogenetic context. Al Bennett, Ted Garland, and I used phylogenetic approaches in papers that extend the Dawson-Licht-type studies of lizard thermal evolution. Jonathan Losos has similarly used phylogenies to reanalyze Schoener's "Size Rules." Further, Losos, Garland, Miles, and Dirk Bauwens and colleagues have independently used phylogenetic approaches to explore patterns of coevolution of morphology, physiology, and locomotor performance.

Although the incorporation of phylogeny into comparative biology has been solidified only within the last decade, it is worth recalling that one of the first papers to bring a phylogenetic perspective to evolutionary ecology was an underappreciated paper by Ernest Williams (1972). In this bold and synthetic paper, Williams integrated phylogenetic information, invoked the Schoener "Size Rules," as well as considered contemporary ecological data to "reconstruct" the history of the radiation of *Anolis* lizards on Puerto Rico. Williams subsequently extended this approach to other island faunas in a philosophically similar paper in Lizard Ecology II.

The comparative method is certainly an active and exciting focus in evolutionary ecology at the present time. It undoubtedly will continue to remain active, precisely because it serves as a crucial link between evolutionary systematics and evolutionary ecology (and many other disciplines). Moreover, if grant funds become increasingly difficult to obtain, many of us will be forced to become comparative biologists—even without grant funds, we can readily mine libraries for vast amounts of existing comparative data.

Nevertheless, comparative methodology will have to address two gaps if it is to remain a leading area of research. First, it will somehow have to transcend the fact that it is by nature an entirely descriptive approach. As Bennett and I have argued, comparative biologists cannot do manipulative experiments in the past, and this inevitably limits our confidence in any comparative conclusions. All is not hopeless, however, as indirect manipulative experiments are sometimes possible. For example, Barry Sinervo and I used "allometric engineering" to test whether interpopulational differences in locomotor performance were merely a consequence of interpopulational differences in egg size. Second, although today's comparative methods attempt to reconstruct the past, they do so by analyzing data only on living, not on extinct, species. In effect, in reconstructing the past, comparative studies actually ignore the past. Given all of the publicity surrounding the movie *Jurassic Park*, should not we encourage attempts to bring in data on fossils? Jon Roughgarden is one of the few evolutionary ecologists who has attempted to use both the fossil record and plate tectonics as a test of specific hypotheses developed from extant species. Interestingly, Roughgarden extracted body size data from fossils to help test his theoretical models (published in *Lizard Ecology: Studies of a Model Organism*) of taxon cycles in body size in *Anolis*.

Phenotypic Selection in Nature: Evolution in the Present

At the time of Lizard Ecology II, comparative studies of traits were common, but studies of phenotypic selection in nature were quite rare. In a sense, this represented a real gap in evolutionary ecology, for studies of phenotypic selection in contemporary populations are the only way one can study evolution in the present.

Why were studies of selection so rare? Based on my own experience, I would generalize that most evolutionary biologists would be skeptical that selection could be detected in the field, simply because of the widespread belief that selection is generally weak. Further, catastrophic events (e.g., Bumpus's famous winter storm) that may cause strong (thus detectable) selection are sufficiently unpredictable that NSF would hardly fund a proposal that depended on the occurrence of a catastrophe.

Against this presumed background of skepticism, Stan Fox's early documentations of selection on individual behavioral traits came as a shock. Then came a stunning series of papers that really made us skeptics recant. Peter Grant (who incidentally published a commentary in *Lizard Ecology: Studies of a Model Organism*) and his students published an inspiring set of field studies on selection on Galapagos finches. Russ Lande and Steve Arnold provided a conceptual and analytical framework for studying phenotypic selection. Steve Arnold developed an elegantly simple but powerful paradigm ("morphology, performance, fitness;" see Miles' chapter) that stimulated many of us. And finally, Arnold and Al Bennett's laboratory studies with garter snakes showed that individual variation was measurable, repeatable, and genetically based.

These studies opened up a new field of studying selection in natural populations. A number of us (Bruce Jayne, Al Bennett, Ted Garland, Don Miles, Art Dunham, Karen Overall, Howard Snell, Joyce Tsuji, Rickie van Berkum, Butch Brodie, Dirk Bauwens) began to look in detail at individual variation—especially in locomotor performance in reptiles. We have used individual variation to explore whether and how selection was occurring on those traits. This remains an active field!

It is important to emphasize that studies of selection in nature require a demographic base–after all, selection is mediated through demographic processes of birth, death, and dispersal. Don Tinkle was one of the first to argue explicitly that demographic data could be used in combination with other perspectives (e.g., performance) to study the dynamics of selection. In a sense, then, the burst of activity in this area really reflects a growing appreciation of Tinkle's legacy–namely, that a demographic perspective is crucial to any analysis of selection in natural populations.

Where will this field go? Four issues immediately come to mind. First, as the work of Peter Grant and his students shows so strikingly, studies of selection must be done in multiple years to be convincing. Otherwise, observed patterns may merely be idiosyncratic. Having personally been involved with Art Dunham and Karen Overall in such a long-term study, I know first hand that this is a sobering prospect, especially given the difficulty of funding as well as the personal effort of sustaining any long-term study. Second, as Tom Mitchell-Olds, Ruth Shaw, and Dolph Schluter argued, studies of selection need to transcend an exclusively descriptive approach. In lizard ecology, Barry Sinervo's use of experimental manipulations of hatchling size in a demographic context stand out as a model. Third, lizard ecologists need to use DNA fingerprinting to explore the subtleties of selection. Some efforts along these lines are currently being made in several laboratories. Finally, lizard ecologists need to bring in metapopulation approaches, as argued herein by Jean Clobert and colleagues. Most lizard populations do not evolve in isolation.

Quantitative Genetics: Evolution in the Future

At the time of Lizard Ecology II, one would be hard pressed to find the words "heritability" or "genetic correlation" in any paper in lizard evolutionary ecology. Gary Ferguson made some pioneering attempts to study the heritability of life-history traits in *Uta*. However, the real stimulus for many of us was again the empirical work of Arnold and Bennett on garter snakes and the theoretical work of Lande and Arnold.

The relevance of quantitative genetics is simply this: if one wants to predict how contemporary populations will respond in the future to contemporary patterns of selection (above), one must appreciate the underlying genetic architecture of the traits of interest. After all, the response to selection is governed by the direction and strength of selection, and of the genetic architecture of involved traits. In effect, then, (quantitative genetics provides a crucial–albeit controversial–window into evolution in the future.

The first major herpetological studies in quantitative genetics were focused on *Thamnophis*, which for present purposes can be considered honorary "legless lizards." Arnold studied the inheritance of feeding preferences, and Arnold and Bennett did the same for defensive behavior and locomotor performance. Their work led to a variety of studies primarily of locomotor performance of snakes and lizards (Rickie van Berkum, Joyce Tsuji, Ray Huey, Ted Garland, Butch Brodie, Paul Hertz, Dirk Bauwens, Raoul van Damme, Aurora Castilla). Garland and colleagues have greatly extended these studies by studying the quantitative genetics of lower levels of physiology.

What about the future? To be honest I am not very sanguine about the future of quantitative genetics in lizard ecology. The reasons are simple yet compelling. First, robust estimates of genetic parameters–especially genetic correlations–unfortunately require sample sizes that are impractical for most lizards. Second, estimates of narrow-sense heritabilities, which are required for predicting evolutionary responses to selection, require complex breeding designs (e.g., half-sib) that are again impractical for most lizards. Third, the option to use laboratory selection experiments (artificial selection, laboratory natural selection), which are an alternative way of generating genetic insights, are again impractical for most lizards.

I see very few easy solutions available to lizard ecologists. Perhaps the best is simply long-term monitoring of populations that have been introduced–intentionally or otherwise. Selander and Johnston pioneered this approach in their studies of the introduced house sparrow; Jon Losos and Tom Schoener are now exploring phenotypic shifts in *Anolis* that were intentionally introduced to small islands in the Bahamas. Such introduced populations offer splendid opportunities–at least to those who are very patient–for studies in evolutionary ecology. Of course, studies of introduced populations will not directly provide information on genetic architecture, but they will provide direct measures of evolutionary responses to selection–assuming of course that a response to selection has occurred.

Phenotypic Plasticity: Change within a Lifetime

Phenotypic plasticity has been a hot issue in recent years in evolutionary biology, but remarkably little has yet been done in lizard evolutionary ecology. At the time of Lizard Ecology II, very few herpetologists, other than physiological ecologists, had even attempted to manipulate phenotypes. [Indeed, the paradigm has been much more towards the "elimination" of phenotypic plasticity by acclimating test animals to a single set of conditions.] Early exceptions were Paul Licht and Gary Ferguson, who had independently used supplementary feeding in their studies with *Anolis* and *Sceloporus*, and Vic Hutchison, who manipulated acclimation to explore phenotypic plasticity in physiological traits of a variety of amphibians and reptiles. Nevertheless, most lizard ecologists have not yet jumped on this bandwagon.

The impetus for studying phenotypic plasticity in natural populations has, however, just received a kick in the pants via an important theoretical model of Steve Adolph and Warren Porter, who studied patterns of geographic variation in life-history traits of *Sceloporus undulatus*. Most evolutionary ecologists–lizard or otherwise–implicitly assume that geographic variation has an evolutionary basis, and hence invoke *ad hoc* adaptive explanations to explain observed patterns of variation. In contrast, Adolph and Porter integrated climatological, biophysical, and physiological data into a nonevolutionary model and found that plastic phenotypic responses–not evolution–might account for much of the observed geographic variation in life-history traits. Although their simple model is readily subject to criticism by proponents of individual-based physiological models, its conceptual foundation serves as a well-reasoned argument for a reevaluation of phenotypic plasticity in the study of life-history evolution.

Phenotypic plasticity can of course be studied in a variety of ways. Field biologists have several options. Peter Niewiarowski's chapter shows how transplants can be used in this regard. [I personally am surprised that transplant experiments are so rare in herpetology, given that "Clausen, Keck, and Heisey" is (or at least was) required reading in most introductory evolution courses.] Licht, Ferguson, and Clobert et al. (herein) successfully used food supplementation in their field experiments. Clearly this type of environmental manipulation is often feasible with lizards, though sometimes very labor intensive! Karen Overall uses manipulations of the biophysical ecology of lizard eggs to show that the developmental environment of an egg can have profound and long-lasting impact on the future size and survival of lizards. Finally, physiologically structured models, such as those of Adolph and Porter or of Art Dunham, will provide theoretical insights into phenotypic plasticity in field studies of populations.

The study of phenotypic plasticity is much more common in laboratory physiology–although physiologists call this phenomenon "acclimation," not phenotypic plasticity. More such studies in physiology or in functional morphology would certainly be welcome, if directed towards testing specific hypotheses, as in Joyce Tsuji's work with acclimation of metabolism in populations of *Sceloporus*.

A special case of phenotypic plasticity is temperature-dependent sex determination, which–thanks largely to Jim Bull–is well documented now in a diversity of reptiles. These "developmental switches," to use Richard Levins' terminology, are nonreversible during an individual's lifetime. Most of us expect that developmental switching is rare in lizards. In one sense, this may not be true. As Barry Sinervo has shown, the amount of yolk a female adds to her eggs can have profound effects on the size of adult lizards. Similarly, Karen Overall finds that the environment of an egg has a profound influence on the future size of lizards. In effect, if lizards mature at a small size, they stay small.

Another underexplored area of phenotypic plasticity involves cross-generational environmental effects (e.g., maternal effects), which can have profound effects on evolutionary trajectories. Chapters by Jean Clobert et al. and Sinervo represent useful attempts to study such cross-generational effects in field context. However, investigations of such effects in lizards are still quite rare.

Concluding Remarks

The most exciting developments (for me personally) in lizard evolutionary ecology are studies that involve systematics or demography or both into the investigation of other areas. Viewed from a distance, this is perhaps not very surprising. Systematics provides a necessary foundation for analyzing historical patterns of evolution. Similarly, demography provides a necessary foundation for understanding patterns of selection in the present, and in so doing for helping to predict evolution in the future. I suspect that lizards will continue to play an important role in such studies because (1) lizard systematics is good and getting better fast and (2) lizards are perhaps the paradigmatic animal for many demographic studies.

I find it refreshing that lizard ecologists are increasingly developing experimental approaches to test various hypotheses. Probably the first experimental studies with lizards were directed toward competition, island biogeography, and life-history traits (Schoener, Dunham, Roughgarden, Pacala, Ferguson); but experimental manipulations (Sinervo, Clobert, Cathy Marler, Henry John-Alder, Mike Moore) are now being directed at a variety of issues.

Of course, lizards are not ideal for all studies in evolutionary ecology. Lizards are hardly the animals of choice for quantitative genetics, for laboratory selection experiments, and for studies of phenotypic plasticity. More seriously, the molecular genetics of lizards is regrettably a non-field. This should be rectified. Otherwise, lizard evolutionary ecologists will be left behind, while many others will move along on a vicariant, intellectual (and funded) event.

Much of my own very recent work has been with *Drosophila* as well as with lizards. I am not the only "herpetologist" who has recently began working with such lower forms of life. For example, Tom Schoener is actively studying the ecology of spiders as well as that of *Anolis* in the Caribbean, Jim Bull is manipulating the laboratory evolution of phage and bacteria, Al Bennett is doing selection experiments to study thermal adaptation in bacteria, Martin Feder is investigating molecular physiology of stress responses in *Drosophila*, and Ted Garland is doing artificial selection on activity and performance in house mice.

Should these herpetological "defections" be seen as a cue that lizard evolutionary ecology is moribund, or at least so perceived by certain people? I personally do not think so. My own rationale may be instructive in this regard. I decided to start working with flies because I was intrigued by certain questions that had evolved specifically from Bennett's and my research with the evolutionary physiology of lizards. Much to my regret, however, I could not easily answer these specific questions with lizards; but I could with flies. Thus lizards motivated the questions, but the questions motivated a switch from lizards. I know that Al Bennett and Ted Garland shared similar motivations.

Consequently, to ask whether lizard ecology is or is not a moribund field is really an inappropriate question. Further, it is also inappropriate to ask whether lizards or flies or bacteria or birds or mice are better models for evolutionary ecology. The reason is that no single animal is the model for all of evolutionary ecology. Lizards are great–and will remain great–for many questions in evolutionary ecology, but not for all.

                                         Raymond B. Huey

CHAPTER 9
DETERMINANTS OF DISPERSAL BEHAVIOR: THE COMMON LIZARD AS A CASE STUDY

Jean Clobert, Manuel Massot, Jane Lecomte, Gabriele Sorci,
Michelle de Fraipont, and Robert Barbault

Habitat fragmentation is a major environmental problem facing conservation biology (Soulé 1986). Theoretical studies have considered the effect of habitat size and isolation on the persistence of a species in networks of habitats (for review, see Hanski 1991a). Numerous empirical examples have been provided to illustrate the importance of the size of a patch (Schoener and Schoener 1983) and of the connectivity among patches (Stamps et al. 1987; Fahrig and Paloheimo 1988; Fahrig 1990; Sjögren 1991). However, the importance of within-patch population size, structure, and dynamics was only recently considered in studies of among-patch dynamics (Hanski 1991b; Hasting 1992; Lebreton and Gonzalez-Davila 1993) and was theoretically investigated more thoroughly in so-called source-sink metapopulations (Pulliam 1988; Howe et al. 1991; Davis and Howe 1992). Specifically, the type of competition within patches was shown to be a potential determinant of the movements between patches (Davis and Howe 1992).

At the population level (within patch), emigration and immigration processes have long been recognized to be major factors of population regulation (Lidicker 1962, 1975; Krebs 1992), and a large body of literature focuses on why, who, and under what circumstances an animal definitively leaves its natal site or its previous breeding place (Swingland and Greenwood 1983; Chepko-Sade and Halpin 1987; Bunce and Howard 1990; Stenseth and Lidicker 1992). The time lag in recognizing the importance of intrapatch dynamics (population studies) at the level of among-patch dynamics (metapopulation studies) could be better understood by determining why individuals move.

Dispersal is generally believed to have evolved in response to three main problems (Fig. 9.1): habitat instability, intraspecific competition, and inbreeding depression (Johnson and Gaines 1990; MacDonald and Smith 1990). However, dispersal can entail costs: (1) changing habitat may reduce food intake by reducing efficiency of food searching due to unfamiliarity with a new habitat, or because the habitat is of poorer quality (Lidicker 1975; Shields 1987); (2) individuals changing social environment may encounter higher aggressiveness from nonfamiliar or nonkin individuals (M'Closkey et al. 1987a; Shields 1987; Anderson 1989; McShea 1990), or will possibly preclude the evolution of helping (Koenig et al. 1992);

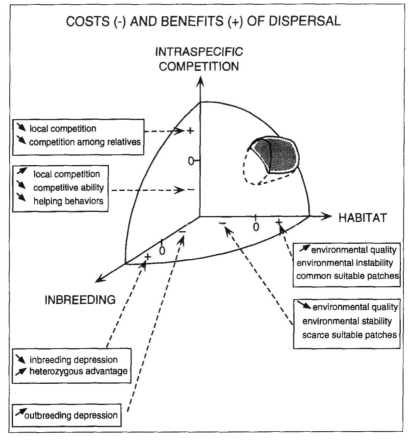

COSTS (-) AND BENEFITS (+) OF DISPERSAL

INTRASPECIFIC
COMPETITION

local competition
competition among relatives

local competition
competitive ability
helping behaviors

HABITAT

environmental quality
environmental instability
common suitable patches

INBREEDING

environmental quality
environmental stability
scarce suitable patches

inbreeding depression
heterozygous advantage

outbreeding depression

Figure 9.1. Costs and benefits of dispersal associated with inbreeding, intraspecific competition, and habitat structure. The curved plane is defined by the points where the costs and benefits of dispersal are at equilibrium. Above the plane, dispersal behavior will be selected; below, philopatry will be selected. The volume represents a species where the area above the plane would include categories of individuals selected for dispersal (for example, juveniles in most species), and the area below the plane would include categories of individuals where philopatry constitutes the selected strategy (for example, adults).

(3) changing habitat or social environment will increase the chance of mating with a genetically distant partner, consequently decreasing local adaptation by breaking combinations of coadapted genes (Shields 1983; Anderson 1989; Dhondt *et. al.* 1990; Blondel et al. 1992); (4) in all cases, individuals may experience an increased risk of predation during the dispersal phase and early in the settling phase (Christian and Tracy 1981; Snell et al. 1988). Thus, dispersal appears to be affected by multiple factors (Dobson and Jones 1985; Marks and Redmond 1987; Shroeder and Boag 1988; Lidicker

and Stenseth 1992). Therefore, the balance between benefits and costs of dispersing symbolized along the three axes of Fig. 9.1 (habitat, competition, and inbreeding) is likely to determine the evolution and importance of dispersal behavior for a given species. Variation in propensity to disperse with age, sex, and individual characteristics can be better understood in this framework: individuals of different ages or sexes have neither the same requirements (e.g., not the same food), nor the same social position (e.g., not the same competitive milieu). Each age and sex will have a different position in the three-dimensional space of Fig. 9.1, and each will experience different selection pressures favoring either dispersal or philopatry. This can be generalized at the individual level since individuals also differ in their ability to cope with their environment (broad sense) because they are in different locations, have undergone different histories, or are not genetically identical.

The differences and relationships between metapopulation and population studies can now be better understood. The metapopulation approach has mainly concentrated on problems of change between units of habitat, while population biologists concerned with the evolution of dispersal have mainly considered the problem of intraspecific competition and of inbreeding avoidance at the within-habitat level (Krebs 1992). Intraspecific competition and inbreeding avoidance do not necessarily cause a change of habitat type, so that the scale (in terms of space and probably time) at which these forces act can vary. Hence, it is crucial to identify the respective importance of each variable driving dispersal and at what scale each is acting (Opdam 1992; Krebs 1992).

Reptiles are good candidates for this challenge (Grant 1983; M'Closkey et al. 1987a; Sinervo and Huey 1990), although very few studies have been carried out on correlates of dispersal (Tinkle 1967; Bauwens and Thoen 1980). The aim of our research on the common lizard (*Lacerta vivipara*) has been to integrate a metapopulation approach within an individual-based approach. This enabled us to tease out the respective roles of habitat, intraspecific competition, and inbreeding avoidance in the evolution of dispersal, and to study the population and metapopulation causes and consequences of dispersal. We studied the influence of habitat structure and population density on dispersal. The effect of individual history and characteristics have also been investigated by assessing maternal lineages, by manipulating maternal environment during gestation (e.g., food availability), and by measuring various individual characteristics. Most of these results have been or will be reported in separate papers in more detail (Massot et al. 1992; Massot et al. 1994a,b; Massot and Clobert unpubl.; Lecomte and Clobert unpubl.). Here, these results will be presented in an integrated, synthetic framework.

Methods

The common lizard, *Lacerta vivipara,* is a small (average adult size: 60 mm snout-vent length), live-bearing lacertid inhabiting peatbogs and heathlands. In the populations studied (Mont Lozère, Cévennes, France, 1420 m elevation), the activity season lasts 5 to 6 months (mid-April to mid-September). Males and females reproduce for the first time at the age of 2 years. Mating takes place in May, and females give birth on average to 5 young (20 mm SVL) two months later (mid-July to mid-August). More details on the species life history can be found in Avery (1975), Pilorge (1982, 1987) and Pilorge et al. (1983).

In 1986, we selected two neighboring areas (A: 3500 m² and B: 4300 m²) 500 m apart and separated by inhospitable habitats (Massot et al. 1992, 1994a,b, for a more complete description). Since the beginning of the study, no individuals were found to move between area A and B. These two areas have similar heterogeneity in terms of (1) habitat structure containing a mixture of grass, heath, trees, and rocks and (2) habitat distribution having a central zone with a higher structural diversity (HSD) surrounded by zones with a more uniform structure (low structural diversity—LSD; Fig. 9.2 shows area B). In area B for example, the difference in habitat distribution was accompanied by a difference in lizard density (HSD: 700 adults/ha, LSD: 430/ha), home range size (significantly greater in LSD than in HSD for most categories of individuals; Massot 1992), length at birth (65% of neonates in HSD vs. 40% in LSD having a SVL higher than the average calculated on all individuals; log-linear analysis males: $X^2 = 8.08$, $df = 1$, $P = 0.005$, females: $X^2 = 11.54$, $df = 1$, $P = 0.001$), and date of birth (HSD 4 days earlier than LSD; ANOVA, $F_{1,95}=14.32$ $P < 0.001$), but not adult-female length (ANOVA, $F_{1,213}= 1.21$ $P = .272$). Although these habitats are adjacent (Fig. 9.2), we consider them as two separate units because interhabitat and intrahabitat dispersal may be different (Stenseth and Lidicker 1992).

Maternal lineage was established by removing a maximum (between 50% and 90%) of gravid females from the populations during July, and by keeping them a few weeks in the laboratory until parturition. During this period, we manipulated the maternal environment (e.g., food, density; more details are given in the section on prenatal effects). After measuring the size, weight, and speed of mothers and young, we released them at the mothers' last capture point.

All individuals were captured and recaptured by hand during the entire activity season (between 20 and 30 days of capture for each area per year). At each capture, we recorded the sex, size, weight, number of external parasites, and color patterns of the individual as well as the time, location, and characteristics (e.g., behavior, substratum) of capture. This enabled us to

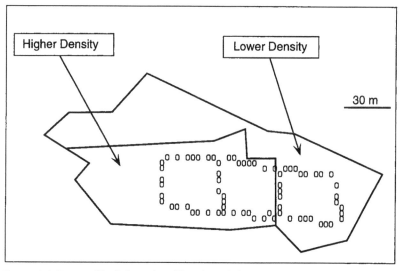

Figure 9.2. Zones of high (HSD) and low (LSD) density in area B (solid lines). The open circles delimit the parts of area B that were fenced. They also symbolize the pitfall traps placed on each side of the openings in the fence. Area A, situated at a distance of 500 m from area B, is not represented here.

estimate average home range size for different categories of individuals (ranging from 20 m to 35 m in diameter, depending on habitat types; Massot 1992). To increase the number of recaptures of juveniles, three adjacent subplots (two HSD and one LSD habitat types) in area B were fenced in 1989, but an opening was left every 4 m (Fig. 9.2). One pitfall trap was placed on each side of each opening, so that lizards could be captured and the direction from which a captured lizard came could be determined (Bauwens and Thoen 1980).

Dispersing individuals may display unique behavioral characteristics (Swingland 1983). The structure of the habitat and the reaction of the species toward the approach of an observer (Bauwens and Thoen 1981) make close observation of a given individual in the field almost impossible. To obtain more detailed descriptions of the behavior of a given individual, we constructed two outdoor enclosures (Fig. 9.3) divided respectively into 12 and 6 boxes (1.5 × 1.5 m). The habitat in each box was standardized, with the most important elements of the natural habitat represented (heath, rock, and grass; for more details see Lecomte et al. 1993). We could therefore remove some individuals with known attributes temporarily from their populations and observe their behavior more intensively in these seminatural enclosures. We assumed that the small size of the box compared to the natural home range was not affecting the comparison of behaviors among individuals, as first observations seem to indicate (Lecomte 1993).

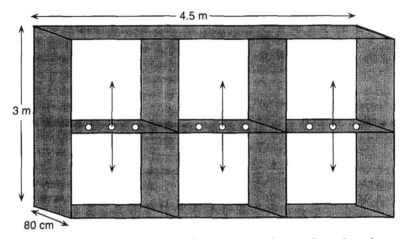

Figure 9.3. A seminatural enclosure of six experimental units where close observations of selected individuals were done. Each unit is a 1.5 × 1.5 m square. The physical structure of each unit was standardized. A piece of rock and of broom was placed in the center of the unit, and each contained grass and heath in as similar a percentage as possible. The holes between units were present only for the juvenile dispersal experiment, where juveniles were allowed to move between two experimental units.

Variance and covariance analyses were performed using SPSS/PC+ (Norusis 1986a,b). Procedure CATMOD (SAS 1985) was used for some log-linear analyses. The type of statistical models used will be specified as required.

Importance of Immigration and Emigration Processes

Many correlative studies have found that immigration—emigration processes were density dependent and were important contributing factors to the regulation of vertebrate populations (Lidicker 1975; Dhondt and Eykerman 1980). Biologists studying small mammals have provided the best experimental evidence of their key role (Krebs 1988), although the results can hardly be generalized since experimental studies based on species with longer generation time are lacking (Sinclair 1989). Population sizes of lizards are thought to fluctuate moderately compared to populations of other vertebrates (Barbault 1975; Schoener 1985; Bauwens et al. 1987), and density-dependent phenomena are consequently suspected to be prominent in lizards. To verify that population sizes were responding to density, and that immigration and emigration processes were important in the common lizard, we experimentally manipulated populations in areas A and B. In August 1986, we removed 500 individuals of various sex and age classes from area A

(HSD and LSD habitats) and transplanted them into the center (HSD habitat) of area B (for a complete description of this experiment, see Massot et al. 1992). After one year, population sizes recovered to the preexperimental levels and many life-history traits responded to the density manipulation (e.g., fecundity, growth rate, juvenile survival rate, survival rate of transplanted individuals; Massot et al. 1992, 1994a,b). The addition of individuals in area B was not followed by increased emigration from HSD habitats to surrounding LSD habitats, and virtually no immigrants from LSD settled into the central zone of B (Fig. 9.4). The readjustment of density in the HSD habitats of the decreased density site A was mainly due to an important immigration of adults, and to a lesser extent yearlings, coming from the peripheral LSD habitats of area A (Fig. 9.4). Marginal subpopulations (individuals inhabiting peripheral LSD habitats in this case) seemed to contribute significantly to the regulation of the central one by buffering its density. In the long term, it would decrease the extinction probability of the central zone and, in turn, of the entire system. This confirmed the predicted role of less demographically productive subpopulations in metapopulation models (Pulliam 1988; Howe et al. 1991). It also emphasizes the importance of adaptability to varying habitat types.

Individual movements were found to be a significant component of the life history of L. vivipara, suggesting that individual variation in this trait could be important and worth examining in greater detail.

Disperser Characteristics and Ability To Settle in Occupied Environments

Dispersal is a three-step process (Anderson 1989; Lidicker and Stenseth 1992) consisting of the departure phase (from the natal site for neonates or from the home range for an adult), the transient phase, and the settlement phase in a new environment. In some species, immigrants are known to differ from residents in morphology (Swingland et al. 1989), life history (Clobert et al. 1988; McCleery and Clobert 1990), and behavior (Pasitschniak-Arts and Bendel 1990). Individual variation in the ability to perform all these steps is therefore likely. We examined the hypothesis that variation was present in our populations by looking for differences in (1) activity pattern between individuals that dispersed and those that did not and (2) the ability of individuals of various age and sex classes have to inhabit a new environment.

In 1988 and 1990, yearlings were temporarily removed and observed individually throughout 5 days in seminatural enclosures (see methods). Only individuals coming from HSD habitats in area B were selected, so that our definition of dispersal is based on a change of social environment rather

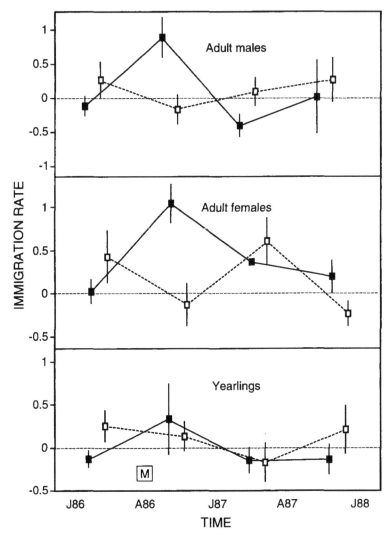

Figure 9.4. Immigration rates into the central zone of area A (reduced population size; solid line and solid square) and of area B (increased population size;
dotted line and open squares). The manipulation of density took place in
August 1986. Immigration rates were the same before the experiment for areas
A and B. Strong immigration compensated for the decrease in density in area
A. Immigration was stopped in area B where the density was increased. Values
of the immigration rates in area A and area B converged again towards the
same values by 1988.

than a change of habitat structure (Cockburn 1992). Individuals were first selected according to the distance between the last capture point and the birth place (last mother capture point): we defined dispersers as individuals having moved more than one adult home range (diameter of 20 m), and philopatric as those that moved less than one adult home range. However, we removed individuals that could not be clearly attributed to a particular class (yearlings that moved 15–20 m). The criteria adopted for the classification of yearlings into residents versus dispersers were slightly different from criteria used in the following sections because the sample size available at this time was small. However, no individual classified as a disperser was observed to return to its natal ground or to include its place of birth as part of the new home range (Massot 1993; Sorci et al. 1994). This removed the possibility that dispersers were just more-active individuals or individuals with a bigger home range than residents. We also verified that there were no length or weight differences between the two types of individuals (SVL: $t = 0.11$, $df = 36$, $P = 0.91$; weight corrected for size: $t = -0.39$, $df = 36$, $P = 0.70$). Broad categories of behavior were distinguished: (1) hiding, basking, and resting in the shade, (2) foraging and walking, and (3) alert and running (see Lecomte et al. 1993). All experiments were performed blind (the observer did not know the status of the observed individual). Each individual was observed two hours per day during 15-minute intervals. The activity pattern was significantly different between dispersers and philopatrics (Fig. 9.5). Dispersers spent more time above ground (basking and walking) than did philopatrics. Philopatrics and dispersers thus differed in their behavior.

The manipulation experiment provided an opportunity to study the settlement phase. By monitoring the fate of the 500 individuals that were introduced in area B, we tried to identify the characteristics associated with successful settlement into the new population. When compared to resident individuals, transplanted individuals that settled successfully were not a random sample of the initial ones. Surviving adult females weighed less, and yearlings and adult males were smaller than nonsurvivors (Massot et al. 1994a). At least for yearlings, this did not result from survivors being younger than nonsurvivors. A small size (controlled for age) was also characteristic of those individuals that naturally immigrated into our study sites (Massot et al. 1994a). Individual variation exists in the probability of settling into a new, already-occupied environment. Smaller individuals may have been permitted by the residents adults to settle, but larger individuals were treated aggressively.

Although both experiments strongly suggest that dispersers differed from philopatric individuals, they suffer from methodological weaknesses that preclude definitive conclusions. The first study pointed out differences in behavior. But we do not know if individuals were different at birth, if

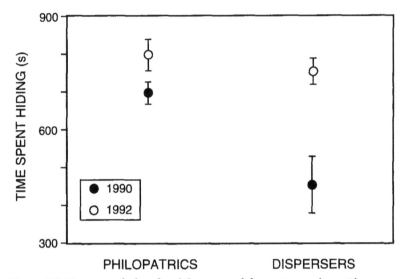

Figure 9.5. Time spent hiding for philopatric and dispersing yearlings. There was a significant effect of year ($F_{1,31}=16.93$, $P < 0.001$), but neither the sex effect ($F_{1,31}=2.25$, $P=.144$) nor the global interaction term ($F_{3,31}=1.39$ $P=.264$) were significant. The status effect was also significant ($F_{1,31}=4.81$, $P = 0.036$.

dispersers and residents experienced opposite selection pressures on their behavior, or if the differences resulted from nonparallel trajectories in their behavioral ontogeny, possibly induced by different past history and social environment. In the second study, individuals that settled successfully in the new population were not necessarily those that would have dispersed in their site of origin (Massot et al. 1994a). In addition, these experiments did not allow us to discriminate between the ability to change social environment and the ability to change habitat structure.

To learn more about factors affecting dispersal, and to investigate the individual characteristics associated with dispersal, we need to follow individuals with known lineage from birth.

Some Pre- and Postnatal Factors Associated with Dispersal

Observed differences in the ability to settle in a new, already-occupied environment, as well as behavioral differences between dispersers and philopatric individuals, can originate at different stages in the life cycle (Ims 1990). They may be caused by the physical structure of the environment or by the social context experienced early in life (postnatal effects, Falconer 1989). They can be present at birth as individual differences in morphology, physiology, or behavior that can be phenotypically induced through maternal past history (long-term prenatal effects), or through the maternal environment during gestation (short-term prenatal effects). Finally, they can be genetically determined.

Area B offered the best opportunity to study postnatal effects on dispersal. It is naturally divided into two zones of contrasting structure (LSD and HSD habitats) that are inhabited by subpopulations with different densities (see methods and Fig. 9.2). Short-term prenatal effects could be investigated independently from postnatal ones by manipulating the maternal environment during gestation. Genetic and long-term prenatal effects are not addressed here.

During gestation, females do not move much and live in a restricted part of their home range (Bauwens and Thoen 1981; Heulin 1984). The last capture point of the female is therefore a good approximation of the future birth place of her offspring. We will define dispersal in two ways (Cockburn 1992): (1) social dispersal: dispersers are those moving more than the diameter of an adult home range (in fact the upper limit of the confidence interval: 30 m). A philopatric individual will be one that moves less than the average diameter of a home range (20 m), those intermediate being excluded; (2) habitat dispersal: dispersers are those moving between zones of different habitat structure. A philopatric individual will be an individual staying in a zone of the same habitat structure. This is not a strict differentiation of dispersal types, since individuals having moved more than one adult home range could also have changed habitat. The distance moved was measured for juveniles by comparing coordinates of the birth place to those of the last recapture point, and for other age and sex classes, by comparing the coordinates of first capture point in year i to the ones of last recapture point in the same year (for those individuals having at least two capture points separated by 10 days). The study started in 1989. From 1990, an average of 80 gravid females were removed per year from area B and gave birth in the laboratory. A total of 7314 captures and recaptures were made by hand or in the pitfall traps of which 1784 were captures of 1311 different juveniles reared in the laboratory.

Age, sex, and habitat effects on dispersal

Social dispersal occurred mainly at the juvenile stage (Fig. 9.6). In densely populated habitats, more than 50% of juveniles dispersed, whereas less then 10% of yearlings and adults of both sexes dispersed ($X^2 = 117.43$, $df = 3$, $P < 0.001$). Males tended to disperse slightly, but significantly more than females when all age classes were combined ($X^2 = 4.67$, $df = 1$, $P = 0.031$).

Social dispersal in juveniles was strongly habitat dependent. In poor and/ or less-populated habitats, juveniles moved less than in densely populated

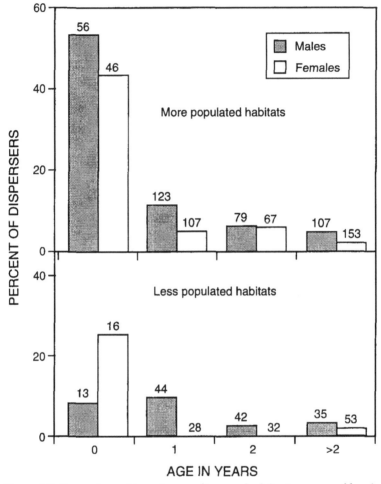

Figure 9.6. Proportions of juveniles, yearlings, and adults (two years old and older) that dispersed. (Upper) Individuals that moved more than one home range in high structural diversity (HSD) habitats (high density). (Lower) The same, but in low structural diversity (LSD) habitats (low density). Sample size appears above bars.

habitats (X^2 = 10.20, df = 1, P < 0.001, and Fig. 9.6). It is difficult to know if this was the result of the habitat structure or of a difference in density induced by the habitat. Social dispersal in juveniles was also year dependent. In 1992, weather conditions during spring and early summer were particularly poor and affected female reproductive success. During this year, only 20% of juveniles dispersed.

Although HSD and LSD habitats in area B had very long boundaries

(Fig. 9.2), habitat dispersal was persistently weak (less than 12%; Fig. 9.7). There was little difference among age classes in this type of movement, although males and young individuals dispersed slightly more than females and older individuals (sex: $X^2 = 8.57$, $df = 3$, $P = 0.036$; age: $X^2 = 4.15$, $df = 1$, $P = 0.042$).

Juvenile dispersal appeared more social than habitat in type. Most mixing between individuals within dense populations in HSD habitat occurred as the result of juvenile movements. However, the amount of mixing was largely habitat- (or density) and year dependent.

Dispersal in yearlings and adults was weak (high philopatry), but had both a social and habitat origin. We will therefore concentrate our effort on understanding heterogeneity in juvenile dispersal.

Resemblance among related juveniles

Observed variation in juvenile dispersal can originate from factors acting either before and/or after birth. One way to investigate this is to examine the resemblance among related juveniles. We restrict this analysis to the more densely populated zone (HSD), because juvenile dispersal was significant in this zone.

Social dispersal was strongly family dependent ($X^2 = 58.08$, $df = 31$, $P = 0.002$), regardless of the sex of the offspring ($X^2 = 16.56$, $df = 30$, $P = 0.977$). Furthermore, brothers tended to disperse in the same direction ($X^2 = 14.00$, $df = 1$, $P = 0.001$), as did sisters ($X^2 = 9.00$, $df = 1$, $P = 0.003$). However, brothers and sisters did not move significantly in the same direction ($X^2 = 2.88$, $df = 1$, $P = 0.090$), but they did not actively avoid each other either (only one angle more than 135 degrees). If inbreeding depression is an important factor, it does not seem that sex differences in dispersal rate or dispersal direction has evolved to systematically avoid it. Very small differences in rate and/or direction may be sufficient to keep close inbreeding at a low frequency, in particular in a high-density situation. However, avoidance of close inbreeding may have evolved through other mechanisms such as kin recognition (Anderson 1989).

Similarity in dispersal resemblance among kin could be caused either by postnatal factors such as sharing a common location or experiencing the same social environment, by prenatal factors such as the short- or long-term history of mothers, or by sharing the same gene pool. Observational data cannot differentiate among these factors, so we conducted the following experiments.

Postnatal effects: Location and adult density at birth

To tease out postnatal from prenatal factors, we performed two experiments in 1991: one designed to study the location effect independently from the family effect (sharing the same mother), the other designed to study the density effect independently from the family effect.

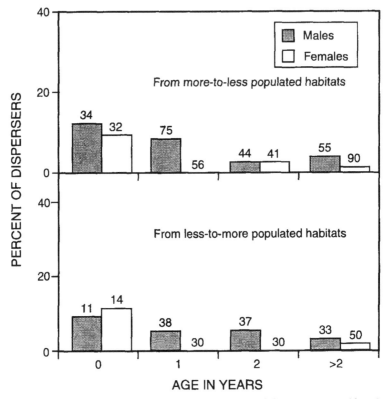

Figure 9.7. Proportions of juveniles, yearlings, and adults (two years old and older) that dispersed. (Upper) Individuals that moved from high-to-low density zones. (Lower) Individuals that moved from low- to high-density zones. Sample size appears above bars.

Family and location effects

For this experiment, we selected 22 females near the end of their gestation period from a population adjacent to area B and maintained them in the laboratory until parturition. At birth, we randomly assigned offspring of the same mother to different terraria. Young (on average 5) from the same terrarium (but of different mothers) were then released at the same location in area B. Ninety-four young were released, of which we recaptured 42, but only 8 different terraria (from 8 different locations) and 6 clutches provided at least 2 captures of different individuals. The family effect was not significant (log linear analysis: $X^2 = 7.84$, $df = 5$, $P = 0.165$). The effect of location was close to significant (log linear analysis: $X^2 = 12.96$, $df = 7$, $P = 0.073$).

Postnatal factors such as sharing the same microhabitat (same location within a habitat type) or the same social environment seemed to influence dispersal. At this stage, prenatal or genetic factors did not seem to be determinant, although we must not dismiss them too early because of the very small sample size on which the analysis was conducted.

Family effect and density

Density and/or habitat are important postnatal factors affecting dispersal of the common lizard. In the field, they were associated with one another and with family, so that we could not separate them in statistical analyses. In 1991, we removed 25 pregnant females from a peatbog near area B (150 m). They were reared in the laboratory until parturition so that it was possible to establish offspring maternal lineage. To examine the role of density at birth, we used the 6-box seminatural enclosure (see methods). Each experimental unit consisted of two boxes connected by small holes that allowed only passage of juveniles (Fig. 9.3). At one end of the unit, we placed adult females at different densities (0, 1, and 2 in a 1.56 m²), the other end remaining empty. After one day, 4 young were put in the adult female side, so that three randomly selected siblings experienced different female density. This enabled us to study density and family effects independently. The variable measured was the number of juveniles that passed definitively to the empty side. Each experiment lasted 8 days and was replicated 4 times with different families. The family effect was significant (log-linear analysis, $X^2 = 27.12$, $df = 15$, $P = 0.028$). Individuals coming from the same mother had a similar tendency to colonize empty boxes, whatever the density level. The effect of density was also significant (log-linear analysis. $X^2 = 3.32$, $df = 1$, $P = 0.034$, one-tailed test). Offspring experiencing high densities dispersed more than those experiencing low densities.

This experiment confirmed the role of postnatal factors, here density. It also demonstrated that what happened before birth (genetic and/or maternal effect) also influenced juvenile dispersal.

Prenatal effects: Maternal feeding during gestation

Many prenatal factors are known to influence the future fate of juveniles (Herrenkohl 1979; Christian and Tracy 1981; Bernardo 1991; Kaplan 1992). Maternal characteristics are known to influence juvenile phenotype (Van Damme et al. 1989; Sinervo and Huey 1990). Maternal size, weight, habitat, and social position are the result of past history (long-term maternal effects) and genetics.

Maternal clutch size, clutch weight, and overall investment in reproduction for a given year are the result of both long-term effects (genetic as well as environmental) and short-term effects (environmental conditions during gestation; Pilorge et al. 1983; Ford and Seigel 1989). Covariances between genotypes and environments restrict the interpretation of correlative studies. The study of short-term maternal effects can be best done through an experimental approach, for example by manipulating female environment during gestation (Ims 1990; Sinervo 1990; Bernardo 1991).

As far as we know, very few studies have investigated the influence of short-term maternal effects on dispersal (Ims 1990; Hanski et al. 1991). Having the pregnant females in the laboratory allowed us to manipulate their rearing environment (Massot and Clobert unpubl.). In 1990 and 1991, females from area B were offered two rates of food delivery (one larva of *Pyralis farinalis* every one or two weeks). At birth, juveniles from mothers experiencing food shortage were not smaller that juveniles from well-fed mothers (body length or SVL: $F_{1,200} = 2.28$, $P = 0.240$). However, they were slightly, but significantly, thinner (body weight corrected for body size: $F_{1,196} = 5.28$ $P = 0.023$). We then released them in the field, at the mothers' last capture point.

The analysis of data on dispersal raised a statistical difficulty. All young could not be included in the analysis because sibs cannot be considered as independent statistical units (particularly if genetic or a phenotypic covariance exists among sibs). Disregarding this could cause important discrepancies in the results (Massot et al. 1994b). The small number of brothers or sisters recaptured per family prevented us from using nested covariance analysis, and using only one randomly selected young per mother would have reduced the amount of information. Massot et al. (1994b) proposed the use of numerical resampling techniques (Efron 1982) to solve this problem. The distribution of the tests (log-linear analysis) calculated on 500 resamplings led to the conclusion that giving more food to the mother during gestation increased the dispersal rate significantly (Fig. 9.8). Juveniles from well-fed mothers dispersed at a higher rate than those from mothers experiencing a food shortage. The maternal feeding effect did not result from the confounding effect of juvenile corpulence, since juvenile corpulence (as well as snout-vent length) did not influence dispersal rate significantly (log-linear analysis: median of the tests distribution done on 500 resamplings for males $m = 0.17$, 95% confidence interval 0.059–0.643, and for females $m = 0.57$, 95% confidence interval 0.301–0.939).

Experimentally, a short-term prenatal factor was found to influence juvenile dispersal. Short-term maternal effects during gestation could be a cue for the juveniles to assess the quality of their future environment (Kirkpatrick and Lande 1989; Sorci et al. 1994). However, a higher rate of maternal feeding increased dispersal. A similar result was also found for the imperial eagle (*Aquila adalberti*) by Ferrer (1993). If the sensitivity of juvenile dispersal to maternal effects had evolved to respond to local environmental variation, we would have predicted the opposite trend: individuals being in a poor local environment should have dispersed further. The competitive displacement hypothesis (Anderson 1989) would have led to the

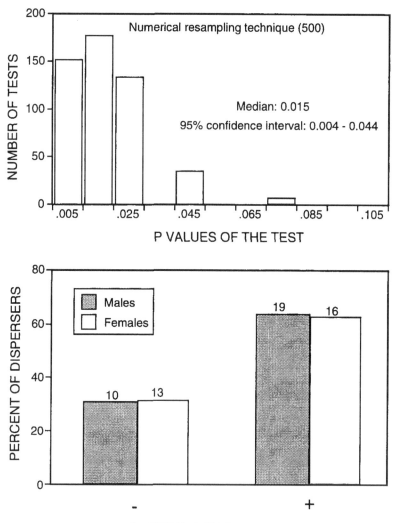

Figure 9.8. (Upper) Comparisons of the percentage of dispersing juveniles born from females experiencing food shortage with juveniles born from well-fed females. Each test was designed using one young per female. Young from each female were selected randomly, and the number of possible resamplings exceeded 1000. The distribution was determined based on 500 resamplings. (Lower) Effect of maternal feeding rate on juvenile dispersal. Both sexes responded in the same way. The total sample size in each category is given above the bars.

same prediction: less fit juveniles forced to disperse further. Since this was not the case, these explanations are precluded (dispersal as a luxury?). The food result must be explained in the light of the other results on habitat and density.

Discussion

Dispersal seems to be determined by multiple factors (Dobson and Jones 1985). Both social dispersal and, provisionally, habitat dispersal appear important in the common lizard based on our experimental results.

Habitat dispersal

Habitat dispersal may be better understood in the framework of metapopulation theory, and particularly in source-sink metapopulations (Pulliam 1988; Howe et al. 1991).

In our case, the two habitat types (LSD and HSD) had very long boundaries in both areas (A and B), so that the use of the concept of metapopulation could have been questionable in this context. However, lizards inhabiting the LSD and HSD habitats differed in several of their life-history traits. Furthermore, few individuals were found to move between habitat types under natural conditions, while the manipulation experiment proved that, when the opportunity is given, individuals from LSD habitats were able to colonize HSD habitats. Each patch could therefore be described as an open population where members have more contact with other members of the same patch than with members of other patches. Each area could thus function as a metapopulation. Although we have not yet calculated the population growth rate of each element of our metapopulation (area B), differences in body size, fecundity, and density are enough to conclude that a difference in demographic regimes exists, and that zones of high density (HSD) are potential "sources," and zones of low density (LSD) potential "sinks." In our case, the definition of source and sink populations is not based on a population growth rate above or below unity, but just that one population is performing better than another. This last feature and the fact that our populations were not geographically separated did not constitute a problem when confronted with theory since recent models of metapopulation have relaxed these assumptions (Davis and Howe 1992).

The density manipulation demonstrated that when numbers were reduced in the source population, regulation of its numbers occurred by habitat dispersal, primarily by yearlings and adults. This was consistent with predictions from some models of source-sink habitats (Howe et al. 1991). However, the exportation of individuals from the source to the sink did not increase when the source population size was experimentally increased

(Massot et al. 1992), as is currently assumed when interpatch density-dependent dispersal rate is considered (Pulliam 1988; Hanski 1991b; Davis and Howe 1992; Hasting 1992). In addition, habitat dispersal is weak in nonexperimental situations. These two features may well be characteristic of a "sink" habitat being capable of self-maintaining its population size and not being a true sink (but see Davis and Howe 1992).

Now, why dispersal by yearlings or adults? Individuals from the sink (LSD habitats) are smaller at birth and when entering the source, they will be less competitive not only with respect to older individuals (yearlings and adults; Pilorge et al. 1987; Massot et al. 1992), but also with respect to individuals of the same age class (immigrant males are smaller-sized than residents; Massot et al. 1994a). Thus, delaying dispersal may reduce the number of individuals (only yearlings and adults) with which they will enter into competition when in the source. Increased social experience may also translate into a behavioral strategy allowing coexistence with dominant adults and larger yearlings. A matrix population model (Caswell 1989) was used to compare the two types of strategies: population change at the juvenile stage and population change at yearling stage (or when yearling compared to when adult). The dominant strategy, in presence of a density-dependent fecundity and migration rate, was to change populations when 1-year old when compared to juveniles (Fig. 9.9), or when adult compared to yearling. At least in our case, the change between populations with different habitat structure or with different densities is not unique to juveniles (see also Tinkle 1967).

Under nonexperimental conditions, however, habitat dispersal was weak, and juveniles moved between habitats at a rate only marginally higher than the other age classes. High juvenile social-dispersal rate probably causes this. Social dispersal can result in change of habitat without habitat dispersal being the principal motivation of the movement.

Social dispersal

The lab experiment suggested that density is one of the postnatal factors inducing departure of juveniles from their natal area. Also, juvenile dispersal was highest in the high-density zones of area B and was weak in low-density parts of the same area. As noted before, this did not entail a change of habitat: most dispersing juveniles stayed in high-density zones. In our case, the importance of the social dispersal in metapopulation dynamics is difficult to assess and at a first glance seemed to have only indirect or delayed effects such as (1) reinforcing the resilience of the source population (HSD habitats), (2) increasing the resistance to immigration from surrounding populations, and (3) keeping a high level of heterozygosity that promotes resistance to fluctuations in the environment. The role of social dispersal will probably be better understood in an evolutionary framework.

In lizards, density is likely to be used as a cue for assessing the quality of a habitat, and the settlement of juveniles is positively related to density

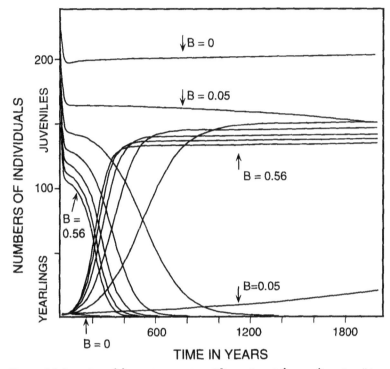

Figure 9.9. Invasion of the mutant strategy "dispersing at the yearling stage" in a
source-sink type of metapopulation where the resident strategy is dispersing at
the juvenile stage. Invasion curves are given for different values of the population growth rate in the sink relative to the source. B is the difference in growth
rates between the source and the sink. The greater the difference in growth rate
B, the quicker the strategy "dispersing at the yearling stage" invades the metapopulation. Fecundity is density dependent both in the source and in the sink.
The migration rate of the sink to the source is dependent on the population size
in the source. The migration rate from the source is also dependent on the density in the source, but it will include only 20% of the individuals (80% being
philopatric whatever the density). Each population was a three-age class Leslie
matrix (without senescence) where survival rates are 0.45 for adults, 0.35 for
yearlings, and potential fecundity multiplied by juvenile survival rate is 2.3.
They correspond to maximum values observed in the field. The demographies
in the sink were varied by changing yearling survival rate, but results are similar
for variation in any other parameters. The ESS were found using ULM program
(Legendre et al. 1993).

(Stamps 1987, 1988). This implies that juvenile dispersal should have more
of a social than a habitat origin, and that density motivates both dispersal
and settlement. This mechanism may explain why juveniles are not changing
habitat structure, but does not explain why they disperse more in high-density populations. Some insight may come from the examination of the family
effects that were detected, and of the theory on social dispersal.

Dispersal rate in low-density zones was too low to test for differences among families, suggesting that postnatal factors dominate over prenatal factors in such situations. Family effects influenced juvenile males and females equally, including the direction of dispersal. An absence of sex differences together with a similarity among brothers and among sisters in the pattern of direction cannot be explained by the inbreeding avoidance hypothesis, although close inbreeding could still be avoided by evolving kin recognition (Fletcher and Michener 1987), as the results of first experiments on mother recognition indicated (de Fraipont et al. unpubl.). The theory of social dispersal currently involves an evolution of dispersal through competition between parents and offspring (Marks and Redmond 1987), competition that will have different evolutionary outcomes depending upon the sex of the juvenile, the social system, and the life-history traits of the species (Greenwood 1980). Theory often advocates that less-competitive juveniles will be driven out of their natal area (social subordination hypothesis: Anderson 1989; Arcese 1989; Hanski et al. 1991; Brandt 1992), implying that dispersers are inferior individuals, particularly in high-density populations (forced dispersal; Lidicker and Stenseth 1992). We have shown that individuals born from females experiencing a food shortage weighed less than individuals born from well-fed females and are therefore likely to be less competitive. However, during the two years the experiment was repeated, juveniles produced by well-fed mothers dispersed more than the other ones. This does not support the hypothesis that dispersers are poor competitors and calls for reexamination of this phenomenon, at least in *Lacerta vivipara*.

Two scenarios for evolution of dispersal in the common lizard

The first scenario is a variation on the previous hypothesis. Superior competitors are produced by mothers living in good local environments during gestation (location effect). Females in such locations should not invest much in reproduction relative to the available resources and should therefore have high survivorship and reproduce the following year (because of decreased costs of reproduction and being in a high resource location). If offspring of these females remain in the same location, they should have a high probability of competing with their mother or father, as well as their sibs. If the environment varies sufficiently at the local scale (within HSD habitats), some females will be in poor local environments and will both (1) have a higher probability of dying because of a higher relative investment in reproduction and (2) produce juveniles that are less competitive. In such a case, it will be advantageous for juveniles produced in good local environments to leave their natal area (competition with the parents and sibs) and, because they are more competitive, to search for other locations where juveniles are of poor quality. For the same reason, individuals produced in poor environments should not move. The probability of competing with parents is weaker (because the probability that the mother will die is higher), and the chance of finding less-competitive juveniles outside of their natal area is small (because they were born in a poorer than average local environment). Some evidence for this comes from a study on parasitism of the parent female and its influence on offspring sprint speed and dispersal (Sorci et al. 1994). In female offspring, philopatry was increased when the mother had a high number of external parasites (indication of a poor local environment). Sprint speed at birth was also increased for females from parasitized mothers, which can be interpreted as a preadaptation to live in an environment with a high number of parasites (Sorci et al. 1994). Philopatric females born from a parasitized mother also had a higher survival rate than philopatric females born from nonparasitized mothers (Sorci et al. 1994). Philopatry could therefore be viewed as an adaptive strategy in poor local environments. However, males did not react in the same way indicating that our scenario is still incomplete.

If the general outline of the above scenario is correct, habitat dispersal can be viewed as a by-product of dispersal caused by local variation in the environment, with high-quality juveniles avoiding colonization of large zones of less-suitable habitat as indicated by their low density (Stamps 1987, 1988). However, this hypothesis implies that environmental quality at the local scale is essentially unpredictable from year to year. Indeed, a high-quality individual leaving a good local environment and becoming established in a predictably poor habitat should be at a selective disadvantage. For example, external parasites in our case had a patchy distribution (Sorci et al. 1994) that varied from year to year. This scenario also implies that the cost of reproduction in this species depends on the environment, although competition among sibs could be sufficient to initiate juvenile dispersal in good locations (Morris 1982).

The second scenario favors more ultimate explanations. It is based on results of the yearly, habitat, and family variation in dispersal and contains some features common to the first scenario. In bad years in less-populated habitats (poorer habitats) and in local environments that are poor during gestation, juvenile dispersal is weak. In good years, densely populated areas provide good local environments during gestation and juvenile dispersal is high. Both juvenile males and females react in the same way, and within families there is no sex difference in dispersal tendency, although males dispersed slightly more than females overall. Therefore, in poor conditions when siblings remain within the home range of their parents, the probability of mating with a kin is higher, whereas under good conditions when siblings disperse in a direction random with respect to the opposite sex, the probability of mating with an unrelated individual is increased. Individuals in poor environments should experience strong selection, and only those individuals with characteristics that increase their fitness in such conditions will survive (e.g., parasitized mothers that produced female offspring with a higher sprint speed than nonparasitized mothers; Sorci et al. 1994). To maintain these adaptations, surviving individuals should selectively mate with individuals bearing the same characteristics, i.e. inbreeding should be an advantage. In other words, inbreeding may not be avoided, or may even be favored, in poor habitat conditions. Inbreeding will increase the retention of coadapted genes that will in turn favor local adaptation (Shields 1983), a particularly interesting strategy when conditions are harsh or severe. In good environments, this strategy is less advantageous (inbreeding depression becoming more costly), and outbreeding can be favored to increase heterozygosity. There is therefore a balance between inbreeding and outbreeding, which should preserve genetic diversity as it relates to environmental variation (Shields 1983; Lynch 1991b). In this framework, habitat dispersal will be the most important force driving the evolution of dispersal, social dispersal being only a by-product, as local genetic coadaptation is probably important at or above the habitat scale. Density at birth will act only as a triggering factor, a good environment during gestation being a good indicator of the overall suitability of the area. If local adaptation (or individual optimization: individuals selecting the local environment most appropriate for their own characteristics) is important at the local level (within a unit of habitat), then variations in environmental quality at the local scale should be weak in any given year. This also implies that dispersers have some fitness advantages over nondispersers under good conditions, but not in poor ones, where the opposite should occur.

Final remarks

If remaining at home or leaving the natal area constitute strategies, we may expect that individuals following these strategies will also differ in other phenotypic traits such as behavior (activity pattern, ways of exploiting the habitat) and life history (e.g., growth rate, age at maturity, age-specific survival), but will have equal overall fitness in the long term. Part of our results are based on the description of a single situation (lack of replication), area B. Habitat dispersal may be more important in other metapopulations, and analysis of data collected in area A will provide a good way to test for the generality of our findings. A better description of the dynamics of populations in different habitat types will be necessary to understand more completely the interplay between habitat and social dispersal and its consequences on population persistence and metapopulation dynamics.

These points and testing the predictions outlined by the two proposed scenarios will constitute our future lines of research.

We have studied movements of the lizard *Lacerta vivipara*, a species that occurs at high enough densities that movements can be examined at the individual and population level. As in other field-oriented studies, it is difficult to quantify rare movements such as long-distance dispersal, or even to document their existence (Stenseth and Lidicker 1992). The study of long-distance movements probably can be best accomplished only by using genetic tools such as DNA fingerprinting (Barton 1992).

Acknowledgments

It is a pleasure to thank all the people who made this study possible. The CNRS (PIR), the Ministry of Environment (SRETIE), and the French Biodiversity Program financially supported this research. The Cevennes National Park and the Office National des Forêts provided most facilities to do our field work. Y. Michalakis, J.H.K. Pechmann, and an anonymous referee critically reviewed a previous draft of this chapter and gave many constructive suggestions and comments. We also thank L. J. Vitt and E. R. Pianka for providing the opportunity to report our work.

CHAPTER 10
COVARIATION BETWEEN MORPHOLOGY
AND LOCOMOTORY PERFORMANCE IN
SCELOPORINE LIZARDS

Donald B. Miles

A common feature among coexisting assemblages of lizards is the tendency to see divergence in habitat exploitation patterns of the constituent species. Coincident with this pattern is the tendency for the species' body proportions to vary in parallel. Similarly, the radiation of a monophyletic clade may lead to the divergence of species with respect to habitat occupancy, substrate use, and morphological attributes (e.g., Williams 1972). In seeking to delineate the interaction between the attributes of an organism with its environment, ecologists exploit a number of approaches, including foraging ecology, behavioral analyses, physiological ecology, and patterns of life-history variation. However, repeated patterns of similar morphological characteristics that correlate with specific habitat types, vegetation configurations, or other ecological attributes of species suggest a strong connection between morphology and ecology. Under these circumstances, ecologists may explore the patterns of covariation among a suite of morphological traits to determine the adaptive significance of morphological diversification (e.g., Lauder and Liem 1989; Losos 1990a).

Questions regarding species—environment correlations, patterns of community organization, and the importance of phylogenetic effects on ecological patterns have been addressed using an ecomorphological approach. Earlier ecomorphological analyses were based either solely on patterns of covariation among morphological traits or the correlation between morphology and ecology (e.g., Collette 1961; Moermond 1979a; Ricklefs et al. 1981; Miles and Ricklefs 1984; Pianka 1986; Douglas 1987; Miles et al. 1987; Scheibe 1987; Losos 1990a). These studies demonstrated that a species' morphological attributes correspond with various ecological and environmental variables. However, the validity of inferences generated from these studies was based on two critical assumptions. First, the correlation between the phenotype and ecology is a result of the action of natural selection operating on the phenotype (Douglas 1987; Ricklefs and Miles 1994). Second, the functional correspondence between morphology and ecology is mediated by performance of the organism at one or more ecologically relevant tasks (Losos 1990a; Ricklefs and Miles 1994). With few exceptions (e.g., Arnold and Bennett 1988; Losos 1990a), the assumption that similar morphological attributes among species have a common functional basis and result in comparable performance capacities has rarely been tested. In part, this may be attributable to the absence of an appropriate theoretical framework in which to express the relationships among morphology, performance, and ecology. In short, ecomorphological analyses are rooted in the assumption that variation in the form-function complex represents adaptations to prevailing environmental conditions (e.g., Bock 1980; Bock and von Wahlert 1965). Consequently, linking variation in morphology to extrinsic selective features of an organism's environment requires understanding how variation in morphology leads to differences in performance.

A number of authors have discussed the need to determine variation in ecologically relevant measures of performance in order to assess the interaction between morphology and various components of the environment (e.g., Bock 1980; Bock and von Wahlert 1965; Huey and Stevenson 1979; Arnold 1983). Variation in performance may affect an organism's ability to exploit specific ecological opportunities (Huey and Stevenson 1979). Conversely, variation in morphology may constrain or enhance performance (Ricklefs and Miles 1994).

A statistical formalization of the linkage between morphology, performance, and fitness was presented by Arnold (1983) who described a model, based on path analysis, that described a framework for studying morphological adaptations. A critical component of the model was that ecologically relevant measures of performance linked morphology to fitness. Thus, the connection between morphology and fitness occurred through an intervening variable, performance. This model may be expanded to include the relationships between morphology and ecology (Fig. 10.1). Here, morphology and some ecological variables are linked by ecologically relevant behaviors or performance capacities. A given point in the morphological space may be connected by a performance path. Specific paths may be precluded due to morphological or energetic constraints. An underappreciated feature of ecomorphological correlations is that performance mediates the strength of the derived relationship (Losos 1990a; Ricklefs and Miles 1994). Variation in morphology may enhance or constrain performance capacities. Therefore, the concordance between the morphological space and ecological space is affected by performance. Consequently, the magnitude of covariation between morphology and performance indelibly affects the relationship between morphology and ecology and yields correlated patterns in the two (Ricklefs and Miles 1994).

Two approaches may be adopted in an ecomorphological analysis: (1) intraspecific studies, where correlations between specific morphological and performance variables are compared to biomechanical predictions; or (2) a comparative approach, where several species that differ in some ecological trait are included in the analysis, and patterns of covariation between

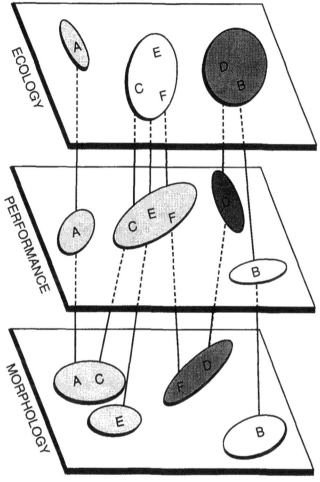

Figure 10.1. Diagrammatic portrayal of the relationships among morphology, performance, and ecology. Each component may be envisaged as a two-dimensional space based on a univariate or multivariate analysis of morphology, performance, or ecology. Taxa within each ellipse represent species or populations that were not statistically distinguishable. Taxa enveloped in separate ellipses were significantly different. As demonstrated in this figure, the lines between each level show that the connection between morphological variation and ecology is accomplished through an intermediate performance or behavior variable, e.g., locomotory performance. The strength of the interaction between morphology and performance may be estimated by a path analysis. In some circumstances, species with similar morphological characteristics (A, C) may exhibit differences in performance which translate into different ecological attributes. Other species may show divergent morphological traits (D, B) and performance measures, but manifest similar ecologies.

morphology and ecology are assessed. An inevitable weakness of the latter
approach is that patterns of concordance among species in morphology-per-
formance-environment correlations may be due to similarity by descent
rather than convergence (e.g., Losos and Miles 1994). Phylogenetic relationships among species pose an additional complication when using inferential
statistics, i.e., the nonindependence of species values may violate assumptions of many statistical procedures. Therefore, ecomorphological analyses
should incorporate a phylogenetic perspective (Felsenstein 1985; Lauder
1990, 1991; Garland and Janis 1993). Recent comparative methods provide
the means for including estimates of phylogenetic relationships among species, such that the influence of history is removed or sequestered in subsequent analyses (see Brooks and McLennan 1991; Gittleman and Luh 1992;
Harvey and Pagel 1991; Losos and Miles 1994; Martins and Garland 1991;
Miles and Dunham 1993).

    For over 20 years, ecologists have attempted to assess the relationship
between morphological divergence and the radiation of species into struc-
turally different components of the habitat. For example, lizards in the
genus *Anolis* in the Greater Antilles exploit a variety of microhabitats that
differ in vegetation coverage, perch type, and perch height. Species of different phylogenetic origins, but similar habitat preferences, converge with
respect to their morphological attributes. Among islands, the patterns of
ecological-morphological association were so repeatable that Williams
(1972, 1983) designated these species categories as "ecomorphs." Additional
analyses coupled morphology and locomotory behavior in the context of differences in habitat structure, as embodied in the "habitat matrix" model
(Moermond 1979a; Losos 1990a,b). Differences in morphology and locomotory mode, e.g., running, jumping, or crawling, were inferred to represent
constraints imposed by the three-dimensional structure of the habitat (e.g.,
Moermond 1979a; Pounds 1989). Variation in the dispersion, length, and
diameter of perches may have selected for the correlated evolution of specific morphological characteristics and locomotory behaviors within the various species of *Anolis*. These results from previous ecomorphological
analyses reveal a strong coupling between habitat structure, morphology,
and locomotory capacity. Presumably, differences observed among species
in various ecological characteristics reflect adaptive differences in morphology which are a consequence of differential performance capacities
(e.g., Laerm 1974). These previous instances of morphological divergence
associated with habitat structure contrast sharply with the apparent lack
of divergence in at least one other group of lizards (e.g., *Liolaemus*; Jaksic
et al. 1980).

    Why should locomotory performance have an association with habitat
use? Sprawling locomotion is a common form of movement for lizards
regardless of taxonomic affiliation (Sukhanov 1968; Rewcastle 1980, 1983).

Specific morphological attributes for running on tree branches may differ from those for climbing vertical rock faces or tree trunks (e.g., Sinervo and Losos 1991). Variation in limb elements and pelvis characters may translate into differences in maximal speed. Furthermore, a variety of studies have demonstrated the ecological and selective significance of various components of locomotion (e.g., Losos 1990a,b; Jayne and Bennett 1991; Garland this volume; Miles unpubl.). Within a population of *Thamnophis sirtalis,* relatively faster individuals had higher survivorship. Species characterized by sit-and-wait foraging patterns tend to be faster than species that forage actively (Huey et al. 1984). Also, movement patterns among lizards tend to vary and may relate to differences in habitat preferences of the species (Avery et al. 1987). As shown by a number of authors, foraging mode covaries with habitat specificity, body proportions, and certain life-history characteristics, such as relative clutch mass (Vitt and Congdon 1978; Vitt 1981; Dunham et al. 1988a). Similarly, evidence is available for certain species that document shifts in leg characteristics that correlate with habitat shifts (e.g., Kramer 1951; Laerm 1974; Pounds et al. 1983). Because leg characters are predicted to correlate with locomotion, habitat shifts in morphology may lead to shifts in performance. Therefore, locomotory performance may be part of a suite of traits related to a species' habitat affinities. An untested aspect of locomotion is whether a single set of morphological traits consistently covaries with performance. Do different morphological features correlate with performance depending on specific movement requirements of the organism? That is, can one document divergent ecomorphological patterns that correspond with differing substrate associations within a monophyletic clade of lizards?

The purpose of this chapter is to extend earlier studies by Laerm (1974), Moermond (1979a), Williams (1983), and Losos (1990a,b) by deriving preliminary estimates of the covariation between morphology and performance among and within sceloporine species (Phrynosomatidae). Here, I seek to ascertain whether species that use similar substrate types have evolved similar form-function complexes. Therefore, I describe a comparative analysis of the relationship between morphology and performance among species that exploit similar substrate types as an initial assessment of convergence in form and function within a clade of lizards that display variation in substrate preference. I make the assumption that species exploiting substrate types of different physical characteristics and dimensions vary in morphology, which consequently results in sprint-speed differences. In this analysis, I restrict attention to skeletal elements of the pelvic girdle and a single measure of performance, maximum sprint speed. This latter trait has been shown to be an ecologically relevant measure of performance. Previous studies demonstrated that speed is a repeatable, heritable trait (Huey and Dunham 1987;

Tsuji et al. 1989; Huey et al. 1990). Furthermore, other analyses revealed significant correlations with behavioral dominance (Garland et al. 1990b; Robson and Miles unpubl.) and survivorship (Jayne and Bennett 1990; Miles unpubl.; see also Bennett and Huey 1990 for a review of the ecological relevance of locomotory performance).

In this study, I have four goals. First, to characterize the patterns of covariation among a suite of morphological traits and assess similarities among species that share common substrate preferences. Second, I test for species-specific patterns of variation in locomotory performance. Although a few studies have demonstrated significant correlations between morphology and locomotory performance among species (e.g., Losos 1990a,b), analyses conducted at the intraspecific level resulted in weak or nonsignificant correlations (Garland 1985; Losos et al. 1989). The third goal is to describe the relationship between hind limb morphology and performance among and within species. Specifically I use Arnold's (1983) performance gradient paradigm to assess the covariation between morphology and performance. Finally, I examine intraspecific ecomorphological patterns in a historical context to determine the extent to which the correlations between morphology and locomotion reflect adaptively-based variation or concordance with phylogeny.

The trends in morphology portray a situation of considerable divergence in morphology and locomotory performance among phrynosomatid species. Both interspecific and intraspecific analyses revealed significant correlations between morphology and locomotion. However, patterns at the intraspecific level were not consistent with those found at an interspecific level. Historical analyses found little evidence for a phylogenetic effect among each morphological and locomotory character. Consideration of the variables when mapped on a phylogenetic tree showed evidence of unidirectional transitions among substrate classes and specific adaptive patterns of ecomorphological correlations.

Methods

Basic biology of phrynosomatid lizards

Species in the family Phrynosomatidae are small, diurnal lizards found throughout large parts of western North America and extending south to Central America (Etheridge 1964; Stebbins 1985). Phrynosomatids occur in a variety of habitats, from low-elevation hot deserts to pine forests, and often are characterized by specific substrate preferences. They are primarily terrestrial, saxicolous, arboreal, or arenicolous; however, some species may exploit more than one substrate type. Because of the diversity in habitat occupancy and substrate preferences, Phrynosomatid lizards provide an ideal system for ascertaining associations among morphology, locomotory performance, and ecology.

The 9 species included in this study are all insectivorous, diurnal lizards found in various parts of southwestern North America. The advantage of using these species is the variety shown in their habitat occupancy and substrate preferences (Table 10.1). Three of the species were strictly saxicolous, or have populations that exploit rocky substrates (*Petrosaurus mearnsi, Urosaurus microscutatus, Sceloporus jarrovi*). Three species have predominately terrestrial habits (*Cophosaurus texanus, Uta stansburiana, Sceloporus woodi*). Finally, three species are found only on trees or have populations that exhibit arboreal preferences (*Urosaurus graciosus, Urosaurus ornatus, Sceloporus clarkii*). Of the 9 species, *U. ornatus* and *U. microscutatus* have populations that may be either arboreal or saxicolous. Other species may exhibit variation in substrate use, such as *S. jarrovi* and *S. clarkii*. For example, the former species is predominantly found on rocky substrates, but will on occasion use fallen dead logs as basking sites (Miles pers. obs.). The arboreal lizard, *S. clarkii*, uses rocky perches (Vitt pers. comm.). Patterns of substrate association in these species are thus complex, depending on the available habitat structure and presence of syntopic species with similar ecological requirements. However, for purposes of this study, the substrate associations of each species recorded at the sample locality will be used in the analyses.

Field studies

Individuals from each of 9 species were captured from various localities in California (*P. mearnsi, U. microscutatus*; see Table 10.1), Arizona (*Cophosaurus texanus, Uta stansburiana, U. ornatus, U. graciosus, S. clarkii*, and *S. jarrovi*) and Florida (*S. woodi*). At each sampling locality, the habitat characteristics and substrate use of each individual were recorded based on observations at the site of capture. Sample sizes for morphological and performance measurements are provided in Table 10.1 along with the species' ecological attributes and collecting localities. Individuals were brought back to a field laboratory for measurement and performance trials. Each individual was color marked and individually identified with a unique toe clip. In addition, snout-vent length and tail length were recorded to the nearest 0.5 mm using a stiff metal ruler. Body mass was measured to the nearest 0.01 g using a portable electronic balance.

Locomotory performance

Estimates of locomotory performance were made using a protocol similar to that described in previous analyses (Huey and Hertz 1982; Garland 1985; Van Berkum 1986; Losos 1990a). Briefly, I induced lizards to run down a

Table 10.1. Lizard species included in this study, sample sizes for the morphology and performance measurements, their general ecological attributes, and collecting localities.

Species	N	Ecological Traits		Sample Locality
		Habitat	Substrate	
Petrosaurus mearnsi	25/75	Desert	Saxicolous	Santa Rosa Mountains, CA
Uta stansburiana	25/57	Desert	Terrestrial	near Lee's Ferry, AZ
Urosaurus ornatus	25/116	Desert/ Oak/ Pine Forest	Arboreal/ Saxicolous	Saguaro National Monument, Tucson, AZ
Urosaurus graciosus	25/128	Desert	Arboreal	NW Phoenix, AZ
Urosaurus microscutatus	10/39	Desert	Saxicolous/ Arboreal	3 mi. north of Jacumba, CA
Sceloporus woodi	15/23	Southern Pine/Oak	Terrestrial	Ocala National Forest, FL
Sceloporus clarkii	15/20	Desert/ Oak	Arboreal	Molino Basin, Santa Catalina Mountains, Tucson, AZ
Sceloporus jarrovi	25/36	Pinyon/ Juniper	Saxicolous	Cochise Stronghold, Dragoon Mountains, AZ
Cophosaurus texanus	6/6	Desert	Terrestrial	Saguaro National Monument, Tucson, AZ

2 m long raceway by repeated taps on the tail. Eight infrared photocell stations were spaced between 10 - 25 cm along the length of the track. Lizards would occlude light beams as they sprinted down the raceway. Elapsed times between each adjacent photocell station were stored in a Compaq portable III computer. Sand was used as a substrate to measure stride characteristics from the spoor (Huey 1982). Lizards were maintained at a body temperature of 36° - 38°C (mean temperature was 37°C) throughout the performance trials, which was within the range of active temperatures measured in the field. Sprint-speed trials were all conducted within 1 day of capture. Each individual was run twice in rapid succession and then allowed to rest at least 1 hour before sprinting again. Eight runs were obtained for all individuals. I took as an estimate of maximum velocity the fastest speed registered by a lizard along any 25 cm segment of the track. Only those runs in which the lizard sprinted down the track were included in the analysis. Instances where an individual jumped, reversed, or shimmy buried were eliminated.

Morphology

A number of functional morphological studies have considered the contribution of various appendicular skeletal elements to sprawling locomotion, with particular emphasis on the pelvic girdle and hindlimb morphology (e.g., Haines 1942; Snyder 1954; Laerm 1974; Brinkman 1980, 1981; Rewcastle 1980, 1983; Peterson 1984; Hildebrand 1985). Many of these characters cannot be reliably obtained from external measurements. Therefore, I used a portable X-ray unit (Min-Xray model 210) to obtain X-rays of lizards. A total of 8 hind limb traits were taken from the radiographs. I measured the hind limb elements to the nearest 0.01mm using a video image analysis program (Measurement TV©, DataCrunch Software). The following morphological traits were digitized directly from the radiographs: width of the pelvis (at the acetabulum), ilium length, pubis length, the angle formed at the pubic symphysis, and the lengths of the femur, tibia (excluding the astragalocalcaneum), fourth metatarsus, and fourth and fifth phalanges.

Statistical methodology

All statistical analyses were performed using "size adjusted" values for each variable. Species included in this study exhibited statistically significantly differences in average adult snout-vent length. Therefore, to detect morphological differentiation attributable to shape variation, I removed the confounding influence of body size by pooling all observations, regressing each morphological and locomotory character against snout-vent length (SVL), and retaining the residuals for subsequent analysis (see Dunham and Miles 1985). Because of the strong correlation between SVL and body mass ($r = 0.98$, $P = 0.001$), similar results were obtained when using body mass. Prior to multivariate analyses, I determined whether each variable conformed to a normal distribution (Dillon and Goldstein 1984). All morphological variables were normally distributed after a log transformation except for maximum velocity, which required a natural log transformation.

Patterns of variation and covariation among size-free morphological characters were initially defined using principal components analysis. A second purpose of PCA was to determine whether levels of morphological differentiation were associated with species-specific variation or suggested ecologically relevant patterns. Principal components were extracted using the covariance matrix based on the 11 log-transformed morphological measurements.

Morphological differences among species were determined using canonical variates analysis (PROC Candisc, SAS Institute 1990). Eleven size-free morphological traits were included in the analysis with eight species (*P. mearnsi, U. stansburiana, U. graciosus, U. microscutatus, U. ornatus, S. clarkii, S. jarrovi,* and *S. woodi*) used as the groups. Interpretation of each

significant canonical axis was facilitated by examining the direction and magnitude of the correlations between canonical and original variables (Dillon
and Goldstein 1984). Extent of morphological differentiation was ascer-
tained by computing Mahalanobis D^2 values and associated critical values.
All possible pairwise comparisons among species were determined; however, significance levels were adjusted using a Bonferroni correction factor
for table-wide tests. I performed this analysis to determine whether position
of species centroids in the morphological space reflected ecological similarity rather than phyletic differences.

   Species-level variation for size-free locomotory performance was evaluated through standard univariate ANOVA procedures. Results using residuals were similar to those obtained from an analysis of covariance with snout-
vent length included as the covariate. No evidence of a nonlinear relationship was evident when plotting residuals against performance. The relationship between morphology and maximum velocity was determined in two
ways. First, I computed standard partial correlation coefficients between
maximum velocity and the 11 morphological variables. This was done at
both the interspecific and the intraspecific level. Next, I calculated the performance gradient by extracting standardized partial regression coefficients
from a multiple regression analysis. Here, maximum velocity was entered as
the response variable, and the 11 size-free morphological variables were
included as predictor variables. Because estimates of coefficients from a
multiple regression are affected by high correlations among predictor variables and outliers among other factors (e.g., Mitchell-Olds and Shaw 1987),
I performed several regression-diagnostic features prior to interpreting
results from the analysis; specifically, I calculated estimates of multicolinearity, influence, and autocorrelation.

Phylogenetic approach

As discussed by several authors, the similarity among closely related species
in any number of phenotypic traits may be a consequence of phylogenetic
relatedness or similar responses to extrinsic selective agents (Miles and
Dunham 1993; Harvey and Pagel 1991; Brooks and McLennan 1991; Losos
and Miles 1994). Use of recently developed comparative procedures
requires availability of a well-supported phylogenetic hypothesis. I drew
upon a recently published and well-supported cladogram of the Phrynosomatidae in general and the *Sceloporus* group in particular (Wiens 1993).
Forty-five characters were included in the analysis and described variation
in soft anatomy, osteology, scalation, coloration, karyotype, and behavior. A
cladistic analysis produced a single tree, with a tree length of 66 steps and a
consistency index of 0.803. Unfortunately, Wiens (1993) did not resolve the

relationships of the species within the genera *Urosaurus* and *Sceloporus* in his overall study. Therefore for two of the comparative analyses, I relied on supplemental data from alternative sources or other procedures (e.g., Maddison 1991) to resolve these relationships. For the three *Sceloporus* species, I used the phylogeny based on karyotype data derived by Hall (as described by Sites et al. 1992). For the species in the genus *Urosaurus*, I used each possible dichotomous branching sequence.

Several analytical procedures are available that incorporate information embedded in a phylogeny into ecological and evolutionary studies. In this analysis, I employed two techniques. First, to determine the association between a phylogenetic hypothesis and morphology or performance, I used the method of phylogenetic autocorrelation (Cheverud et al. 1985). Given a matrix of phylogenetic relatedness among species, in the form of patristic distances, branch lengths, or genetic distances, phylogenetic autocorrelation analysis calculates the concordance between a phenotypic trait and branching structure within a clade. In the present analysis, I used a patristic distance, or the number of bifurcations necessary to connect any two species to a common ancestor, as a measure of phylogenetic relatedness. The analysis proceeds through a partitioning of the total variation in a trait into two components: a phylogenetic value and a specific value. The former value represents current variation in a trait attributable to descent from a shared ancestor. Association between a trait and measures of phylogenetic relatedness are given by the phylogenetic autocorrelation.

A second approach for understanding evolution of the association between form and function, e.g., morphology, performance, and ecology, involves a protocol outlined by Lauder (1991). In this analysis, Lauder divides morphology, performance, and ecology into separate levels and determines whether any level may be modified independently. A test of the integration of complex systems consists of comparing changes at each level. Highly integrated systems should show strong dependencies with change at one level followed by change at others. Mapping changes on a phylogeny allows historical influences to be included in the analysis. Modifications at any level that follows the branching structure of the tree would provide evidence of historical congruence in the evolution of ecomorphological associations. Patterns of change that follow ecological modifications rather than phylogeny could be interpreted in terms of adaptively based shifts in response to a change in the selective environment (e.g., Greene 1986; Baum and Larson 1991).

Results

Morphological variation

Principal components analysis based on the 11 size-free variables required five axes to explain 92% of the total trait variation. The first axis, which explained 41% of the total variation, had high loadings for the relative lengths of the femur, tibia, metatarsus, and fifth toe (Table 10.2). This axis largely described a gradient of increasing relative size of the distal elements of the hind limb. Two *Sceloporus* species (*S. clarkii* and *S. jarrovi*) were characterized by relatively short distal limb elements, and *P. mearnsi* had elongated limb elements. The former pair of species are primarily climbers, whereas the latter species is restricted to rocky habitats. The terrestrial species, *Uta stansburiana* and *S. woodi*, also had relatively long distal limb elements. The second axis, comprising 20.4% of the total variation, had high positive loadings for the tail and fourth toe, and negative values for pelvis width. On this axis, lizards with long tails and fourth toes, but narrow pelves were at one end, e.g., *Urosaurus graciosus*, and species with wide pelves and short toes and tails at the other, e.g., *U. stansburiana, S. woodi*. Axis three, which extracted 19% of the size-free variation, portrayed a gradient that contrasted relative lengths of the fourth toe against the tail. All three *Sceloporus* species had relatively long fourth toes but short tails. Relative body mass, which may be interpreted as body mass per unit length, had the highest loading on the fourth axis; which I interpret to represent the relative stockiness of the lizards. Finally, the fifth axis contrasted length of the ilium against the angle formed by the fusion of the pubis bones at the symphysis. This axis provides a measure of pelvis shape. Species at one end of the axis have a long, narrow pelvis, e.g., *U. graciosus*, and species at the other a box-like, short, broad pelvis, e.g., *U. stansburiana*.

Inferences regarding the ecological significance of patterns of morphological variation may be obtained by examining the position of species in the space defined by each axis. As an example, in Fig. 10.2, PC axis 2 is plotted against PC axis 3. This plot shows considerable overlap among species, which would suggest that similarity in substrate preferences does not necessarily lead to similarity in morphology. Yet, consider the position of saxicolous species in the morphological space. These species provide an example of morphological convergence. Individuals of *P. mearnsi* and *S. jarrovi* occur within the central portion of the cloud of points. In contrast, the arboreal species, *U. graciosus, U. ornatus*, and *S. clarkii*, do not show coincident positions in morphological space. *Urosaurus graciosus* is actually positioned at the far end of axis 2 and quite distant from the latter two species.

Despite the apparent inconsistency between morphology and substrate association, principal component analysis revealed morphological patterns

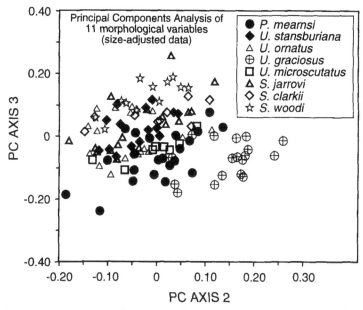

Figure 10.2. Position of species in the morphological space defined by principal components axes 2 and 3. PCA was based on size-free variables.

Table 10.2. Loadings from the first 3 principal component axes based on size-adjusted variables.

Variable	Axis 1	Axis 2	Axis 3	Axis 4	Axis 5
Tail	0.29	0.63	-0.67	0.21	0.07
Mass	0.08	-0.05	0.47	0.87	-0.64
Pelvis Width	0.64	-0.60	0.12	0.09	0.11
Ilium Length	0.33	-0.37	0.03	0.07	0.79
Pubis Length	0.63	-0.47	0.13	-0.01	-0.01
Angle of Pubis	0.12	-0.50	0.07	0.13	0.29
Femur	0.95	-0.21	-0.06	0.01	-0.03
Tibia	0.96	-0.05	-0.08	-0.05	-0.14
Metatarsus	0.91	-0.07	0.10	-0.05	-0.21
Fourth Toe	0.35	0.66	0.64	-0.11	0.09
Fifth Toe	0.85	-0.14	-0.19	-0.08	0.13
Eigenvalue	0.01	0.009	0.007	0.003	0.001
Percent Variance	41.0	20.4	18.8	7.0	4.0

that may be related to ecology. For example, the arboreal species, *U. graciosus*, is characterized by a relatively long tail, narrow pelvis, long fourth toe, and slender body. The terrestrial species, *U. stansburiana* and *S. woodi*, were characterized by relatively long tails, wide pelves, broad pubic angles, and long fourth toes.

Evidence of interspecific morphological differentiation was determined using canonical variates analysis. The first five axes were highly significant ($P < 0.001$). However, only three axes were necessary to explain 95% of the total variation. Correlations between the canonical variables and the original variables are presented in Table 10.3. The first axis, which explained 57% of the variation, separated species based on the relative lengths of the femur, tibia, metatarsus, and fifth toe. This axis partitioned *P. mearnsi*, *S. jarrovi*, and *S. clarkii* from the remaining species. The former species had relatively long elements of the hind limb, while the latter two had shorter elements. Axis 2, which accounted for 24% of the variation, partitioned species based on the relative pelvis width and angle of the pubic symphysis against relative tail length. This axis separated *U. graciosus*, which has a long tail but narrow pelvis and short femora, from the remaining species. The third axis accounted for 14% of the variation and characterized differences among species based on positive loadings for tail length and negative loadings for metatarsus and pubis. Both *U. ornatus* and *U. microscutatus* were positioned at the positive end of the axis, and *S. woodi* and *U. stansburiana* were found at the negative pole. Evidence for species-specific patterns of variation versus ecological patterns may be discerned by examining species positions in the canonical variates space. Positions of species along axes 2 versus 3 reinforce patterns from the PCA (Fig. 10.3). The second axis may be interpreted as a gradient that separates the primarily arboreal species. *Urosaurus graciosus* is at one end of the morphological space, followed by *S. clarkii* and then *U. ornatus*. The position of this latter species is found within a cluster of points at the positive pole of axis 2. Individuals of *U. ornatus* tend to show an alliance with *U. microscutatus* and *P. mearnsi*. Species-specific differences tend to emerge along axis 3. Terrestrial species tend to be at the negative portion of axis 3 (*U. stansburiana*, *S. woodi*), while saxicolous species (*P. mearnsi*, *U. microscutatus*, *S. jarrovi*) occur at the more positive portion of the axis.

Above patterns are qualitative, however, and may reflect distortions attributable to portraying a multidimensional space in a two-dimensional plot. Therefore, I computed Mahalanobis D^2 distances between all species and associated tests of significance (Dillon and Goldstein 1984). The D^2 values are a multivariate generalized distance which controls for covariation among morphological variables. Table-wide significance levels associated with pairwise comparisons were adjusted using a Bonferroni correction. Distances between *P. mearnsi* and all other species ranged from 29.3–100.7

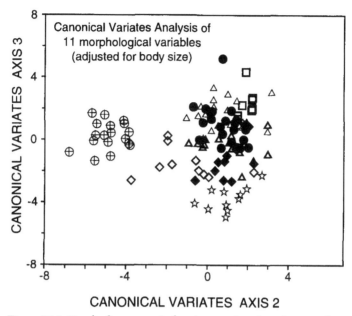

Figure 10.3. Results from canonical variates analysis based on size-free data. Plot of positions of species along CV axis 2 and CV axis 3. Species symbols are the same as in Fig. 10.2.

Table 10.3. Summary of a canonical discriminant analysis based on the 11 size-adjusted morphological variables.

Variable	CV Axis 1	CV Axis 2	CV Axis 3
Tail	0.56	-0.59	0.51
Mass	-0.39	0.53	-0.61
Pelvis Width	0.42	0.90	-0.08
Ilium Length	0.51	0.83	0.21
Pubis Length	0.67	0.56	-0.37
Pubis Angle	-0.18	0.92	0.31
Femur	0.86	0.49	0.02
Tibia	0.95	0.29	-0.08
Metatarsus	0.86	0.29	-0.39
Fourth Toe	0.54	-0.29	-0.68
Fifth Toe	0.84	0.32	0.15
Eigenvalue	11.4	4.8	2.8
Percent Variance	57.0	24.0	13.9

and were highly significant (Table 10.4), suggesting that *P. mearnsi* occupied a unique part of the morphological space. The arboreal species, *U. graciosus*, exhibited large distances, 39.7–59.3, from the other species, again indicating that this species occurred in a distinct and separate portion of morphological space. The two terrestrial species, *U. stansburiana* and *S. woodi*, were not significantly different from one another. Yet they were significantly different from the other species. No differences were found between *U. ornatus* and *U. microscutatus*, although both were significantly different from the other species. Similarly, *S. jarrovi* and *S. clarkii* were very similar in morphological characteristics.

Variation in locomotory performance

Body size and maximum velocity were positively and significantly correlated among the eight phrynosomatid species ($r = 0.90$, $P < 0.001$; Fig. 10.4). In general, larger species were either arboreal (*S. clarkii*) or saxicolous (*P. mearnsi, S. jarrovi*) and had the fastest velocities (all were greater than 2.40 m/s). Terrestrial species were within a cluster of small-sized species (average SVL = 50.2 mm). The smallest species, *U. microscutatus*, also had the slowest maximum velocity.

As shown in Table 10.5, variation in size-adjusted maximum velocity ranged from 1.77 m/s (*U. microscutatus*) to 2.48 m/s (*S. woodi*). An ANOVA based on the size-free values of maximum velocity revealed significant differences among species ($F_{7,142}$, $P < 0.001$). A post hoc comparison test revealed that the eight species fell into four groups. They are listed in descending order of maximum velocity. The first group consisted of a single species, *S. woodi*; the second contained only *P. mearnsi*; the third comprised *S. clarkii* and *U. graciosus*; whereas the fourth consisted of *U. microscutatus*, *U. ornatus*, and *S. jarrovi*. One of the species, *U. stansburiana*, did not have a unique association with any particular group of species. Instead, the post hoc comparisons found it to overlap with the third and fourth groups.

Covariation between morphology and locomotory performance

As an initial assessment of the correlation between morphology and locomotory performance, I calculated partial correlations among hind limb elements and maximum velocity, with effects of body size removed (Table 10.6). Significance levels of correlations were adjusted using a Bonferroni correction prior to interpretation of table-wide associations. Interspecific correlations revealed a strong association between hind limb elements and maximum velocity (Table 10.6). Relative lengths of the metatarsus, tibia, femur, and fifth toe had the highest association with maximum velocity. Pelvis width was negatively associated with maximum velocity as expected under predictions from functional studies (e.g., Snyder 1954).

Table 10.4. Mahalanobis' generalized distance D^2 based on size-free morphological measurements. Abbreviations for species are: Pm = $P.$ *mearnsi,* Uta = *Uta stansburiana,* Ug = *Urosaurus graciosus,* Um = $U.$ *microscutatus,* Uo = $U.$ *ornatus,* Sc = *Sceloporus clarkii,* Sj = $S.$ *jarrovi,* Sw = $S.$ *woodi.* Probability values are $^{\circ}P < 0.05,$ $^{\circ\circ}P < 0.01,$ $^{\circ\circ\circ}P < 0.001.$

Species	D^2						
	Pm	Uta	Ug	Um	Uo	Sc	Sj
Pm							
Uta	29.3°°°						
Ug	59.3°°°	40.3°°°					
Um	61.2°°°	23.9°°	57.6°°				
Uo	67.8°°°	22.7°°°	46.4°°	7.8			
Sc	100.7°°°	34.8°°°	39.7°°°	32.9°°°	16.7°°		
Sj	91.3°°°	23.8°°°	52.6°°°	14.2°°°	8.3°°°	11.5	
Sw	36.6°°°	11.9	54.8°°°	53.6°°°	47.9°°°	43.6°°	45.8°°°

Correlations at the intraspecific level present a different pattern (Table 10.6). Two species, *U. microscutatus* and *S. woodi,* had no significant correlations between morphology and performance. Although some correlations had large values, the failure to detect significant correlations may be attributable to low sample size for these two species. For the remaining species, a consistent pattern was the high correlations between maximum velocity and lengths of the metatarsus, tibia, and femur. Note that the correlations for these traits were similar in *U. microscutatus.* Two species, *S. jarrovi* and *S. clarkii,* had significant correlations involving pelvis width and ilium length with speed. Only four species, *P. mearnsi, U. graciosus, S. jarrovi,* and *S. clarkii,* had significant correlations between speed and fifth toe length. Interestingly, all these species climb on either rocks or trees.

Unfortunately, the ecomorphological associations generated in the above analysis may be confounded by indirect correlations among morphological traits. That is, the correlation between femur length and speed may reflect the direct effect of femur length as well as the indirect effects of the other variables through their correlations with femur length. Therefore, I entered the 11 morphological traits in a multiple regression analysis as predictor variables and maximum velocity as the criterion variable (Table 10.7). Regression coefficients were standardized to make them analogous to path coefficients. Thus, each coefficient describes the effect of a given morphological variable on maximum velocity while holding the influence of the other variables constant. The interspecific analysis revealed only four

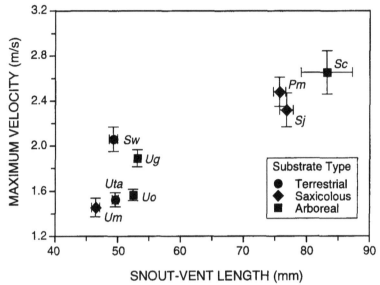

Figure 10.4. Relationship between snout-vent length and maximum velocity. Values plotted are species means ± 1 *SE*. Abbreviations for species are: *Pm* = *Petrosaurus mearnsi*, *Uta* = *Uta stansburiana*, *Ug* = *Urosaurus graciosus*, *Um* = *U. microscutatus*, *Uo* = *U. ornatus*, *Sc* = *Sceloporus clarkii*, *Sj* = *S. jarrovi*, *Sw* = *S. woodi*. Observed substrate affinities of species are given by different symbols: circles denote terrestrial species, diamonds denote saxicolous species, and squares are arboreal species.

Table 10.5. Variation in maximum velocity among species. Results based on size-free values of maximum velocity.

Species	Mean Maximum Velocity (m/s)	Standard Error	Sample Size
P. mearnsi	2.35	0.03	22
U. stansburiana	1.85	0.02	15
U. graciosus	1.77	0.04	8
U. microscutatus	1.79	0.02	21
U. ornatus	2.11	0.02	18
S. clarkii	1.89	0.02	13
S. jarrovi	1.73	0.03	18
S. woodi	2.48	0.04	10

variables associated with maximum velocity: lengths of the metatarsus, pubis, and fifth toe, and the width of the pelvis (Table 10.7). The regression diagnostics revealed significant multicolinearity for two of the species, *U. microscutatus* and *S. clarkii*, and these were excluded. No significant coefficients were derived for *U. stansburiana* or *S. woodi*, although the largest values are presented for heuristic purposes (Table 10.7). All of the remaining species had unique patterns of covariation. Two morphological variables were significant for *P. mearnsi*: tibia (0.79) and pelvis width (-0.95). Only one variable was significant for *U. ornatus*: femur length (1.35), and *S. jarrovi:* ilium length (0.62). The strongest ecomorphological association was shown by *U. graciosus*, which had significant coefficients for relative mass, metatarsus length, tibia length, pubis length, and pelvis width. These results parallel predictions generated from earlier functional morphological studies (e.g., Snyder 1954; Hildebrand 1985).

Historical analysis

A phylogenetic autocorrelation analysis was performed to assess the impact of phylogeny on morphological and performance trait values. In this application of phylogenetic autocorrelations, I used a modified patristic distance as a measure of phylogenetic relatedness. For each species, I determined the number of bifurcations necessary to connect two species to a common ancestor (Miles and Dunham 1992). Overall, neither the morphological traits nor maximum velocity exhibited significant phylogenetic autocorrelations. Such a result would lead to the conclusion that historical effects were not important in ecomorphological associations derived in the earlier analyses. The absence of a phylogenetic effect may be in part attributable to the choice of species included in the analysis. Two genera, *Petrosaurus* and *Uta*, are represented by a single species and substrate association. Species in the genus *Urosaurus* are found on only two of the three substrate categories, while species of *Sceloporus* were chosen to cover each substrate category. The absence of a phylogenetic effect within the phrynosomatids as a general phenomenon must be tested in an expanded analysis, which should include additional species of *Urosaurus* and *Sceloporus*.

An alternative method is to map character changes in ecology on a phylogeny and determine whether morphological and performance differences occurred in synchrony with the inferred evolutionary transitions (e.g., Greene 1986; Lauder 1990, 1991; Baum and Larson 1991; Reilly and Lauder 1992). The phylogeny for the Phrynosomatidae as resolved by Wiens (1993) is presented in Fig. 10.5. Under this hypothesis, I treated *Cophosaurus texanus* as the outgroup to the taxa included in this study. This outgroup is part of a larger clade that also includes the genera *Callisaurus, Holbrookia,* and *Uma* (collectively known as the "sand lizards"). In Wiens' tree

Table 10.6. Partial correlations between maximum velocity and skeletal elements of the hind limb with the effects of body size removed. Probability values are as follows: °P < 0.05, °°P < 0.01, °°°P < 0.001, and +P = 0.1.

Species	Tail	Mass	Fourth Toe	Meta-tarsus	Tibia	Femur	Pelvis Width	Ilium	Pubis	Angle	Fifth Toe
Interspecific	0.16	0.002	0.33°°°	0.81°°°	0.76°°°	0.70°°°	0.34°°°	0.24°°°	0.49°°°	-0.16°	0.69°°°
Pm	0.01	-0.29	0.03	0.32	0.55°°	0.46°°	0.02	0.20	-0.16	0.07	0.49°°°
Us	0.43°	-0.02	-0.32	0.47°	0.55°°	0.53°°	0.08	-0.01	-0.10	0.38+	0.37+
Ug	0.27	0.41°	-0.17	0.66°°°	0.44°	0.54°°	0.05	0.05	0.33	-0.20	0.49°°
Um	-0.17	0.09	0.27	0.46	0.54	0.57+	-0.38	0.09	-0.29	0.21	0.61+
Uo	-0.04	0.44°	-0.05	0.53°°°	0.47°°	0.58°°°	-0.24	0.08	-0.16	-0.35+	0.11
Sc	-0.01	0.42	0.06	0.92°°°	0.91°°°	0.93°°°	0.89°°°	0.77°°°	0.84°°°	0.25	0.84°°°
Sj	0.09	0.19	0.19	0.65°°°	0.60°°°	0.72°°°	0.43°	0.52°°°	0.09	0.16	0.68°°°
Sw	0.32	0.03	0.21	-0.12	0.41	-0.01	0.02	-0.38	-0.30	0.46	0.18

Table 10.7. Summary of a multiple regression analysis using the size-free morphological variables. The coefficients presented in this table are standardized regression coefficients. These may be interpreted as path coefficients. Results from both the interspecific and intraspecific analyses are presented. ° $P < 0.05$, °° $P < 0.01$, °°° $P < 0.001$, and + $P = 0.1$.

Species	Tail	Mass	Fourth Toe	Meta-tarsus	Tibia	Femur	Pelvis Width	Ilium	Angle	Fifth Toe
Interspecific				0.61°°°			-0.25°°		0.16°	0.34°°°
Pm					0.79°	0.66+	-0.95°°°			0.35+
Us			-1.34+	-1.13+		-1.76+				
Ug		0.71°°°		0.54°°						
Uo					-0.75°	1.36°	-0.58°°	-0.52°°	0.82°°	
Sj		-0.48+						0.62°		

Phrynosoma has a sister taxa relationship to the sand lizards. Species within these genera tend to exhibit terrestrial habits. Of the remaining species, *P. mearnsi* is the sister taxa to *Uta*, *Urosaurus*, and *Sceloporus*. The latter two genera are sister taxa. Among the *Sceloporus* species, *S. merriami* is a basal species, and *S. woodi* is the sister group to *S. jarrovi* and *S. clarkii*. Although presented as a fully resolved clade, the relationships among the *Urosaurus* species are still unknown.

I began the historical analysis by optimizing substrate preference on the cladogram (Miles and Dunham 1993; Losos and Miles 1994). The phylogenetic tree was entered into McClade 3.0 (Maddison and Maddison 1992), and character states of interior nodes were determined by minimizing the number of character transitions (maximum parsimony). Six evolutionary transitions in substrate preference were detected (Fig. 10.5). Based on the outgroup, terrestriality was inferred to be the ancestral substrate type. Four transitions were from terrestriality to rock dwelling. An additional two transitions were from rock dwelling to arboreality. It appears that at least within phrynosomatids, arboreality arose from a saxicolous rather than a terrestrial ancestor. In no instance did I detect evolutionary reversals in substrate association. Thus, shifts in substrates appear to be unidirectional.

Patterns of historical congruence (sensu Lauder 1991) may be inferred by comparing results of analyses of morphological and performance differentiation in a phylogenetic context (Fig. 10.5). Evidence of historical congruence would occur when changes in morphology, performance, and ecology were coincident with branching patterns of the phylogeny or are unvarying within a clade. Noncongruence, and therefore putative evidence of adaptive shifts, may be suggested when changes in morphology, performance, and ecology are independent of phylogenetic relatedness of species.

Because detailed information was unavailable about morphology and performance for *C. texanus*, *Sator*, and *S. merriami*, I excluded them from the historical analysis. The differences in morphology, as derived from the canonical analysis, and locomotory performance, as revealed by the ANCOVA, may be compared in a phylogenetic context with the evolutionary transitions in substrate association.

Ecomorphological patterns of the terrestrial species

The two terrestrial species exhibited similar patterns of covariation in morphology based on the D^2 values, which may suggest phylogenetic conservatism. Because *S. woodi* was found to be relatively faster than *Uta*, their performance capacities were not congruent.

Ecomorphological patterns associated with the shift from terrestrial to saxicolous habits

Each transition from terrestriality to rock dwelling was accompanied by a morphological shift, e.g., *P. mearnsi*, *Urosaurus*, and *S. jarrovi*. Only one

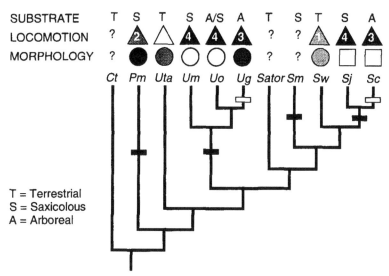

Figure 10.5. Historical analysis of the association between morphology, locomotory performance, and ecology. The phylogenetic tree for Phrynosomatidae was based on 45 characters, which included osteology, squamation, soft anatomy, coloration, and karyotype traits (Wiens 1993). The tree had a length of 66 steps and a Consistency Index of 0.803. The fully resolved *Urosaurus* clade is one possible solution to the polytomy. Branching pattern for *Sceloporus* is based on the karyotype data of Hall (see Sites et al. 1992). The outgroup for the analysis was *Cophosaurus texanus*. Closed bars denote evolutionary transitions from terrestrial to saxicolous habits. Open bars denote transitions to arboreality from a saxicolous ancestor. Triangles represent variation in locomotory performance. Species with similar numbers and shading within triangles were found to be not statistically different with respect to locomotory performance (based on results from Table 10.3). Circles and squares designate variation among species in morphology as summarized by canonical discriminant analysis. Species with circles of the same shading had similar morphological characteristics (see Table 10.4). *Sceloporus jarrovi* and *S. clarkii*, represented by open squares, were shown to be statistically similar. Species without data are represented by question marks.

common morphological trait changed with the transition in substrate preference, body size. In each transition, the saxicolous descendent species was larger than the terrestrial ancestor. Beyond this trait, little evidence of morphological convergence related to saxicolous habits was manifested. The morphological patterns for *Petrosaurus* were unique and significantly different from the remaining species (Table 10.4). All the *Urosaurus* species were statistically different from the *Sceloporus* species (Table 10.4). The pattern of generic differentiation suggests that the transitions in ecology were coincident with clade (intergeneric) specific morphological variation. The three saxicolous species exhibited similar locomotory performance capacities, although there was no evidence of a concordance between the rock dwelling morphology and locomotory performance.

Ecomorphological patterns associated with the shift from saxicolous to arboreal habits

Two patterns of differentiation are evident in the transition from rock dwelling to arboreality. In *U. graciosus*, a shift in morphology and performance accompanied the shift in substrate. This result held regardless of the various outcomes possible when resolving the polytomy. However, in *S. clarkii*, there is evidence of phylogenetic conservatism in morphology, but a shift in performance, because *S. clarkii* and *U. graciosus* had similar values for maximum velocity. In contrast, the two *Sceloporus* species, *S. clarkii* and *S. jarrovi*, exhibited very similar patterns in morphology. The remaining arboreal species, *U. ornatus*, was similar to *U. microscutatus* in morphology and maximum velocity. Again, the similarity between these species is possibly an instance of phylogenetic conservatism.

Discussion

Species in the lizard family Phrynosomatidae exhibit a diverse range of substrate associations. Furthermore, species vary in external morphological characteristics, some having relatively long tails and flattened body form, such as *P. mearnsi*, other species with long tails and narrow body form, e.g., *U. graciosus*, and others displaying relatively short tails and long limbs, e.g., *S. woodi*. Do differences in morphology among species represent adaptations to specific ecological differences in habitat occupancy or substrate association? How does variation in morphology translate into ecological differences? One possible answer is that locomotory performance acts as a transducer, which links morphological differences to ecology.

Results from this study demonstrated marked differentiation in two trait complexes: (1) skeletal elements of the hind limb and pelvis and (2) locomotory performance within a single clade of lizards. Furthermore, among species, pelvis width, metatarsus length, and fifth toe length significantly predicted maximum velocity. Intraspecific correlations between morphology and locomotion varied considerably, but a consistent pattern was the association between distal limb elements with maximum velocity (Tables 11.6 and 11.7).

Similarity in the relationship between form and function may be a consequence of phylogenetic relatedness. However, an analysis designed to detect associations between a trait and phylogenetic relatedness failed to demonstrate a correlation between either locomotion or morphology with phylogeny. However, a potential weakness of the phylogenetic autocorrelation approach is the inability to include more than one variable. Hence, phylogenetic patterns of covariation cannot yet be resolved.

Optimization of the species differences in morphology, performance, and ecology onto a phylogeny provides an alternative method for unraveling historical effects versus evidence of adaptive shifts. At least 6 evolutionary transitions in substrate association were detected by optimizing the patterns of substrate use onto phrynosomatid phylogeny. Four shifts involved transition from a terrestrial ancestor to a saxicolous descendent, whereas two involved a saxicolous ancestor to an arboreal descendent. A simultaneous comparison of changes in morphology and locomotion with substrate shifts among species provided evidence for both specific adaptations in response to a shift in ecology as well as phylogenetic conservatism. The saxicolous *P. mearnsi* and arboreal *U. graciosus* were clear examples of species displaying unique adaptations for a derived substrate association. Phylogenetic conservatism may be characteristic of two species pairs: *U. microscutatus* and *U. ornatus*, and *S. clarkii* and *S. jarrovi*.

Many earlier studies that detailed patterns of morphological variation in an ecological or functional context were performed using species other than those examined here. Therefore, patterns extracted from Phrynosomatidae may be compared with earlier functional and ecological morphology results to determine the generality of conclusions.

Patterns of morphological variation

Phrynosomatid lizards exploiting different substrate types did not exhibit clear patterns of convergence in size-free morphological variation. Nevertheless, constituent species included in this analysis displayed a remarkable degree of morphological divergence. Such a pattern suggests potential responses at the generic or specific level to perhaps other more detailed ecological characteristics. This contrasts with previous analyses of *Anolis*. For example, Losos (1990a) found that *Anolis* species inhabiting similar substrate types or microhabitat categories, e.g., crown, trunk, or grass, tended to converge in various aspects of morphology, such as lamellae number and length of the tail and fore- and hind limbs. Levels of divergence among the phrynosomatids also fail to follow a pattern of conservatism exhibited by *Liolaemus* lizards in Chile. Jaksic et al. (1980) found little evidence of differentiation in body proportions among 12 species with respect to substrate associations. Three exceptions to this pattern were noted: a ground-dwelling species, *L. lemniscatus*, was characterized by relatively short limbs, and two shrub-climbers, *L. schroederi* and *L. chiliensis,* had longer tails (Jaksic et al. 1980). In an analysis consisting of various assemblages of lizards occurring in 20 different habitats in the southwestern U.S. habitats, Scheibe (1987) used morphology to predict habitat use. He found a strong correlation between variation in head length and width, and front leg and rear foot length with canopy structure shrub density. His analysis suggested that large lizards occurred in densely vegetated areas, and relatively smaller, slender species exploited more open habitats. Unfortunately, the functional connection between morphology and habitat use was not presented.

Among phrynosomatids studied here, the described gradients of morphological variation in relation to substrate use provide a strong fit with expectations based on biomechanical and functional studies (e.g., Snyder 1954; Hildebrand 1985). For example, terrestrial species tended to have relatively longer hind limb elements than did the arboreal species (e.g., Pianka 1986). Also, arboreal species were characterized by a slender body plan, long tail, and long hind toes for grasping and stability during climbing (e.g., Snyder 1954; Ballinger 1973b; Ricklefs et al. 1981; Pianka 1986). Two genera within Phrynosomatidae (*Urosaurus*, *Sceloporus*) have evolved arboreal habits, yet patterns of morphology are not coincident. For example, the population of *U. ornatus* in this study was strictly arboreal, yet it was characterized by a relatively wider pelvis and shorter limbs and tail than *U. graciosus*. This dichotomy between *U. graciosus* and *U. ornatus* is a general pattern regardless of the population from which the latter species is drawn. Comparisons among populations of *U. ornatus* reveal morphological and locomotory differences between arboreal and saxicolous populations. However, I note that this is a preliminary study and many other species of *Sceloporus* exhibit arboreality and so should be included in future analyses. Furthermore, species of phrynosomatids may use similar substrate types, yet exhibit markedly divergent exploitation patterns. For example, some species of *Sceloporus* may be found on rocks, yet certain species may have crevice-dwelling habitats, and others may favor talus slopes. In each example, we would classify those species as saxicolous, but the use of large boulders rather than talus may involve different patterns of movement and hence favor divergence in morphological characteristics.

Morphological differentiation among species and substrates

Discriminant functions provided evidence of morphological differentiation among species. Nearly all variation was explained by the first three axes, suggesting considerable redundancy in the morphological data. Two species were immediately separated from the others, *P. mearnsi* and *U. graciosus*. Both have unique ecological characteristics. The former species is restricted to rocky outcroppings or canyons with granitic walls (Miles pers. obs.) and has a flattened body *bauplan* for exploiting narrow crevices and cracks interspersed throughout the habitat. These lizards have a fairly distinctive morphology relative to other species. *Urosaurus graciosus* is also restricted to a particular habitat type, trees and shrubs. Morphological attributes that allow *U. graciosus* to exploit desert trees and shrubs set it apart from remaining species. Terrestrial species were quite similar in morphology, with rather broad pelves and long limbs. There was some overlap between the remaining *Sceloporus* and *Urosaurus* species. Remaining species in each of these genera exploited either arboreal or rocky substrates. However, Mahalanobis D^2 values revealed that genera were significantly different in morphology. Thus, these results fail to document a consistent pattern of covariation in morphology that corresponds with exploitation of particular substrate types.

Covariation between morphology and maximum velocity

Many functional and biomechanical analyses stressed the importance of the hind limb and pelvic girdle elements for generating propulsion and support during locomotion in lizards. Despite the strength of association found in the analysis of form, many functional studies found little or no correlation between locomotory performance and morphology (e.g., Garland 1985; Losos et al. 1989; but see Losos 1990a,b; Huey et al. 1990). ecomorphological associations presented here demonstrated strong correlations between hind limb and pelvis elements and speed. Partial correlation analysis yielded far more correlations of greater magnitude, but this may reflect indirect effects of uncontrolled intercorrelations among morphological variables. Regression analysis, which controls for the influence of other predictor variables still arrived at significant associations. A characteristic feature of the intraspecific analysis was the lack of consistent variables correlating with speed. In fact, considerable heterogeneity was evident with respect to the correlation between speed and morphology. This result reinforces Bock's (1959) conclusions about multiple pathways of adaptive change.

Historical analysis

Finally, I employed a phylogenetic approach to determine whether derived patterns were a reflection of similarity due to ancestry or a response to varying selective regimes. Phylogenetic autocorrelation analysis found no significant correlations between trait values and phylogeny. Hence, many patterns may be representative of independent evolution since divergence from a common ancestor. However, implementation of Lauder's protocol of a multilevel analysis of form-function complexes revealed a different conclusion. Evidence of both adaptive change (sensu Baum and Larson 1991) and phylogenetic conservatism was present. The shift in morphology and performance, as seen in *P. mearnsi* and *U. graciosus*, suggests probable adaptations to saxicoly and arboreality, respectively. The absence of a morphological shift after an evolutionary transition to arboreality, as seen in *S. clarkii*, provides an example of morphological conservatism. Based on these data, the relationship between morphology, performance, and ecology is complex within Phrynosomatidae. Additional data are needed for *S. merriami*, *C. tex-*

anus, and *Sator* to obtain a more detailed estimate of how morphology and performance may interact to affect a species' ecological characteristics.

Summary and Future Directions

Eight species of phrynosomatid lizards in this study demonstrated marked patterns of divergence in morphology, locomotory performance, and intraspecific correlations between trait complexes. Hind limb morphology and locomotory performance were strongly correlated, but no consistent pattern emerged regarding the specific morphological components that tended to predict performance. A historical analysis of substrate use revealed at least 4 transitions between terrestrial to saxicolous habits, and 2 transitions between saxicolous and arboreal habits. Examination of derived patterns of morphological and locomotory differentiation in a phylogenetic context revealed evidence of adaptation to substrate type as well as instances of phylogenetic conservatism in morphology and performance.

Patterns documented here provide an initial analysis into assessing relationships among morphology, performance, and ecology. Three extensions are immediately evident. First, a rather coarse description of ecology was used in this study, i.e., substrate preference. In fact, the ways that lizards use habitats vary considerably, and a more empirical description is necessary. Other habitat-related factors should be included, for example escape behavior or climbing ability (e.g., Jaksic and Nunez 1979; Losos 1990a). Thus, lack of evidence for convergence with respect to substrate use in this study may reflect the narrow and categorical nature of this trait. For example, Sinervo and Losos (1991) describe variation in arboreality within a single species, *S. occidentalis*. Exploitation of tree habitats can vary, for some individuals or species may use vertical perch types exclusively, while others may only exploit primarily horizontal surfaces. Sinervo and Losos (1991) found that differences in arboreal tendencies of individuals within a population correlate with leg length and ability to run along perches of varying diameters. Thus, the manner in which species exploit the environment will have selective consequences leading to differences in the patterns of morphological covariation. Consequently, additional studies should be performed to examine whether intraspecific variation in habitat use leads to variation in morphology and locomotory performance. Preliminary studies comparing *U. ornatus* populations suggest that arboreal and saxicolous populations are characterized by divergent ecomorphological patterns (Miles unpubl.). Second, only one performance measurement was included in the analyses. Other ecologically relevant tasks should also be incorporated, such as jumping performance, sprint performance on various substrates, and others. Third, the number of species and individuals included in the analysis was rather limited. Analysis should be expanded to incorporate as many species with as large a sample size as possible. Differences in intraspecific ecomorphological patterns may reflect low statistical power rather than emergent biological properties related to habitat use. Finally, a greater use of multi-level historical analyses (e.g., Reilly and Lauder 1992) may lead to refinements in understanding the evolution of ecomorphological associations. That is, each level in an analysis, from morphology, performance, and ecology should be evaluated in light of a phylogenetic hypothesis. Such a protocol should distinguish derived ecomorphological associations that emerge in response to a shift in the ecological milieu from ancestral associations.

Acknowledgments

Various aspects of this research were supported by grants from NSF (BSR 86-16788, IBN 92-07895), National Geographic Society, and Ohio University (OU Research Committee and Ohio State Board of Regents Research Challenge Fund). I wish to thank the many field assistants without whose help this research would not have been completed. The assistance and advice of Laurie Vitt, William Gutzke, Scott Moody, Steve Reilly, and Ray Huey substantially improved field, laboratory, and analytical aspects of the study. I also thank Ray Huey, Laurie Vitt, and an anonymous reviewer for providing critical comments of earlier versions of this chapter.

CHAPTER 11
PHYLOGENETIC ANALYSES OF LIZARD ENDURANCE CAPACITY IN RELATION TO BODY SIZE AND BODY TEMPERATURE

Theodore Garland, Jr.

The causes and consequences of variation in locomotor costs and capacities have received considerable attention from physiological ecologists and comparative physiologists during the last 20 years (reviews in Taylor et al. 1982; Bennett 1983; Bennett and Huey 1990; Bennett 1991; Full 1991; Gatten et al. 1992; MacMillen and Hinds 1992; Djawdan 1993; Garland 1993; Garland and Losos 1994; Miles this volume). Two main reasons for this attention are apparent. First, locomotor performance is known or thought to be causally related to success in many activities that affect fitness in nature, including foraging, courtship, and escape from predators (reviews in Bennett 1983; Hertz et al. 1988; Bennett and Huey 1990; Garland et al. 1990b; Jayne and Bennett 1990; Garland and Losos 1994; Bulova in press). Second, selection is thought to act more directly on whole-animal performance abilities (e.g., speed, stamina) than on lower-level morphological or physiological traits (e.g., limb proportions, muscle contractile properties, enzyme activities; see Fig. 11.1). Thus, many recent studies of locomotion by organismal biologists have focused on or at least included direct measures of performance abilities, rather than only isolated morphological, physiological, or biochemical characters.

Much of the work quantifying locomotor capacities has involved reptiles (citations above and Brodie and Garland 1993; Miles this volume). For several reasons, reptiles and lizards in particular are attractive groups for studying locomotor abilities. Even ignoring birds, extant Reptilia exhibit a great range of locomotor modes, including swimming (e.g., turtles and crocodilians), burrowing (e.g., amphisbaenians), and limbless crawling (e.g., snakes) (Zug 1993). Even excluding snakes, squamate reptiles offer an impressive array of locomotor types, also including limbless forms, arboreal forms with specialized toe pads, gliding forms, and bipedal species with toe fringes that aid locomotion on water or sand (references in Garland and Losos 1994). As compared with terrestrial mammals, lizards have limited stamina but not necessarily limited speed or acceleration, which raises both physiological and ecological questions (Bennett 1983, 1991). Lizards also experience relatively variable body temperatures when active and have a large range of body sizes (both within and among species), both of which may be expected to affect absolute locomotor abilities. Finally, lizards are relatively

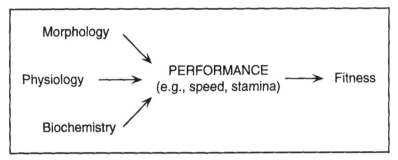

Figure 11.1. Conceptual relationships among individual variation in lower-level traits (morphology, physiology, biochemistry), organismal performance (e.g., various measures of locomotor abilities), and Darwinian fitness (modified from Arnold 1993; Garland 1994; Garland and Losos 1994; see also Huey and Stevenson 1979; Bennett 1980; Garland et al. 1990a). More complicated versions of this paradigm have also been proposed, sometimes inserting behavior between performance and fitness; it is also recognized that many of the relationships between levels are context dependent, and strongly affected by environmental circumstances (e.g., temperature; see Garland and Losos 1994). When considering interspecific variation, Darwinian fitness can be replaced by ecological factors and/or by measures of evolutionary success. The present comparative study is one of the first to begin with a survey of an organismal performance trait (treadmill endurance capacity) as the starting point for future studies of either the mechanistic bases of interspecific variation in performance or the behavioral and ecological correlates or consequences of variation in performance.

cooperative subjects for many measures of locomotor abilities (Bennett and Huey 1990; Garland and Losos 1994).

The most commonly measured aspects of lizard locomotor performance have been maximal sprint running speeds (e.g., Bennett 1980; Garland 1985; Huey et al. 1990; Losos 1990a; Bauwens et al. 1994; Miles this volume) and energetic costs of locomotion (Bennett and Gleeson 1979; John-Alder and Bennett 1981; John-Alder et al. 1983, 1986; Dial et al. 1987; Walton et al. 1990; Secor et al. 1992; Autumn et al. 1994). Far less information is available on variation in stamina (review in Garland and Losos 1994) or, at the other extreme, acceleration or jumping abilities (Huey and Hertz 1984; Carothers 1986; Losos 1990a). The purpose of the present study is to survey and quantify variation in treadmill-endurance capacities among species of lizards. The data set is one I have been accumulating since 1982, and the species included are opportunistically eclectic rather than a representative sample of all extant lizard clades. Nevertheless, the present study is the first broadly based comparative study of reptiles to begin with a survey of an organismal performance character as a starting point for future physiological or ecological studies (Fig. 11.1; see also Djawdan and Garland 1988; Garland et al. 1988; Djawdan 1993 on mammals). Chapters in each of the first two volumes on lizard ecology mentioned the general relationships between activity capacities and behavior or ecology, but did not directly address quantitative interspecific variation in locomotor abilities (Tucker 1967; Bennett 1983).

In addition to documenting interspecific variation, I address whether stamina is related to body size and/or body temperature, using a recently developed statistical method–phylogenetically independent contrasts–that explicitly uses information on phylogenetic relationships. This analysis indicates that stamina has evolved in a positive fashion with both size and temperature. Further studies will be required to elucidate other factors underlying interspecific variation in stamina, as well as the selective forces that have shaped its evolution.

Methods

Data collection

Data presented herein were gathered from March 1982 through August 1993. Most animals were captured by hand or by hand-held slip noose and returned to the laboratory within 1–2 or at most a few days of capture. When possible, animals were tested immediately; otherwise, they were maintained with access to incandescent lamps for thermoregulation during the day and given ad libitum access to water and fed (mealworms, crickets, fruits, vegetables, dog food) approximately every other day.

All animals tested appeared to be in good health; obviously gravid females were excluded, because they may show reduced locomotor abilities (Shine 1980; Garland 1985; Garland and Else 1987; Sinervo et al. 1991; review in Garland and Losos 1994). Sample sizes ranged from one to 58 individuals per species (Table 11.1: $\bar{x}$ = 10.2 for 54 species, excluding *D. dorsalis, A. cristatus,* and *V. salvator*). For some species (especially *Pogona, Gambelia, Sauromalus, Phrynosoma*), some individuals tested refused to walk to exhaustion on the treadmill, instead biting or inflating their bodies (cf. Crowley and Pietruszka 1983; John-Alder et al. 1986; Mautz et al. 1992; refs. therein). Some *Phrynosoma* squirted blood from their suborbital sinuses. For some other species, such as *Cophosaurus texanus,* nearly one-half of the individuals tested exhibited frantic activity during each trial, resulting in rapid exhaustion. All such individuals were excluded from the analyses. All species' mean values are treated equally in statistical analyses, with no attempt to weight for sample size. As this is the first broad-scale comparative study of endurance capacities of lizards, I considered it appropriate to include all available data (e.g., including the four species represented by a single individual) and hence possibly err on the side of completeness.

Table 11.1. Treadmill-endurance running times at 1.0 km/h for 57 species or subspecies of lizards. Listing order of species follows that of Figure 11.3.

Species	N	Temp. (°C)	Body Mass (grams)		Endurance (minutes)		Within-species log-log regression		Predicted $\log_{10}$ Endur. at Max. Mass
			Min	Max	Min	Max	intercept	slope	
Agaminae (5 species)									
Physignathus lesueuri	15	35.8	3.6	559.2	0.848	9.7	0.295	0.182	0.79486
Ctenophorus nuchalis	58	40	1.3	48.3	1.673	169.5	0.809	0.615*	1.84578
Ctenophorus fordi	9	37	0.7	3.9	1.825	9.4	0.402	0.713*	0.81999
Pogona vitticeps	10	37.7	4.1	421.4	8.053	189.9	0.692	0.370	1.66381
Pogona barbata	3	37.7	100.8	317.9	9.952	31.1	-0.024	0.510	1.25126
Polychridae (1 species)									
Anolis carolinensis	12	31.9	2.3	5.9	1.082	2.1	0.080	0.159	0.20255
Iguanidae (5 species)									
Dipsosaurus dorsalis		40		65					1.117609†
Amblyrhynchus cristatus		35		2885					1.30103†
Ctenosaura similis	17	40	21.4	760.1	3.643	35.7	0.402	0.270*	1.17887
Sauromalus obesus	1	39.5		192.0		7.8			0.88947
Sauromalus hispidus	2	38.4	443.5	530.5	10.9	16.8			1.13239†

	n								
Crotaphytidae (2 species)									
Gambelia wislizenii	7	`38.7	15.4	31.8	3.3	53.0	-1.323	1.720	1.26058
Crotaphytus collaris	12	37.3	14.2	33.6	3.4	29.8	0.555	0.254	0.94287
Phrynosomatidae (17 species)									
Uta stansburiana	21	37.2	1.8	4.8	1.4	3.7	0.294	0.054	0.33050
Urosaurus ornatus	21	36.9	1.8	4.4	0.8	1.9	-0.164	0.602*	0.22313
Sceloporus undulatus	7	37.5	3.0	11.7	1.1	2.8	-0.141	0.477*	0.36763
Sceloporus virgatus	13	36.4	2.3	5.1	0.9	2.5	-0.239	0.795	0.32244
Sceloporus occidentalis	19	34.7	7.2	14.2	1.3	3.8	-0.716	1.023	0.46192
Sceloporus olivaceus	1	36.4		30.6		1.6			0.19875
Sceloporus jarrovi	8	37.6	5.5	23.4	0.9	1.7	-0.243	0.336*	0.21713
Sceloporus clarkii	1	37.6		43.1		3.0			0.47596
Sceloporus magister	2	39.4	46.9	65.8	5.8	77.5			1.32726‡
Uma inornata	3	38	7.3	17.3	1.3	7.7	-1.439	1.895	0.90897
Callisaurus draconoides	20	40	8.2	19.2	3.4	18.3	-0.022	0.748	0.93720
Cophosaurus texanus	8	40.3	4.2	14.3	2.8	4.7	0.672	-0.152	0.54184‡
Holbrookia maculata	11	37.9	2.1	6.4	0.9	3.3	-0.131	0.545	0.30724
Phrynosoma coronatum	3	37.7	28.0	38.0	2.4	8.7	-5.920	4.335	0.92862
Phrynosoma cornutum	13	37.0	21.2	49.9	1.7	11.2	-0.200	0.468	0.59553
Phrynosoma platyrhinos	1	39.3		19.1		3.0			0.47900
Phrynosoma modestum	11	36.4	3.7	16.6	1.6	6.3	0.384	-0.022	0.36510‡

Table 11.1 (cont.)

Species	N	Temp. (°C)	Body Mass (grams)		Endurance (minutes)		Within-species log-log regression		Predicted log$_{10}$ Endur. at Max. Mass
			Min	Max	Min	Max	intercept	slope	
Gekkota (5 species)									
Coleonyx brevis	2	35	1.8	1.8	0.9	1.3			0.04336†
Teratoscincus przewalskii	25	15		9.1					-0.10791†
Lepidodactylus lugubris	7	34	0.8	1.8	0.5	1.1	-0.148	0.681	0.01852
Hemidactylus turcicus	9	35	2.3	3.6	0.7	1.2	0.1463	0.223	-0.02144
Hemidactylus frenatus	10	34	2.5	4.3	1.0	1.7	-0.228	0.608*	0.15889
Scincidae (7 species)									
Ctenotus regius	10	36.8	1.0	5.7	2.1	19.7	0.350	1.036*	1.13192
Ctenotus leonhardii	5	36.8	1.3	8.4	6.3	38.8	0.983	0.203	1.17002
Egernia cunninghami	25	35.7	7.0	313.6	3.4	9.3	0.579	0.102	0.83277
Tiliqua rugosa	6	36.6	471.5	652.3	11.5	45.4	-1.388	0.982	1.37494
Tiliqua scincoides	6	36.1	16.3	552.0	5.6	96.3	0.032	0.678*	1.89014
Eumeces laticeps	9	35.6	23.2	37.8	6.1	44.7	-2.941	2.893	1.62406
Eumeces skiltonianus	3	35	1.4	7.8	2.0	3.3	0.286	0.205	0.46897
Teiidae (7 species)									
Cnemidophorus t. tigris	24	40	10.3	25.8					2.07918
Cnemidophorus t. gracilis	12	40.0	8.4	22.1	13.0	71.2	0.089	1.218*	1.72483
Cnemidophorus t. marmoratus	8	40.8	6.8	18.8	17.2	79.8	1.766	-0.122	1.62477‡

Cnemidophorus sexlineatus	5	38.7	4.1	6.8	27.6	77.8	1.448	0.189	1.60501
Cnemidophorus uniparens	9	38.9	2.6	8.0	2.8	9.4	0.672	0.101	0.76295
Cnemidophorus exsanguis	2	39.9	13.0	14.5	11.0	37.5	3.486		1.30709‡
Cnemidophorus gularis	3	40.4	13.0	14.4	18.7	22.9		-1.913*	1.31472‡
Lacertidae (3 species)									
Pedioplanis lineoocellata	11	38.6	2.7	4.3	2.1	22.4	0.661	0.002	0.66181
Lacerta agilis	9	33.3	3.1	19.7	1.3	2.9	-0.069	0.401*	0.45008
Lacerta vivipara	19	30.3	1.1	4.7	0.8	1.8	0.260	0.144	0.12203
Anguidae (2 species)									
Gerrhonotus multicarinatus	6	35	5.9	31.3	1.4	3.3	-0.230	0.489*	0.50181
Gerrhonotus coeruleus	4	35	0.8	8.8	0.7	2.1	-0.086	0.427*	0.31821
Varanidae (1 species)									
Varanus salvator	35			505					0.84510†
Helodermatidae (2 species)									
Heloderma suspectum	6	31	24.5	673	2.9	17.4	0.168	0.360	1.18580
Heloderma horridum	7	31	194	1220	5.08	45.3	0.597	0.227	1.29644

* Slope significant at $P < 0.05$ (2-tailed test).

† Data for four species are taken from the literature: *Amblyrhynchus cristatus* data estimated from Gleeson (1979, 1980); *Dipsosaurus dorsalis* data estimated from John-Alder and Bennett (1981), John-Alder (1983, 1984a,b)(see also Cannon and Kluger 1985; Gleeson 1985); *Teratoscincus przewalskii* prediction from Autumn et al. (1994); *Varanus salvator* data estimated from Gleeson (1981; see also Gleeson and Bennett 1982). For these four species, body mass listed under "maximum" column is actually mean of animals tested.

‡ Mean of $\log_{10}$ endurance reported and analyzed (see text).

Endurance capacity was measured as the length of time lizards could maintain pace at 1.0 km/h on a motorized treadmill. Consistent with previous studies, trials were terminated when lizards failed to maintain pace following 10 consecutive taps or pinches (at < 1-s intervals) about the tail and hind limbs; this protocol yields repeatable measures of stamina (John-Alder and Bennett 1981; Garland 1988 unpubl.; John-Alder et al. 1986; Garland and Else 1987; Tsuji et al. 1989). The belt surface varied somewhat among laboratories, but was generally a type of rubberized cloth that provided good traction. The size of the area within which lizards could walk was varied in relation to lizard size (see Garland 1984; John-Alder et al. 1986; Garland and Else 1987).

Some species and individuals maintained pace more steadily than others, but every attempt was made to minimize any tendency to sprint forward and ride back on the belt, because this behavior may affect both the energetic cost of locomotion and stamina (cf. Thompson 1985; Full pers. comm.). At all times during a trial I used the mildest stimulation possible to keep a lizard moving but avoid having it become excessively frightened and engage in frantic activity, which usually leads to rapid exhaustion. All individuals were tested on two consecutive days, and the longer time was used as the measure of stamina. Notes were recorded at the end of each trial, and trials were repeated on an additional day if frantic activity occurred. If an individual never cooperated, its endurance time was excluded. Within species, individual differences in endurance time were generally repeatable from trial to trial. Some of the endurance data have been reported previously (Garland 1984, 1993; Garland and Else 1987; John-Alder et al. 1986; Beck et al. unpubl.).

All species were measured at or near the mean body temperature of animals when active in the field, as indicated in the literature (references in John-Alder and Bennett 1981; Avery 1982; Garland 1984; John-Alder et al. 1986; Garland and Else 1987; Hertz et al. 1988; Tsuji et al. 1989; Beck et al. unpubl.). Temperature was maintained by (1) placing the treadmill in an environmental chamber or (2) placing incandescent lights above the treadmill or by blowing warm air onto the belt surface with a portable hair dryer, controlled by a digital thermocouple temperature controller. In most cases, body temperatures of individuals were taken with a quick-registering mercury thermometer immediately after exhaustion; either the mean of these values or the maintained ambient temperature was used in data analysis. Body mass was recorded at the end of each trial.

For *Cnemidophorus tigris tigris* from Dale Dry Lake, San Bernardino Co., California, a mean endurance time of 2 hours was estimated from Garland (1993). Data for four additional species were taken from published sources. Three of these species were measured in the laboratory of

A. F. Bennett at the University of California-Irvine: *Amblyrhynchus crista-tus* (20 min, estimated from Gleeson 1979, 1980), *Dipsosaurus dorsalis* (15 min, estimated from John-Alder and Bennett 1981; John-Alder 1983, 1984a,b; see also Cannon and Kluger 1985; Gleeson 1985), and *Varanus salvator* (seven min, estimated from Gleeson 1981; see also Gleeson and Bennett 1982). Endurance data for *Teratoscincus przewalskii* (0.78 min) were from a predictive equation for animals tested at 15°C (Autumn et al. 1994). Data on field-active body temperatures of these geckos range from 9.9°–21.5°C, with a mean of 15.3°C; endurance was measured at 15° and at 25°C (Autumn et al. 1994). I have therefore used the endurance value for 15°C rather than for 25°C. For all of the foregoing species, mean body masses for individuals tested were substituted for "Maximum" body mass in Table 11.1.

Statistical analyses

Lizards exhibit relatively indeterminate growth, at least as compared with birds or mammals. For many species, I tested a broad ontogenetic size range, and in 40 of 44 cases endurance correlated positively with body mass and/or age, although few of these correlations were statistically significant at $P < 0.05$ (Table 11.1; see also Garland 1984; Garland and Else 1987; Garland and Losos 1994). Using mean body mass or mean endurance would therefore be unrepresentative of typical adults for many species. Moreover, my sampling of juveniles versus adults was not consistent across species. For comparative analyses, I therefore analyzed the logarithm of the maximum body mass within my sample and the predicted log endurance at that mass, computed from a least-squares linear regression equation for each species (right-most column of Table 11.1). All regression analyses were performed using SPSS/PC+ Version 5.0 (Norusis 1992). For *Cophosaurus texanus*, *Phrynosoma modestum*, *Cnemidophorus tigris marmoratus*, and *Cnemidophorus gularis* ($N = 3$), regression slopes were negative (see Table 11.1). For these four species and for *Sauromalus hispidus*, *Sceloporus magister*, *Coleonyx brevis*, and *Cnemidophorus gularis* (all $N = 2$), I analyzed the log of the maximum mass and the mean log endurance.

Examination of univariate and bivariate distributions and of residuals from regressions indicated that transformations were appropriate. Endurance and body mass were $\log_{10}$ transformed prior to all analyses, as is common in allometric studies. Body temperature was strongly left-skewed, with a single species measured at 15°C and all other values falling between 30° and 40°C. Thus, values for this independent variable were not evenly spaced on the raw scale. Body temperature was therefore transformed by raising it to the 10th power (and then divided by 10^{14} to simplify labeling of graphs). This transformation achieved a more-or-less even spacing of body temperature (see Fig. 11.6), which should improve statistical power while avoiding undue statistical influence by the single species measured at a very low (although ecologically realistic: Autumn et al. 1994) temperature.

Felsenstein's (1985) method of phylogenetically independent contrasts was applied using the PDTREE program of Garland et al. (1993; this and other PC-based comparative method programs are available from the author on request in exchange for a formatted 3.5-inch, 1.44-megabyte disk). This method uses independently derived information on phylogenetic relationships (topology and branch lengths) of the species being analyzed in an attempt to transform the species' mean values to be statistically independent and identically distributed, thus permitting the use of conventional parametric statistics to address a variety of evolutionary questions (Felsenstein 1985; Harvey and Pagel 1991; Martins and Garland 1991; Garland 1992; Garland et al. 1992; Pagel 1993; Martins this volume).

An example of the computation of independent contrasts is presented in Fig. 11.2. The full phylogeny used for analyses is depicted in Fig. 11.3. A complete description of the sources used to construct this "best current compromise" phylogeny is available from the author on request (e.g., see Murphy et al. 1983; Wyles and Sarich 1983; Kluge 1987; Montanucci 1987; Estes and Pregill 1988; Good 1988a,b; Dessauer and Cole 1989; Mindell et al. 1989; de Queiroz 1992; Sites et al. 1992; Zug 1992; Wiens 1993; refs. therein). Relationships of major lineages within Iguania are uncertain. Thus, following recommendations of Frost and Etheridge (1989), these relationships are represented and analyzed as an unresolved polytomy including the subfamily Agaminae plus the families Polychridae (represented here only by *Anolis carolinensis*), Iguanidae, Crotaphytidae, and Phrynosomatidae.

In the face of the unresolved node for Iguania, recommendations of Purvis and Garland (1993) were followed for computing independent contrasts and for estimating statistical relationships between traits. Specifically, from the 57 taxa represented as the tips in Fig. 11.3, the full $N-1 = 56$ independent contrasts were computed. Unresolved nodes lead to uncertainty as to the degrees of freedom available for hypothesis testing; in the worst case, one degree of freedom is lost for each unresolved node (Purvis and Garland 1993). The polytomy for Iguania in Fig. 11.3 represents four unresolved nodes in comparison with a fully resolved, dichotomous tree. Thus, to be most conservative, four degrees of freedom (df) should be subtracted when testing for relationships between endurance contrasts and body mass and/or temperature contrasts. However, these relationships were highly significant (all $P < 0.001$), so reducing the df had no qualitative effect on the conclusions.

Checks of branch lengths as described in Garland et al. (1992), and again using the PDTREE program, indicated that for all three traits the absolute values of standardized contrasts showed significant negative relationships

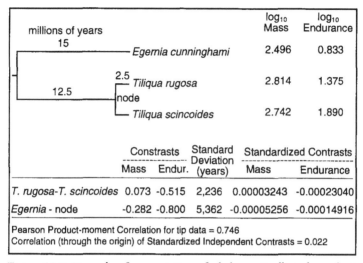

					log₁₀ Mass	log₁₀ Endurance

Figure 11.2. Example of computation of phylogenetically independent contrasts (Felsenstein 1985), using data for three species of Australian skinks (from Table 11.1). "Contrasts" for mass and endurance are differences between species' mean values at the tips of the phylogeny; the value for the node is computed as a weighted average of its two descendants (in this example, the branch lengths from the node to the two tips are equal, so a simple average is the same as an average weighted by the reciprocal of the branch lengths). A "standard deviation" of an independent contrast is a square root of the sum of the branch lengths of the contrast; branch lengths leading to estimated nodal values are lengthened to reflect uncertainty, as described in Felsenstein (1985). "Standardized contrasts" are contrasts divided by their standard deviations; these values are used in statistical analyses, such as multiple regression through the origin. Note that branch lengths can be transformed prior to computation of contrasts (Garland et al. 1992); in the present analysis, branch lengths were log transformed for endurance and for body temperature, but were raised to the 0.6 power for body mass (computations in this figure use the untransformed branch lengths).

with their standard deviations ($r = -0.333$ for log body mass, $r = -0.449$ for log endurance, $r = -0.483$ for body temperature raised to the 10th power). Such relationships must be eliminated before testing for correlations between sets of independent contrasts. Independent contrasts are ratios representing an amount of difference for a given trait (divergence at the phenotypic level) divided by branch lengths (see Fig. 11.2 and Garland 1992). As is commonly the reason for computing ratios, division by the denominator is intended to standardize or "scale" the numerators so they can be compared directly. Thus, the ratio should not show any correlation with its denominator, or spurious correlations between sets of ratios will be introduced. Analogously, it is inappropriate to test for a correlation between, say, resting and maximal metabolic rates by correlating simple mass-specific

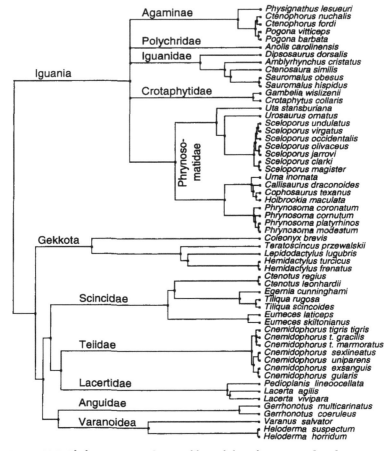

Figure 11.3. Phylogenetic topology and branch lengths estimated as divergence times for the 57 taxa of lizards studied herein. Basal split is at 190 million years. The unresolved polytomy at the base of the major lineages within Iguania (taxonomy follows Frost and Etheridge 1989) is arbitrarily placed at a depth of 100 million years.

values, because mass-specific metabolic rates typically show a negative relationship with body mass. Instead, metabolic rates can be divided by body mass raised to some appropriate power, determined empirically, or residuals from regressions on body mass can be computed (e.g., see Garland 1984, 1985; Garland and Else 1987; Garland and Losos 1994).

Given the correlations noted above, various transformations of branch lengths were applied in an attempt to reduce or eliminate them. For log body mass, branch lengths raised to the 0.6 power yielded a correlation of –0.065; for log endurance, log transformed branch lengths yielded $r = -0.108$; for body temperature raised to the 10th power, log branch lengths yielded $r = -0.081$: all of these correlations are statistically insignificant. The foregoing transformations were therefore used for computing independent contrasts (see Fig. 11.4). Residuals from the independent contrasts multiple regression of endurance on mass and temperature were well behaved, further justifying the transformations. It is important to note that the decisions to transform branch lengths (or the phenotypic traits) are empirically based for the purpose of satisfying assumptions of parametric statistical tests and do not represent arbitrary attempts to induce correlations between variables (Garland et al. 1992).

Results

Species varied tremendously in their endurance capacities (Table 11.1; Figs. 11.5–11.7). For example, range of endurance running times in the three largest species of Scincidae (*Egernia cunninghami*, *Tiliqua* [*Trachydosaurus*] *rugosa*, *Tiliqua scincoides*) was 10-fold (see also table 1 of John-Alder et al. 1986), even though they are of similar body size and body temperature. The three largest species of Agaminae (*Physignathus lesueuri*, *Pogona vitticeps*, *Pogona barbata*) showed a similar range in stamina with minor differences in size and temperature. Even populations of *Cnemidophorus tigris* exhibited substantial differences in stamina (Table 11.1).

The foregoing comparisons demonstrate that body size and body temperature do not explain all of the intra- or interspecific variation in stamina. Nonetheless, Figs. 11.5–11.7 do indicate an overall positive interspecific relationship of $\log_{10}$ treadmill endurance with both $\log_{10}$ body mass ($r = 0.531$) and body temperature[10] ($r = 0.512$). A conventional nonphylogenetic, stepwise multiple regression was highly significant ($F_{2,54} = 31.14$; $P < 0.0001$; multiple $r^2 = 0.536$), with the partial regressions for both $\log_{10}$ body mass ($F_{1,54} = 31.75$; $P < 0.0001$; partial $r^2 = 0.282$) and body temperature[10] ($F_{1,54} = 29.48$; $P < 0.0001$; partial $r^2 = 0.254$) being highly significant. Residuals from this multiple regression were reasonably well behaved, and none of the standardized residuals was more than three standard deviations from the mean. In this multiple regression (Fig. 11.7), the partial regression coefficient for $\log_{10}$ body mass estimates the allometric scaling exponent: 0.364 ± 0.130 ($\pm$ 95% confidence interval). Such nonphylogenetic analyses generally yield unbiased, although inefficient, estimates of scaling relationships; however, the significance tests and confidence intervals must be treated with caution owing to the nonindependence of species' mean values (Harvey and Pagel 1991; Martins and Garland 1991; Pagel 1993).

The phylogenetic analysis indicated that standardized independent contrasts in endurance were positively correlated (Fig. 11.8) with contrasts in both body mass ($r = 0.501$) and body temperature ($r = 0.440$). A stepwise

A.   B.

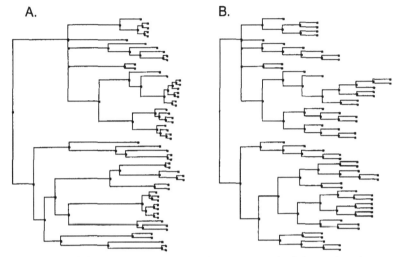

Figure 11.4. Phylogeny of Figure 11.3 with branch lengths (A) raised to the 0.6 power and (B) $\log_{10}$ transformed (cf. Garland et al. 1992). These transformations were used when computing phylogenetically independent contrasts (see text).

multiple regression through the origin showed that both contrasts in mass (partial $F_{1,54}$ = 19.78; P < 0.0001; partial r^2 = 0.251) and contrasts in body temperature (partial $F_{1,54}$ = 14.53; P = 0.0004; partial r^2 = 0.159) were highly significant predictors of contrasts in endurance (multiple $F_{2,54}$ = 18.75; P < 0.0001; multiple r^2 = 0.410). Note that r^2 values for regressions through the origin are not directly comparable to regression models including an intercept, as were used for the nonphylogenetic analyses above. With regression through the origin, as must be used with independent contrasts (Garland et al. 1992), r^2 indicates the proportion of variability in the Ys about the origin that is explained by regression (Norusis 1992).

Residuals from the multiple regression using independent contrasts were approximately normally distributed. Only one of the standardized residuals was more than three standard deviations from the mean, that for the contrast between *Cnemidophorus sexlineatus* (with relatively high stamina) and *Cnemidophorus uniparens* (with the lowest stamina for any measured *Cnemidophorus*; see Table 11.1). As the inclusion of unisexual species of *Cnemidophorus* (here, *uniparens* and *exsanguis*) in a bifurcating phylogeny with the sexual species and subspecies is an extreme oversimplification at best (e.g., see Dessauer and Cole 1989), I would refrain from interpreting anything about this. Other checks of the residuals (e.g., see Norusis 1992) did not indicate any notable outliers, leading to the general interpretation that after accounting for divergence in body mass, divergence in body temperature, and estimated divergence times, no unusually rapid evolutionary divergences in stamina are apparent.

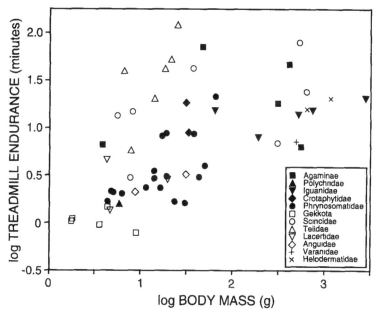

Figure 11.5. Positive log-log relationship between treadmill endurance running capacity and body mass for 57 species or subspecies of lizards (data from Table 11.1). Note that all solid symbols represent species within Iguania, whereas open symbols, +, and x represent species within Scleroglossa (represented here by Gekkota, Scincidae, Teiidae, Lacertidae, Anguidae, Varanidae, and Helodermatidae).

Discussion

Results presented herein indicate, unsurprisingly, that species of lizards with larger body size tend to have greater absolute stamina. After accounting for the effect of body size, body temperature also accounts for a significant amount of the remaining variance in treadmill endurance (all species were measured at temperatures representative of active lizards in the field). Multiple-regression analyses using phylogenetically independent contrasts confirm that stamina has evolved in a correlated fashion with both body size and body temperature; thus, in an evolutionary sense, it appears that bigger is better and warmer is better (cf. Bennett 1987b, 1990). The latter result indicates that evolution has not been able to compensate fully for the general Q_{10} effect on biochemical and physiological rate processes. An alternative approach would have been to measure each species at its optimal temperature for stamina (cf. Bauwens et al. 1994 on sprint speed), but this information is available for only a handful of species (see Bennett 1990). Moreover, measuring each species at its physiologically optimum temperature would have been ecologically unrealistic for those species (e.g., *Gerrhonotus*,

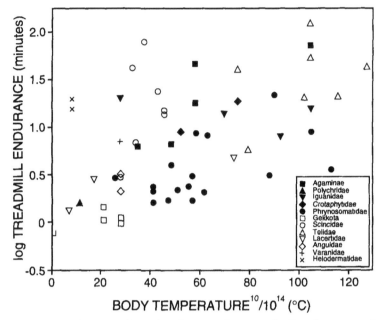

Figure 11.6. Positive relationship between $\log_{10}$ treadmill endurance and body temperature[10].

Teratoscincus przewalskii; Bennett 1980; Autumn et al. 1994) which are typically active at body temperatures considerably below their optimum for stamina.

The present comparative data suggest that if selection favors increased stamina in a lineage, one consequence may be increased body size and/or increased body temperatures while active. This hypothesis would be further supported by data indicating that within-population genetic correlations of stamina with size and with temperature tend to be positive in lizards (references in Brodie and Garland 1993; Garland and Losos 1994). Because stamina is probably mechanistically linked to both body temperature and body size (through its differential effects on limb cycling frequencies in relation to muscle mass and fuel and waste product storage capacities), positive genetic correlations seem likely to be both present and persistent over evolutionary time.

If size and/or temperature evolved as a correlated response to selection on stamina per se, this would have ramifications for food requirements, because both size and temperature affect total metabolic rate. Given that stamina might initially be subject to natural selection because of its relevance for foraging abilities (see Bennett 1983, 1991; Garland 1993; Cooper

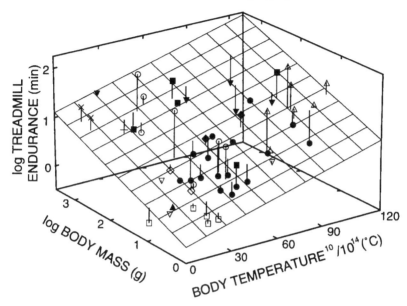

Figure 11.7. Three-dimensional plot showing positive relationship of $\log_{10}$ tread-mill-endurance running capacity with both $\log_{10}$ body mass and body temperature[10] (data from Table 11.1; symbols as in Fig. 11.5). Plane represents the multiple regression of endurance on both mass and temperature (see text). Note that $\log_{10}$ body mass and body temperature[10] are uncorrelated ($r = 0.017$).

this volume), among other things, one can envision complicated scenarios involving positive feedback of selection, correlated response, and further selection. Moreover, stamina might be linked mechanistically with maximal sprinting abilities (Djawdan 1993; Garland and Losos 1994; but see Garland 1988; Garland et al. 1988), because these two aspects of locomotor abilities share some common morphological and physiological bases. Given that sprinting abilities can also affect foraging abilities, the situation becomes even more complicated. One can imagine, for example, that the high body temperatures of active *Cnemidophorus* both enhance stamina and increase food requirements; the former permits foraging widely, whereas the latter may necessitate it. Wide foraging in turn diminishes territoriality and leads to a mating system in which males may pursue females for hours during courtship, thus creating a predisposition for sexual selection to act on stamina (see also Stamps 1983; Garland 1993; Martins this volume). Wide foraging and conspicuous courtship (e.g., in open desert habitats characteristic of many *Cnemidophorus* species) would place a premium on antipredator mechanisms that include high sprint speed and/or the ability to run long distances when pursued by a predator (i.e., high stamina; for distance running

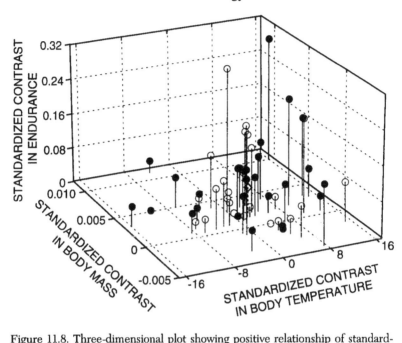

Figure 11.8. Three-dimensional plot showing positive relationship of standardized independent contrasts in $\log_{10}$ treadmill endurance capacity with contrasts in both $\log_{10}$ body mass and body temperature[10] (see text for statistics: contrasts in body mass and in body temperature were uncorrelated [$r = 0.086$], so multicollinearity was not a problem). Solid circles represent contrasts within the Iguania; open circles are contrasts within Scleroglossa (the bottom half of the phylogeny shown in Fig. 11.3); open diamond represents the basal contrast between these two clades. The fact that the basal contrast is not unusual indicates that these two clades do not differ in mean endurance, after accounting for effects of body mass and temperature (see Garland et al. 1993). Note that all contrasts in endurance have been set to positive while switching signs for mass and temperature contrasts accordingly (see Garland et al. 1992).

capacities, see Fig. 2 in Garland 1993); indeed, the *Cnemidophorus* studied herein also possess relatively high maximal sprinting abilities (Garland unpubl.). Phylogenetic comparative studies of *Cnemidophorus* and its relatives might be used to test such ideas (Brooks and McLennan 1991; Harvey and Pagel 1991). In any case, the foregoing scenarios emphasize the need to consider the evolution of whole organisms–not just single characters–in the appropriate behavioral, ecological, and phylogenetic context (e.g., Bennett and Huey 1990; Losos 1990a; Bauwens et al. 1994). Charles Darwin clearly appreciated this view, and it has been emphasized by many recent physiological ecologists, although some laboratory-oriented comparative physiologists and biochemists seem occasionally to have forgotten it.

Might the positive correlation with body mass be an artifact of the measurement protocol, which involved testing all species at the same absolute speed of 1.0 km/h? Probably not; consider other possible measures of stamina. One alternative would be to determine the maximum speed that each individual could maintain for some arbitrary length of time, perhaps 5 or 15 minutes. This would require measuring each individual at a series of speeds, and thus maintaining animals for extended periods of time in the laboratory, given that only one run to exhaustion per day is appropriate to allow for sufficient physiological recovery between trials (see Gleeson 1991). Such a protocol was logistically impossible in the present study. Moreover, even if it were undertaken, it would almost certainly show that larger individuals and species could maintain higher speeds for a given length of time, i.e., that stamina correlates positively with body size. Another index of stamina is the maximal aerobic speed, defined as the speed at which the maximal rate of oxygen consumption ($\dot{V}O_{2max}$) is attained (John-Alder and Bennett 1981; Taylor et al. 1982). Although ontogenetic data are limited to a single species (Garland and Else 1987), and comparative data for adults are limited to about 15 species of lizards (Fig. 23 in Garland 1993), in both cases the correlation with body mass seems to be positive. Thus, irrespective of what protocol might be used to measure the running endurance of lizards of different sizes, I am confident that it would show species with larger body sizes to have generally higher absolute stamina.

Why choose 1.0 km/h as the speed at which to measure stamina? This speed was chosen for two primary reasons. First, preliminary studies (e.g., John-Alder and Bennett 1981; Garland 1984) had indicated this to be a speed at which most individuals of most species would exhaust within a reasonable length of time, e.g., several minutes. This makes the measurement of many individuals per day feasible. But more importantly, most lizards do not engage in locomotor activity for more than a few minutes at a time, except in some unusual circumstances (Hertz et al. 1988; Garland 1993). Second, a survey of normal walking speeds of undisturbed lizards in nature showed that 1.0 km/h is not atypical for many species (appendix I of Garland 1993); certainly, it is only about 1/10 of the maximal sprint speed of many species (references in Garland and Losos 1994).

What about expressing locomotor performance on a per unit body length basis, as is commonly done in the literature on fishes? Might such an expression of stamina for lizards, or perhaps testing lizards at a speed that was a constant number of body length per second, be more "ecologically relevant?" I think not. What evidence have we that performance relative to body size—as opposed to absolute performance—is the more important determinant of success in foraging, courtship, or escaping from predators? None of which I am aware. Rather, absolute locomotor abilities (e.g., sprint speed in meters per second or stamina in minutes) would seem to be the behaviorally and ecologically relevant measure of performance. In my opinion, a potential predator or prey organism does not "care" about a lizard's relative sprint speed, only how fast it can move in absolute terms! I do not believe that animals live on a per unit size basis (cf. Garland 1983). In fact, however, we have essentially no direct empirical evidence on this point (Hertz et al. 1988; Bennett and Huey 1990; Garland et al. 1990b; Jayne and Bennett 1990; Garland and Losos 1994).

Although locomotor abilities per unit size may have no special ecological relevance, it is of interest to compare the scaling of locomotor abilities with indices of movement in nature. In terrestrial mammals, daily movement distances (actual distances walked) scale approximately as body mass$^{0.3-0.6}$ (Garland 1983; Goszczynski 1986; Altman 1987). A positive scaling of daily movement distance probably also applies to lizards, although the scaling exponent has not been described (see Fig. 1 in Garland 1993). If daily movement distance scales to an exponent similar to that for the scaling of endurance at 1.0 km/h (estimated as about 0.36 in the present nonphylogenetic multiple regression), then endurance relative to typical daily movements would be similar in lizards of different size. Comparisons of the scaling of stamina with the scaling of home range area or of typical walking speeds should also prove interesting (see Stamps 1983; Christian and Waldschmidt 1984; Hertz et al. 1988; appendix I of Garland 1993).

Any measure of locomotor performance can be affected by motivational factors in addition to morphology and physiology (Garland and Losos 1994). The recent emphasis on direct measures of locomotor performance (see Introduction; Fig. 11.1), however, is generally based on the assumption that the latter are more important than the former. A number of previous studies have been successful at identifying mechanistic correlates of individual variation in locomotor abilities in reptiles (e.g., John-Alder 1983, 1984a,b; Garland 1984; Garland and Else 1987; Tsuji et al. 1989; Garland et al. 1990a; Huey et al. 1990; reviews in Bennett and Huey 1990; Bennett 1991; Garland and Losos 1994)–performance gradients sensu Arnold (1983)–but comparable studies of interspecific variation have lagged behind (Bennett and Huey 1990; Losos 1990a; Bennett 1991; Garland 1993; Garland and Losos 1994). Apparently, motivation alone cannot account for all of the interspecific differences discovered in the present study because body size and temperature statistically explain about one-half of the variation among species. From a physiological perspective, future studies should attempt to determine what other factors are associated with variation in stamina (e.g., maximal oxygen consumption, relative heart size or muscle mass, blood oxygen carrying capacity; see Tucker 1967; Bennett 1983, 1991; Gleeson 1991; Garland 1993). Examination of the present data set would allow one to choose for study species which vary relatively widely in stamina but which are relatively closely related, of similar size, and active at similar body temperatures (e.g., within Agaminae, Phrynosomatidae, *Cnemidophorus*, or Scincidae; see Table 11.1). Also of importance will be determining whether constraints or trade-offs may exist that affect the joint evolution of different aspects of locomotor abilities, such as speed versus stamina (Garland 1988; Garland et al. 1988; Losos 1990a; Djawdan 1993; Garland and Losos 1994).

Direct inspection of species' mean values indicates that stamina can vary by 10-fold even among related species that have similar body sizes and similar active body temperatures (e.g., the three largest species of Scincidae and the three largest species of Agaminae). Both body size and body temperature, as independent variables, fall on the left side of Fig. 11.1; they represent proximate, mechanistic determinants of stamina, assuming that the present correlations actually reflect causal relationships. What might be the ultimate, evolutionary causes of interspecific variation in stamina?

One might hypothesize that interspecific or interpopulation variation in stamina would correlate with typical walking speeds, general movement rates, or perhaps home range area (cf. Bennett 1983; John-Alder et al. 1983, 1986; Hertz et al. 1988; Autumn et al. 1994), which might reflect such environmental factors as habitat heterogeneity, availability of cover, or predator or prey abundance (Bulova in press; on mammals see also Garland et al. 1988; Djawdan 1993). Hertz et al. (1988) compiled data for nine lizard species suggesting a positive correlation between treadmill endurance at 1.0 km/h and average daily movement distance, but quantitative field data on such traits are scarce (see Stamps 1983; Christian and Waldschmidt 1984; Pietruszka 1986; McLaughlin 1989; Garland 1993; Garland and Losos 1994) and presently limiting to further comparative analyses. Some anecdotal observations may be relevant here. The Australian water dragon, *Physignathus lesueuri*, lives in forested areas and rarely strays far from water, into which it often retreats when approached by humans and presumably by other predators. This species also shows the lowest stamina of the five Agaminae tested, the other four of which occur in more open and more arid habitats. Similarly, the Australian skink, *Egernia cunninghami*, seems rarely to stray far from the cover of exfoliating granite boulders; except for the much smaller *Eumeces skiltonianus*, it possesses the lowest stamina of the seven Scincidae tested. Finally, *S. magister* has the highest stamina of the seven species of *Sceloporus* tested, also occupies relatively open, arid habitats, and probably moves relatively great distances (Lowe pers. comm.).

Even in the absence of quantitative and comparable data on field activities, conventional wisdom identifies certain entire clades of lizards as generally having high activity levels (e.g., Scincidae, Teiidae, Lacertidae, Varanoidea in Fig. 11.3; Cooper this volume). The genus *Cnemidophorus* is a case in point (Garland 1993). Present data suggest that the genus *Cnemi-dophorus* appears to be characterized by generally high stamina, even after accounting for their relatively high body temperatures (see Fig. 11.7; Garland 1993; Cullum, pers. comm.; but see Bennett and Gleeson 1979 on *C. murinus*). Scincidae also have a reputation for being relatively active, widely foraging animals, and most of the species tested herein do have relatively high stamina; the same applies to the two species of *Heloderma* (Fig. 11.7).

Comparing the stamina of entire sets of species, such as species within different clades, is typically addressed by analysis of covariance. Garland et al. (1993) discuss several alternative phylogenetically based methods for ANCOVA. From the analytical perspective of phylogenetically independent contrasts (see "Methods"), hypotheses about whether clades show significant differences in some aspect of the phenotype are tested by asking whether the contrasts at their bases are unusual. For a priori hypotheses about differences between two clades, the (residual) independent contrast connecting the bases of the two clades can be tested for statistical significance. In the present study, I had no strong a priori hypotheses about clade differences. Also, the clades with reputations for being relatively highly active also are relatively closely related, i.e., the Scincidae, Teiidae, Lacertidae, and Varanoidea (see Fig. 11.3). Therefore, I simply examined the residuals from the independent contrasts multiple regression of endurance on body size and body temperature (see Fig. 11.8). Using standard criteria (e.g., Norusis 1992), none of the residual contrasts was a strong outlier (see last paragraph of "Results" section).

Therefore, based on the present data, after accounting for phylogenetic topology, estimated divergence times, evolutionary divergence in body size, and divergence in body temperature, I see no strong evidence that any of the clades discussed above experienced an unusual event during the evolutionary history of its stamina. By unusual event I mean a substantial change in stamina underlaid by change in something other than body size or body temperature, and perhaps caused by selection. Of course, general examination of all residual contrasts is not a very powerful test as compared with a priori hypotheses about specific contrasts (e.g., at the bases of clades). Larger sample sizes and more thorough sampling of clades may thus alter the foregoing conclusion. For example, the only species of *Varanus* in the present data base is *V. salvator*, a riparian species, which is almost certainly less active than are species which dwell in more open, arid habitats. However, a more powerful way to test hypotheses about the evolution of endurance capacity will be to obtain quantitative data on behavior and ecology, thus allowing each species to be fully used in a multiple correlation, regression, or path analysis, rather than having to lump and compare cladistically or ecologically defined groups of species (cf. Garland et al. 1993). In such an endeavor, differences among populations within species should not be overlooked (e.g., *Cnemidophorus tigris* in the present study; see also Bulova in press). Studying population as opposed to species-level differences allows a more direct examination of microevolutionary processes (Huey et al. 1990; Garland and Adolph 1991; Sinervo et al. 1991; Garland et al. 1992; Garland and Losos 1994; Niewiarowski this volume).

Acknowledgments

Animals were collected under a variety of permits, including California Scientific Collector's Permit #0746 (1983), New South Wales Dept. of Parks and Wildlife (1983–1984), Arizona Game and Fish Dept. Scientific Collecting Permit #GRLND000184 (1990), and New Mexico Dept. of Game and Fish Scientific Collecting Permit #1901 (1990 and 1992). I thank A. F. Bennett (U.C.-Irvine), A. J. Hulbert (Univ. of Wollongong), and R. B. Huey (Univ. of Washington) for laboratory space and support in years past; B. A. Adams, D. D. Beck, A. F. Bennett, S. J. Bulova, L.-L. ChangChien, J. Clobert, C. J. Cole, W. E. Cooper, M. R. Dohm, P. L. Else, D. L. Gregor, R. B. Huey, A. J. Hulbert, C. H. Lowe, T. O'Leary, A. Ramírez-Bautista, T. W. Reeder, J. R. Sayce, W. C. Sherbrooke, G. Sorci, J. D. Stein, C. R. Townsend, J. S. Tsuji, J. A. Wilkinson, and the students of Zoology 469 (Spring 1990) for assistance in capturing or loans of animals; K. Autumn and R. B. Weinstein for access to unpublished information; L.-L. ChangChien, K. de Queiroz, and D. L. Gregor for assistance in compiling phylogenetic information; J. A. Jones for computer programming and helpful discussions; and A. F. Bennett, R. Diaz-Uriarte, R. E. Jung, E. R. Pianka, L. J. Vitt, and two anonymous reviewers for comments on the chapter. Supported in part by National Science Foundation grants IBN-9157268 (Presidential Young Investigator Award) and DEB-9220872, the Wisconsin Alumni Research Foundation (administered by the University of Wisconsin Graduate School), and Australian-American Educational Foundation through a Fulbright Grant (1983–1984).

PART IV
POPULATION AND
COMMUNITY ECOLOGY

While lizards are special in many ways, those interested in population and community ecology would hope that they are not unique. The same factors and principles that influence the assembly, dynamics, and persistence of lizard populations should be generalizable to some degree to other organisms. If this were not the case, we lizard ecologists would have little importance outside our own parochial endeavors.

Lizard ecology is blossoming in several ways and leading to generalizable insights far beyond lizard ecology. Some obvious trends in the arena of population biology, community ecology, and biogeography are well illustrated by even a casual comparison of the first volume of Lizard Ecology (1967), with the second (1983), and the present volume. I briefly highlight from my personal perspective some of these trends and a few areas where further blossoming might be needed.

When one sees a building, a bridge, or a mitochondrion, the spatial organization of the parts, the architecture, suggests order and forces that must act in concert to produce that structure. When ecologists speak of "community structure," it is much less tangible. The term community structure is probably one of the most overused but underdefined terms in ecology. Losos (this volume) admirably focuses on more clearly stated and analyzed subquestions: What determines the set of species present in a place, their abundances, and the differences or similarities between communities? One of the most controversial questions in community ecology concerns the extent to which species affect distributions and abundances of other species. Are the species present on a given island, for example, a random assemblage with respect to species identity and ecological characteristics, or has competition acted as a filter, through extinction or failed colonizations, allowing only certain pairs to coexist and certain patterns of resource partitioning? One way, albeit circumstantial, to determine if competition or predation is filtering out prohibited species pairs is by testing statistically to see whether or not two suspected species co-occur on islands less often than expected by chance. But a much better way is to conduct controlled experiments when that is possible.

One striking trend in the history of lizard ecology is a dramatic shift from descriptive natural history to bold, large-scale experiments aimed at understanding the mechanisms of populational dynamics. Tom Schoener, for example, uses entire islands as replicates in experimental introduction and removal experiments. My coworkers and I utilize 20 identical aircraft hangars in Oahu to perform experiments to understand the mechanism of competitive interactions between two geckos. Sinervo (this volume) and colleagues experimentally construct populations of hatchling lizards to have body sizes either larger or smaller than the norm, and Niewiarowski (this volume) describes reciprocal transplant experiments designed to reveal the sources of variation in growth rates of *Sceloporus undulatus*. Such experiments were unheard of only ten years ago.

In this volume Andrews and Wright describe a large-scale and fascinating experiment where large plots of mature rain forest are experimentally watered with a massive and elaborate sprinkler system. The aim is to determine effects on *Anolis* population dynamics. Results underscore the importance in multispecies communities of variables affecting dynamics of one species, often in complicated and nonadditive ways. Addition of water affects egg survival differently than it does adults, affects each differently in wet years than in dry years, and the effects are not additive across years, as plants and other species adjust to alterations and feed back new and indirect effects at varying time scales.

One of the goals of ecology has been to understand community dynamics by decomposing the community into smaller and more manageable subsets of species than are found in the whole community. These complications, amply illustrated by Andrews and Wright, are not unique and should serve as a warning that the dynamic behavior of the full community may not be predicted using only observations of interactions between pairs of species in isolation. Another example of this growing appreciation is the important study by Schoener and Adler (1991) analyzing island distribution patterns of selected sets of bird and lizard species in the Bahamas. Their work suggests a role for diffuse competition. After controlling for differential habitat affinities of the species, they found frequent negative 3-species interactions often in the face of significant (or even positive!) pairwise interactions. Such higher order interactions and indirect effects are important because they imply that simple models of species interactions may be inadequate to describe and predict the behavior of complex communities.

The implications of individual behavioral plasticity to community dynamics of lizards or other taxa is largely unexplored territory. One species can modify the interaction between two other species by simply altering their behaviors, habitat preferences, or sensory abilities. In a large multispecies system, direct effects, indirect effects, higher-order interactions, unknown physical parameters, and individual behavioral adjustments of the component species will play out in complicated ways.

Another refreshing trend in community ecology is the incorporation of history, not just of the environment but of the taxa. Once we allow for the possibility that not all taxa are equivalent in their evolutionary lability and pliancy, then we have to accept, as Losos does in his chapter, that taxonomic affinities must shape community function and the distribution and abundance of species.

A milestone in the ability to address these issues is the development of molecular techniques in systematics. Since fossil preservation in most places is spotty, unraveling the origin of endemics has been fraught with uncertainty. Historical explanations of biogeographic patterns often demand concordance between suspected geological events and resulting speciation events. Without a fossil record, the biological timetable is missing, and comparisons with the geological timetable are impossible. Techniques like DNA annealing rates and protein and DNA sequencing allow not only assessments of relationships between species, but often allow construction of a rough timetable for diversification. Where phylogenies have been constructed using different molecular techniques they are often in good agreement with one another as well as with estimates based on classical techniques and the fossil record, giving us confidence in their accuracy.

The ability to construct reasonable phylogenetic hypotheses helps get to the core of some central questions in ecology and biogeography. It is very difficult to critically distinguish alternative theories for biogeographic processes. Consider the controversy surrounding MacArthur and Wilson's equilibrium theory of island biogeography. The discovery of extinctions on islands is consistent with any number of island biogeographic theories. What is critical in testing equilibrium theory is whether extinction rates appear relatively constant (as predicted by equilibrium theory) or are pulsed and associated with major geological or climatic changes as predicted by non-equilibrium theories. Similarly, dispersal may come largely in pulses as climates change, making old unfavorable habitats now accessible or sea levels bridging or separating land masses. Thus while both historical and ecological theories predict extinctions and dispersal, the temporal phasing is different. Phylogenetic reconstruction gives us an additional tool to examine the temporal modality of faunal diversification.

Another important trend in lizard ecology is the recognition that lizards are part of extended food webs, whose structure influences lizard abundances, brought to our attention by Schoener and his colleagues. The saurocentric ecologist can divide the organic world up into lizards, lizard food, things that eat lizards, things that eat the things that eat lizards, and everything else. Some of the "everything else" category includes organisms that compete with lizards for lizard food and so they indirectly have a bearing on lizard abundance. More and more, we will be seeing studies that relate lizard abundance to features of trophic structure. Gary Polis and colleagues are now conducting studies utilizing the islands in the Sea of Cortez, where lizard biomass can vary by two orders of magnitude from island to island, to assess the relative importance of bottom-up effects (lizard food) versus top-down affects (lizard predators) in accounting for this variance. Schoener and, independently, Jon Roughgarden have verified the impact of lizard exploitation on lizard food (insects) through carefully designed experiments on island systems in the Caribbean. My colleagues and I have observed nearly a 100-fold decrease in diurnal lizard abundance on islands in the mid-Pacific where mongooses were introduced near the turn of the century compared to islands that still lack mongooses.

One area that still needs more study is the effect of "things that eat the things that eat lizards" on lizard abundance. In the tropical Pacific, at mainland sites like tropical Australia and Southeast Asia, Borneo, and the Philippines, lizards can also be strikingly abundant in spite of a rich diversity of lizard predators. Perhaps here, the presence of predators on lizard predators allows lizards to reach high numbers. Consistent with this hypothesis is the casual observation, worthy of further quantification, that large, near-Asian islands like Palau, which lack the very top-level carnivores, such as large felids and snake- and monkey-eating eagles, have uncommonly low lizard numbers. Further out in the Pacific, where lizard predators are now absent, lizard abundance returns to high levels, unless predators like the mongoose are introduced. If such trophic effects indeed explain at a broad-scale the geographical variability in lizard abundance, then we should be able to find complementary patterns in the abundance of lizard food. For example, does mongoose introduction lead to higher ground-dwelling insect numbers in the mid-Pacific islands? Do matched study sites with and without top carnivores differ consistently in lizard food? These are exciting opportunities for comparative ecological work.

How frequent is coevolution within lizard assemblages? Does it influence the structure of communities? If it does shape communities, how does it shape them? These questions have been debated intensely for decades. The development of molecular techniques and new theoretical tools for phylogenetic reconstruction will lead to insights into these questions.

The biggest barrier to testing models of character displacement for specific situations, e.g., the anoles of Lesser Antilles or the *Cnemidophorus* of the Sea of Cortez, is the large number of parameters in the models that are nearly impossible to quantify. For example, quantitative genetic models of character displacement require knowledge of the competitive impact of each specific phenotype on each other plus the heritability of those phenotypes. Yet in practice it is rarely possible to find identical sets of phenotypes so that competitive impacts can be experimentally determined and replicated since in sexual species each individual is genetically unique. Clonally reproducing lizards (e.g. several geckos, unisexual *Cnemidophorus*, parthenogenetic lacertids) offer the experimentalist an important bridge to testing models dealing with the coevolution of competitors. Molecular genetic techniques (allozymes, genetic fingerprints, mt-DNA restriction fragment maps) allow resolution of genetic clones. One can then measure phenotypic variability within a clone, measure ecological differences correlated with genetic differences, and experimentally combine different clonal combinations to measure specific competitive phenotypic interactions. In this way we should be able to test the underpinnings of evolutionary niche theory as applied to the problem of character displacement. Is it true that greater genetic variability leads to greater ecological variability and wider species niche widths? Do larger phenotypes have a competitive advantage in terms of competitive impacts on smaller phenotypes? The tantalizing glimpses of answers to these questions seen in a handful of published studies warrant vigorous pursuit, both theoretical and empirical.

Finally, an area not surprisingly absent from all three Lizard Ecology volumes is missing, not because it is unimportant, but rather intractable experimentally without major commitments of resources. There remains a noticeable lack of landscape-scale studies on lizards–studies that relate local abundance and fluctuations to not only local ecological processes but also to regional processes–i.e., metapopulational phenomenon like patch colonization and extinction rates and the physical arrangement of patches in space. There is probably no more practical question today in conservation biology than whether elimination of a particular habitat patch in one spot (via development) will cause significant "downstream" effects on populations in unaffected patches elsewhere, and how these "effects at a distance" are shaped by the geometry of the landscape. In a provocative new book, Pimm (1991) often points out the large scatter about most attempts relating various populational stability measures to suspected individual characteristics of species (e.g. body size, reproductive rate, average population density) or to the way a species is linked to others (number of trophic levels, number of prey species). Perhaps we have not identified some of the most pivotal influences of temporal fluctuations. One possibility, emerging as a contender, is that the stability properties of a particular species' population are affected at least as much by the abundance of habitat space and the arrangement of habitat patches in space, as by the population densities of species with which it interacts or the temporal variability of the climate. The study of patch dynamics in lizards and other taxa is ripe for further exploration. Eric Pianka's new studies of large-scale fire dynamics and the complex landscape patchwork fires created for lizards in arid Australia hold exciting potential for addressing these issues.

Ted Case

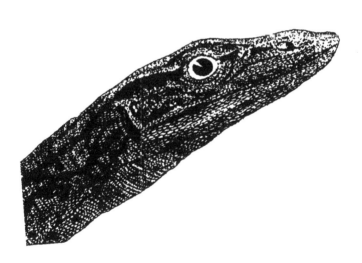

CHAPTER 12
LONG-TERM POPULATION FLUCTUATIONS
OF A TROPICAL LIZARD: A TEST OF
CAUSALITY

Robin M. Andrews and S. Joseph Wright

A central problem in ecology concerns population stability. Connell and Sousa (1983) point out that the assumption of constancy through time is implicit to a wide range of theoretical and empirical studies on ecological systems. Given this assumption, observations of resource partitioning, for example, can be used imply to past and present competition for resources. If, on the other hand, populations fluctuate widely, present patterns of resource use may have little relationship to past interactions among species (Wiens 1977). Actually, in a survey of real populations, magnitude of population fluctuation varied widely (Connell and Sousa 1983). No taxonomic patterns were apparent in this survey; terrestrial plants, terrestrial insects, aquatic invertebrates, birds, and mammals all included populations with low and with high variability in density. Lizards, however, were not included in Connell and Sousa's study.

In fact, observations of 29 populations of 21 species over relatively short periods suggest that constancy through time is characteristic of lizard populations (Schoener 1985). This idea is supported by long-term observations published more recently (Abts 1987; Jones and Ballinger 1987; Strijbosch and Creemers 1988; McLaughlin and Roughgarden 1989). However, concluding that lizard populations are unusually stable from this data set may be premature. One limitation of the data concerns geography; most observations were either for species inhabiting temperate zone regions or small islands in the West Indies. These locations are unlikely to be representative of all habitats occupied by lizards. Another limitation concerns life histories; most observations were for species in which adults are relatively long lived. In this situation, a persistent adult population buffers the effect of variable reproductive success. For example, chuckwallas (*Sauromalus obesus*) in the southwestern U.S. breed infrequently; reproduction is associated with years of high rainfall when food is abundant (Abts 1987). Still, the population fluctuates little because of high adult survival from one year to the next.

We use long-term data on populations of the polychrid (sensu Frost and Etheridge 1989; formerly in the Iguanidae) lizard *Anolis* limifrons to address the issue of stability. The geographic range of *A. limifrons* is mainland neotropical, and individuals are relatively short lived. Thus, this species provides a geographical and life-history contrast to the majority of population studies on lizards. In addition, we will go beyond a description of population fluctuations and discuss the results of an experimental study that tests causal mechanisms behind changes in population density.

In comparison with other lizard species, populations of *A. limifrons* are far from stable. This conclusion is the result of observations for 22 years at Barro Colorado Island (BCI) in Panama (Fig. 12.1). During this period, the population has exhibited large fluctuations in density from one year to the next and even larger variation over decades (Andrews 1991).

Fluctuations in density are correlated with variation in the amount and timing of rainfall (Andrews and Rand 1982, 1990; Andrews 1991). Thus, rainfall appears to drive the large fluctuations in density, although food intake and recruitment are density dependent. *Anolis limifrons* individuals are most abundant at the end of years with wet seasons that are relatively dry, and with dry seasons that are relatively short, as measured by the amount of rainfall during the preceding December and April. December and April are critical months in defining the length of the dry season. For example, if much rain falls in December and April, the dry season is short (3 mo), and if little rain falls in December and April, the dry season is long (5 mo). In contrast, the months of January through March are not relevant (statistically) because rainfall is consistently low.

The two components of the correlation between population density and rainfall are the amount of rain in the wet season and the length of the dry season. Of these two, the association between the length of the dry season and population density makes the most sense. The dry season is a period of very low reproductive activity (Andrews 1979a). For example, at the beginning and end of the dry season, December and April, the number of eggs in the forest is 24% and 4%, respectively, of that during the middle of the wet season (July-September) when egg production is maximal (Andrews unpubl.). Low reproduction in the dry season can be related to population density at the end of the year because recruitment commences when the rains begin in April or May, and the sooner the rains begin, the longer the period in which individuals are added to the population and the larger the population build-up by the end of the year.

The second component of the correlation between rainfall and population density is the amount of rain during the wet season. At the time the experimental component of this study was initiated, we did not know why a wet season with low rainfall was associated with high population density. As we shall discuss, the consequences of manipulations of moisture availability in the dry season suggest a mechanism that relates the amount of rainfall during the wet season to population density.

*Editors' note: Nomenclature of anolines remains controversial; *Anolis* was split by Guyer and Savage (1986), reconsidered by Cannatella and de Queiroz (1989) and Williams (1989), and further reconsidered by Guyer and Savage (1992).

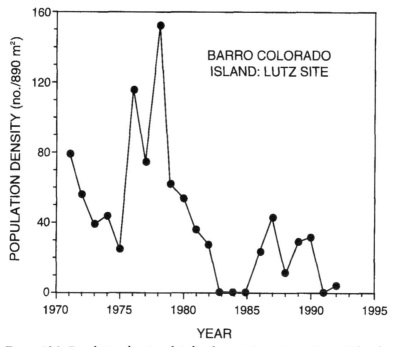

Figure 12.1. Population density of *A. limifrons* at Lutz site at Barro Colorado Island, Panama from 1971 through 1992. Data for 1971–1989 are from Andrews (1991) and data for 1990–1992 are from Andrews and Rand (unpubl.).

The opportunity to evaluate the mechanism or mechanisms by which seasonal variation in rainfall affects population density was provided by a major experimental perturbation at BCI. Water was provided to large areas of rainforest during three successive dry seasons. The amount of water was sufficient to maintain soil moisture and relative humidity near the forest floor at wet-season levels (Wright and Cornejo 1990). Because lizards typically perch less than 1.5 m from the ground, and eggs are laid on or near the ground, dry season conditions were ameliorated by the watering treatments for both lizards and eggs.

Our general expectation was that the density of lizards on watered plots would increase relative to that on control plots over the course of the experiment. Our reasoning was that females would produce eggs during the dry season on watered plots. Reproduction during the dry season, in addition to the wet season, would increase population density because females would produce more eggs during their lifetimes. Furthermore, whereas one generation per year is typical for this species, two generations per year would be possible if reproduction were possible year round.

Materials and Methods

Life history

The life history of A. *limifrons* in Panama is well known (Sexton 1967; Sexton et al. 1972; Andrews 1979a, 1982b; Andrews and Rand 1982, 1983; Andrews et al. 1983; Wright et al. 1984). Pertinent attributes are summarized here. *Anolis limifrons* is a small lizard with a maximum snout-vent length (SVL) and mass (M) of 50 mm and 2 g for both females and males. Females and males become sexually mature at a minimum of 35–36 mm SVL; we considered individuals of 36 mm or more to be adults. The modal size of sexual maturity for females is 41–42 mm SVL and all females are sexually mature at 44 mm SVL.

Reproduction is seasonal, with the highest rates of egg production in the early and middle wet season and the lowest rates during the dry season. Eggs are first laid when rains begin in April or May. Hatching occurs 6 weeks after oviposition and hatchlings are most abundant in August and September.

The reduction of egg production in the dry season seems more related to the effect of dryness per se on eggs than to limited food availability for lizards. Eggs laid in the dry season are subject to desiccation, at least in some years (Andrews and Sexton 1981). In contrast, lizards appear to accommodate to relatively dry conditions (Sexton and Heatwole 1968) by utilizing relatively mesic microhabitats during the dry season; lizards perch closer to the ground and are associated with larger trees during the dry than the wet season (Andrews unpubl.). Despite the fact that arthropod abundance reaches annual lows late in the wet season and during the dry season (Smythe 1982; Levings and Windsor 1982), food availability during the dry season cannot be severely limiting as juveniles grow to maturity during this period (Andrews et al. 1983), and adults increase their fat reserves (Sexton et al. 1971).

Study site and experimental procedures

Studies were conducted at Poacher's Peninsula on Barro Colorado Island (Wright and Cornejo 1990). The forest is among the oldest on BCI and has not been cleared for at least 500 years (Piperno 1990). The terrain is flat or gently sloping and the canopy is approximately 30 m high. Four 150 × 150 m plots were established in late 1984. Water availability was manipulated by watering two of the plots (plot numbers 1 and 2). The remaining plots (plot numbers 3 and 4) served as nonwatered controls. Water was delivered by PVC pipe to 80 sprinklers per plot, mounted 1.8 m above the ground. The objective was to maintain soil water potentials at –0.04 MPa at depths of 25 and 45 cm (Wright and Cornejo 1990). Thus, the schedule for watering was adjusted relative to target levels; watering was initiated in December and continued into April for the three years of observation reported here (1986, 1987, and 1988). During a typical week in the dry season, each manipulated plot received 675,000 kilograms of water during 1.5 h between 1100 and 1400 h on each of five days. Water was taken from Gatun Lake where nutrient concentrations are lower than in rainwater (Gonzalez et al. 1975).

Demographic studies

Lizards were censused on two 30 × 30 m areas (replicates) within each of the four plots. These permanent subplots were picked randomly. Censuses were of two types. During the first type, an observer walked a set path on the perimeter and the two midlines of each 30 × 30 m area and searched for lizards 2 m either side of the transect line. During the second type of census, the 30 × 30 m areas were carefully searched for lizards. Censusing began in January 1986 and ended in October 1988. Twenty-three censuses were conducted: 4, 5, 3, 5, 3, 3 in the 1986 dry and wet seasons, the 1987 dry and wet seasons, and the 1988 dry and wet seasons, respectively. Plots were censused within a few days of one another and comparable effort was spent on all plots within each census period. Lizards captured during both types of censuses were processed and released at the site of capture. Lizards were weighed to the nearest 0.1 g using Pesola® scales and measured from the tip of the snout to the vent (SVL) to the nearest mm. Lizards were individually marked by clipping the terminal phalange of toes (not more than two toes per foot and five toes per individual). Capture-recapture data were used to estimate population density, survival, seasonality of recruitment, an index of condition, and body growth.

Data analyses

Overall, the objective of the experiment was to determine the effects of watering during the dry season on lizard populations. Because observations were made for three years, differences among years were also of interest as a main effect and as an interaction with time of year (January-April: dry season, and May-December: wet season) and treatment (watered or control). Pairwise z-tests follow White et al. (1982). All other statistical analyses were conducted using SAS statistical software (SAS 1982).

    Standard Jolly-Seber models (Pollock et al. 1990) were used to estimate population density and survival. Data for the two subplots within each plot were pooled to ensure that sample sizes were sufficiently large for reliable estimates. Separate analyses were run using the entire data set and for adult data only. Censusing protocols differed somewhat from census to census. The first three and last three censuses were line-transect censuses while during the middle of the study, line-transect censuses were alternated with more intensive searches of entire subplots. The lower intensity of censusing early and late in the study should tend to inflate estimates of density and reduce estimates of survival at these times. However, as the same protocols were used at all sites within any one census, estimates of numbers and survival are consistent relative to one another across sites. For statistical comparisons, we used the means of estimates of numbers and survival for each season and year combination per plot as observations in an ANOVA with repeated measures on season and year.

The proportion of individuals less than 30 mm SVL (< 2 mo old) captured during each season in each year was used an index of recruitment. Recruitment in this sense has two components: the rate at which eggs are produced and their rate of survival to hatching. In a previous study, Andrews (1988) determined that the proportion of young individuals in *A. limifrons* populations at the end of the wet season was directly related to the survival of eggs during that wet season. For statistical comparisons, we used the proportion of individuals less than 30 mm SVL in each of the four subplots per treatment (watered vs. control plots) as observations in an ANOVA with repeated measures on season and year. Proportions were based on the first capture of an individual in each season and year, that is, each individual was represented only once per season and year combination.

An index of condition [IC = $(Mass^{0.3}/SVL) * 100$] was used as a general measure of food intake by adult males and females and of reproduction by adult females (Andrews 1991). Observations for females and males were analyzed separately because of known sexual differences in this index: IC for adult females is relatively high because of the mass of follicles and oviductal eggs (Andrews et al. 1983). Repeated observations on the same individuals in successive censuses were treated as independent observations because (1) size and reproductive status of individuals change with time and (2) recaptures were not common; the probability that a lizard alive at any one census would be captured during that census was relatively low (24–27%, results of Jolly-Seber analyses). The four subplot means per treatment for each season in each year were used as observations in an ANOVA with repeated measures on season and year.

Growth data were fitted to the logistic growth equation because it provided the best description of growth for *A. limifrons* in previous analyses (Andrews et al. 1983). The linear form of this equation is

$$SVL2 = SVL1 * a/[SVL1 + (a - SVL1) * e^{-rD}]$$

where SVL1 and SVL2 are the snout-vent lengths at the first and second capture, respectively, D is the interval in days between captures, a is the asymptotic SVL, and r is the rate constant of the logistic equation (Schoener and Schoener 1978). Intervals < 15 d were not used. Estimates of a and r were determined by nonlinear regression procedures. To avoid bias, records for individuals that had reached their asymptotic size (SVL > 43 mm) at first capture were not used. Preliminary analyses showed that males and females did not differ in their asymptotic size or in their rate constants ($P > 0.5$, t-tests; see also Andrews et al. 1983). Therefore, data for males and females were combined.

To estimate a and r accurately, nonlinear regression analysis requires a relatively large number of records that span the range of possible sizes (see comments by Andrews in Schoener and Schoener 1978). Too few growth records were available for simultaneous comparisons of plots within treatments, years, and seasons. For example, most (70%) dry season records were from 1987, while 95% of wet season records were from 1986 and 1987. Therefore, a single estimate of a and r were made for the watered and control plots by season.

As maximum SVL for A. *limifrons* individuals is 50 mm, many individuals with SVLs of 44 mm or more were still increasing in size. To further evaluate possible differences between the treatments, growth rates (mm/d) for individuals with SVL > 43 mm were used in a three-way ANOVA with treatment, year, and season as main effects.

Egg survival was evaluated in July and August 1986 on the four plots. Details of this study were presented by Andrews (1988).

Results

Population density

Watering was initiated at the beginning of the 1986 dry season. At this time, mean estimated population density (standard errors) of adults was of 117 (53), 64 (52), 114 (161), and 29 (14) for plots 1–4, respectively. Densities on the four plots were similar (all $Ps > 0.05$, pairwise z-tests).

Adult densities during the 1986 dry season represent pretreatment densities because the great majority of these individuals would have been present when watering began. Thus, differences in recruitment or subsequent differences in density between treatments should reflect treatment effects and not preexisting differences in density.

Watered and control plots had very different patterns of population density over the course of the study (Fig. 12.2). Population densities on control plots were stable. In contrast, while densities were similar initially on watered and control plots, during the 1986 wet season, densities on watered plots exploded. At that time, all four possible pairwise comparisons between watered and control plots were statistically significant ($zs = 1.96, 2.19, 2.82, 2.85, Ps < 0.05$). Following 1986, densities on watered plots declined monotonically, and were again similar to control plots toward the end of the study. No pairwise comparisons at periods other than the 1986 wet season were statistically significant ($P > 0.05$). Overall, total population densities on watered plots were significantly higher than on control plots (Table 12.1). Density did not differ as a function of season, although the significant season * year interaction presumably reflects the increase in density from the dry to wet season in 1986 and the decrease from dry to wet season in 1987 and 1988. Density did not differ as a function of year, but near significance ($P < 0.10$) of this main effect and of the year * treatment interaction reflect the very different patterns of density change on the watered and control plots.

Any sampling biases (see "Materials and Methods: Data Analyses") would not have affected results of comparisons between treatments. Bias would have affected all plots, and would have resulted in relatively high estimated densities for the 1986 dry season and the 1988 wet season.

Survival

Survival of adults was not affected by watering and did not differ among years or between seasons (Table 12.1, Fig. 12.3). As predicted by the low sampling effort for the 1986 dry season and the 1988 wet season (see "Materials and Methods: Data Analyses"), estimates for these periods were relatively low. However, results of an ANOVA which excluded these two periods were the same as those utilizing all periods (Table 12.1); the only significant term in both analyses was the season * year interaction ($P < 0.05$).

In general, survival rates of adult *A. limifrons* on the watered and control plots were similar to those observed at other sites in central Panama (Wright et al. 1984; Andrews and Nichols 1990). Andrews and Nichols (1990) did not detect differences in survival rates between adults and juveniles at two other BCI sites. Similarly, overall estimates of survival in this study for analyses that included all individuals and that included only adults were virtually identical for all plots.

Growth

Watering during the dry season affected growth rates but not asymptotic size (Table 12.2). The rate constant of the logistic growth equation in the dry season on the watered plots (0.023) was higher than on the control plots (0.019). Given the a priori assumption that lizards should grow faster under moister conditions (Stamps and Tanaka 1981), this difference was statistically significant ($P < 0.05$, $t = 1.8$, one-tailed t-test). Lizards had similar growth rates on the watered and control plots in the wet season, and asymptotic sizes were similar at all times ($P > 0.05$, t-tests). At these growth rates, a hatchling would grow to sexual maturity (36 mm SVL) in 68 days in the dry season on the watered plots, but this growth would take 81–83 days in the

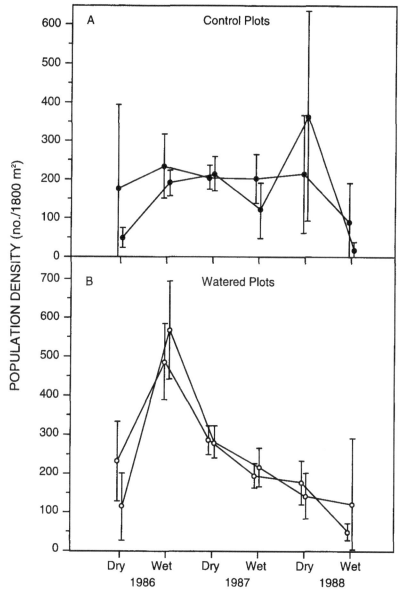

Figure 12.2. Estimates of population density (± 1 SE) of A. *limifrons* on control plots (top panel; closed circles) and watered plots (bottom panel; open circles). Estimates are arithmetic means of time-specific estimates for each season and year combination. Standard errors are pooled standard errors for these periods.

Table 12.1. Repeated measures analyses of variance for population density and adult survival. Where appropriate, Greenhouse-Geisser estimates of probability levels are given. Means are shown in Figs. 12.2 and 12.3, respectively.

Source	df	Density		Survival	
		F	P	F	P
Between subjects					
Treatment	1	21.2	< 0.05	1.0	0.41
Error	2				
Within subjects					
Season	1	0.0	0.92	0.0	0.88
Season*Treatment	1	3.6	0.20	0.3	0.66
Error	2				
Year	2	11.2	< 0.10	0.2	0.74
Year*Treatment	2	12.6	< 0.10	0.1	0.76
Error	4				
Season*Year	2	21.2	< 0.05	26.5	<0.01
Season*Year*Treatment	2	2.7	0.18	1.3	0.37
Error	4				

wet season on all plots and in the dry season on the control plots. In contrast, growth rates of adults with SVLs > 43 mm did not differ as a function of treatment, season, or year ($P > 0.05$, $F_{11,386} = 1.2$, 3-way ANOVA).

Index of condition

Results of the analyses of IC were similar for females and males (Table 12.3). As expected, however, absolute values of IC were higher for females than males (Fig. 12.4). IC did not differ between watered and control plots. Except for the dry season of 1988, values of IC were highest during the dry season and lowest during the wet season, and these seasonal differences were significant. IC declined during the 1986–1988 period, and differences among years were significant as well. The season by year interaction was significant, presumably because of the relatively low values for IC in the 1988 dry seasons.

Juvenile recruitment

The proportion of juveniles did not differ between treatments (Table 12.4, Fig. 12.5). However, seasonal variation was highly significant as was the season * treatment interaction. This interaction reflects the very different pattern of recruitment on watered and control plots. Recruitment on control

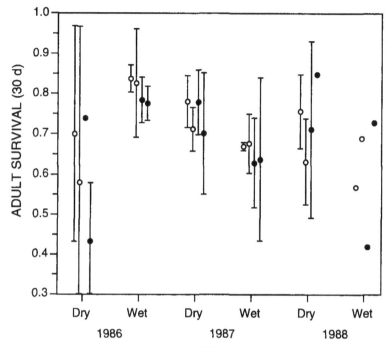

Figure 12.3. Estimates of 30-day probabilities of survival (± 1 *SE*) for adult *A. limifrons* on two control (closed circles) and two watered (open circles) plots. Estimates are arithmetic means of time-specific estimates for each season and year combination. Standard errors are pooled standard errors for these periods (Nichols pers. comm.). Standard errors could not be calculated for some estimates early and late in the study period.

plots was strongly seasonal–relatively few juveniles were present during the dry season and relatively many were present in the wet season. In contrast, on watered plots the proportion of juveniles did not change seasonally. This pattern was consistent across years.

Egg survival

Egg survival from oviposition to hatching was higher on the two watered plots (25% and 42%) than on the control plots (15% and 8%)($P < 0.01$, 2×2 X^2 test) in July and August 1986 (Andrews 1988).

Discussion

Population dynamics

Barro Colorado Island supports a semideciduous tropical forest. Rainfall is seasonal, but only a small minority of tree species are deciduous during the

Table 12.2. Fitted parameters of the logistic growth equation for *A. limifrons* populations in dry and wet seasons on watered and control plots. Estimates of *a* (asymptote) and *r* (rate coefficient) are followed by *n*, asymptotic *SE* (in parentheses). N. S. = not significant.

	Season	
	Dry	Wet
Asymptote, *a*		
Watered	47.1 (63, 0.56)	47.3 (76, 0.44)
	N. S.	N. S.
Control	47.5 (59, 0.62)	47.0 (64, 0.59)
Growth Constant, *r*		
Watered	0.023 (63, 0.0018)	0.019 (76, 0.0011)
	$P < 0.05$	N. S.
Control	0.019 (59, 0.0015)	0.019 (64, 0.0013)

Table 12.3. Repeated measures analyses of variance for the index of condition for females and males (mean values shown in Fig. 12.4). Where appropriate, Greenhouse-Geisser estimates of probability levels are given.

Source	df	Females		Males	
		F	P	F	P
Between subjects					
Treatment	1	0.2	0.67	1.0	0.36
Error	6				
Within subjects					
Season	1	26.5	< 0.01	57.6	< 0.01
Season*Treatment	1	3.0	0.13	1.1	0.34
Error	6				
Year	2	5.9	< 0.05	24.3	< 0.01
Year*Treatment	2	0.8	0.48	0.4	0.67
Error	12				
Season*Year	2	12.7	< 0.01	12.2	< 0.01
Season*Year*Treatment	2	0.4	0.62	1.8	0.22
Error	12				

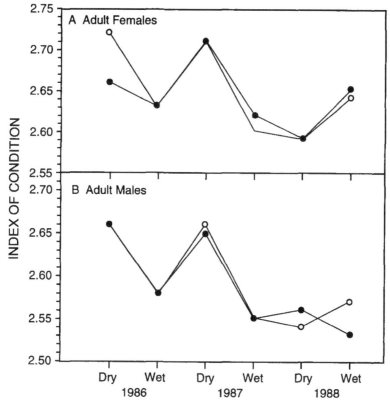

Figure 12.4. Index of Condition (IC) for adult females and males on control (closed circles) and watered (open circles) plots. Because of extensive overlap, SEs are not shown.

dry season (Foster and Brokow 1982). Still, variation in rainfall affects phenology and reproductive success of both plants and animals (Leigh et al. 1982). Some species of plants and animals found in wetter habitats only a few kilometers to the east are not present on BCI, and others more typical of wetter habitats are also found on BCI. For example, *A. limifrons* and its congener *Anolis humilis* are sympatric in the aseasonal rainforests of the Atlantic lowlands of Panama and the rest of Central America. *Anolis humilis* does not occur on BCI or in other places with pronounced seasonality. On the other hand, the range of *A. limifrons* does extend into relatively seasonal habitats, but its distribution as a whole is largely coincident with that of aseasonal rainforests. What features of the dry season at BCI are inimical to *A. limifrons*–a species more typically associated with less seasonal habitats? Our experimental manipulations address this question. Watering during the dry season resulted in a population outbreak of *A. limifrons* during the 1986

Table 12.4. Repeated measures analysis of variance for the proportion of juveniles < 39 mm SVL captured during each season in each year (values shown in Fig. 12.5). Proportions were arcsine transformed before analysis. Where appropriate, Greenhouse-Geisser estimates of probability levels are given.

Source	df	F	P
Between subjects			
Treatment	1	0.3	0.61
Error	6		
Within subjects			
Season	1	85.5	<0.01
Season*Treatment	1	46.9	<0.01
Error	6		
Year	2	0.0	1.00
Year*Treatment	2	2.8	0.12
Error	12		
Season*Year	2	0.3	0.74
Season*Year*Treatment	2	1.6	0.25
Error	12		

wet season. Population densities on watered plots were 500 or more individuals per 1800 m², which was higher than has been previously recorded on BCI (Andrews et al. 1983). Densities on watered plots during the outbreak were more like those of West Indian islands than mainland sites (Andrews 1979b). In contrast, during the 1986 wet season, population densities on control plots were less than half those on watered plots.

Adult lizards were apparently unaffected by the experimental maintenance of wet-season levels of moisture during normally dry periods. Adult survival did not vary between watered and control plots. Adults on watered and control plots had virtually identical patterns of their index of condition. In general, high values of IC are associated with relatively high food intake, while for females, high values of IC are associated with enhanced rates of egg production (Andrews 1991). Thus, food availability and egg production should have been similar on the watered and control plots. Why did watering during the dry season not affect adult lizards? To address this question, A. limifrons populations must be placed in the context of their position in a very complex food web (Pimm 1991). The response of the forest to watering in the dry season was limited. For example, most species of trees were unaffected in terms of seasonal patterns of leaf flush, leaf drop, and fruit production (Wright and Cornejo 1990). In the understory, the phenology of some

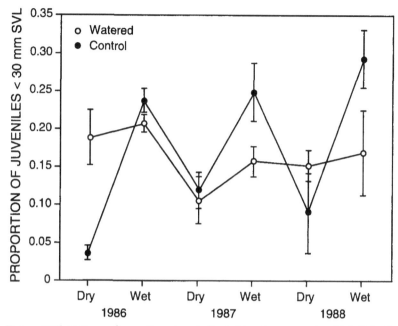

Figure 12.5. Pattern of recruitment as indicated by the proportion (± 1 *SE*) of individuals less than 30 mm SVL on control (closed circles) and watered (open circles) plots.

shrubs was shifted by watering, but productivity overall was not altered (Wright 1991). Thus, for the plant community, production and phenology were little affected by alleviation of water stress during the dry season. Overall, the abundance of arthropods, including prey taxa eaten by *A. limifrons*, did not differ between the watered and control treatments, although arthropod abundance was generally higher in 1986 than in 1987 or 1988 (Wolda and Wright 1992). Observations on arthropod abundance are thus compatible with the patterns of variation in the index of condition, and both indicate that prey availability to *A. limifrons* individuals was not affected by the watering treatment. Judging by the similar rates of survival for *A. limifrons* on the watered and control plots, overall predation rates were also unaffected by the watering treatment. This suggests that predator populations were unaffected by watering or that net predation on *A. limifrons* populations did not change. Literally scores of species are predators on *A. limifrons*. These include mammals, birds, snakes, and other lizards, as well as invertebrates such as spiders and mantids. Responses by such diverse taxa to watering could be quite variable without any obvious change in the mortality of one prey species. Given that the food web was not disrupted by watering during the dry season, it is perhaps not surprising that survival and food intake for *A. limifrons* individuals were not affected by what, at first sight, was a major environmental perturbation. We assume that the lack of response to the watering treatment by adult lizards can be extended to juveniles as well. Adults and juveniles have similar rates of survival (Andrews and Nichols 1990) and eat the same types of prey (Sexton et al. 1972).

In contrast to the "lizard" stage, the egg stage was strongly affected by the watering treatment, and the population process most closely linked to the outbreak in 1986 and the subsequent population decline on the watered plots was variation in egg survival. Watering had a marked effect on recruitment (Fig. 12.5). Normally, the proportion of juveniles is low in the dry season and high in the wet season when most egg laying occurs. This was the pattern of recruitment on the control plots. In contrast, on watered plots, recruitment into the population was initiated early in 1986; young juveniles made up an unusually large proportion of the population in the 1986 dry season. Because recruitment started early, more individuals were added to the populations on watered than control plots. Enhanced survival of eggs during the 1986 wet season (Andrews 1988) also would have contributed to the increase in population density on the watered plots in 1986. Moreover, given observed growth rates, hatchlings produced in the dry season on the watered plots would have matured, and their offspring would have been in the censused population during the following wet season. However, in 1987 and 1988, the proportion of young juveniles on watered plots was the same as on control plots in the dry season and considerably lower in the wet season. As adult survival did not vary with treatment, we conclude that the population decline in 1987 and 1988 was the result of relatively low recruitment on the watered plots.

Causality?

If recruitment were enhanced by amelioration of dry season conditions on the watered plots in 1986, why was it not enhanced in the following two years? One explanation is that the response to watering in the dry season was a function of prevailing ambient conditions.

Population density is correlated with rainfall. Thus, the overall response to watering during the dry season should reflect, what, on the basis of rainfall, are good and bad years for population growth. *Anolis limifrons* individuals are more abundant at the ends of years when rainfall during the wet season is relatively low and when the dry season is relatively short. In this regard, 1986 should have been a particularly favorable year for populations of *A. limifrons*. In 1986, annual rainfall at BCI was 217 cm with December and April rainfall totalling 49 cm. In contrast, in 1987 and 1988, annual rainfall was relatively high, 295 cm and 260 cm, respectively, while December and April rainfall was low, 40 cm and 21 cm, respectively. Given the amount and pattern of rainfall during 1986 through 1988, population densities in the 1986 wet season should have been higher than in the 1987 and 1988 wet seasons. Furthermore, in 1986, the combination of watering during the dry season plus low rainfall during the wet season should have provided ideal conditions for an increase in population density on watered plots relative to the control plots. On the other hand, in 1987 and 1988, high rainfall in the wet season was not favorable for population growth, and differences in density between watered and control plots should have been lower during these years than in 1986. This is exactly the pattern observed. Densities on watered plots were higher than those on control plots in 1986 but not in the other two years. Apparently, watering during the dry season did not compensate for the effect of high rainfall in 1987 and 1988.

Year-to-year changes in IC support the idea that environmental conditions deteriorated over the 1986–1988 period. Lizards were in better condition in 1986 than in the following years irrespective of treatment. This suggests that food intake and reproductive output would have been highest in 1986, and lower in 1987 and 1988. Moreover, as IC was similar on the watered and control plots, the decline in IC must have been associated with changes in environment that were unrelated to effects of watering.

While changes in population density and IC of A. *limifrons* on the watered and control plots were consistent with the hypothesis that responses were a function of ambient weather conditions, the pattern of recruitment was not. Watering during the dry season should have enhanced recruitment irrespective of rainfall, and recruitment during the wet season should have been similar on the watered and control plots irrespective of rainfall. This was observed in 1986, but not afterwards. Instead, in 1987 and 1988, recruitment during the dry season was similar on the watered and control plots, and recruitment during the wet season was depressed on watered plots relative to control plots.

This problem is resolved if we consider that watering the second or third years, may not have been a simple replication of the first year. Watering in successive years could have changed the environment so that some factor or factors became detrimental to either the rate of egg production or to egg survival. The similar values of IC for females indicate that the rate of production of eggs probably did not differ between watered and control treatments. If this is correct, then continued watering must have had an inimical effect on egg survival. We speculate that continued wet conditions were associated with the build-up of pathogens or predators normally reduced in numbers by dry conditions. Such an effect would be consistent with the relatively low recruitment on the watered plots in 1987 and 1988, and the (apparently) monotonic decline of population density on the watered plots. It is also consistent with known interactions between *Solenopsis* ants, the major predator on eggs (Andrews 1988), and moisture. *Solenopsis* are more abundant in the wet than the dry season, are more abundant in wet than dry microhabitats at any time of the year, and increase in numbers when plots are watered experimentally in the dry season (Levings 1983; Kaspari pers. comm.).

Results of the watering experiment suggest causal mechanisms behind the correlation between population size and rainfall. First, because egg production by *A. limifrons* is a facultative response to moisture, short dry seasons enhance the rate of population growth by simply increasing the time during the year available for reproduction. A longer reproductive season means that more individuals can be recruited into the population. More eggs are produced by individual females over their lifetimes, and eggs produced early in the breeding season can give rise to a second generation that season. Second, because egg survival is enhanced if the wet season is relatively dry and reduced if the wet season is relatively wet, the amount of rainfall during the wet season is negatively related to population density by the end of the wet season.

The first of these mechanisms is actually a description of known features of the demography of *A. limifrons*. The second is speculative, but plausible, and would be relatively easy to test. The critical result at this point is, however, that abundance of this tropical lizard is critically related to variation in the physical environment. Variation in rainfall affects when and how long eggs are produced and, through the effect of rainfall on egg predators or pathogens, how many eggs survive.

Conclusions

Results of this study provide insights that should be helpful with regards to one of the critical issues in the coming decades of research on lizards. This issue is the recognition and conservation of endangered species or populations. The first insight is that lizard populations normally fluctuate in size, and some normal fluctuations are of high amplitude. We have documented wide temporal variation in the density of a tropical lizard. This, and observations on other species (e.g., Wolda and Foster 1978; Rand et al. 1983), indicate that species of tropical forests fluctuate just as widely as species of temperate regions. Widely fluctuating populations, irrespective of habitat, create two problems. First, distinguishing between natural declines in density and declines related to anthropogenic changes in the environment may be difficult even with studies lasting a decade or more (Pechmann et al. 1991). The second problem is that reserves set aside for conservation purposes must be sufficiently large to accommodate both the high and low points of natural fluctuations. This point is not novel, of course, as it is one of the general guidelines of conservation biology (Soulé 1987).

Another insight, which is perhaps novel, is that ecological studies that do not explicitly include the egg stage are not likely to develop a useful understanding of how both normal and anthropogenic environmental changes affect population dynamics. The reason is that lizards in eggs are more affected by variation in physical and biotic environment than lizards after hatching. Eggs are relatively small and immobile—they do not have the homeostatic abilities of lizards; eggs are stuck with their nest site while posthatching individuals have the ability to seek out favorable microclimates and to avoid predators. Thus, similar changes in the physical and biotic environment will be harsher from the viewpoint of the egg than from the perspective of posthatching lizards. Such differences between eggs and posthatching lizards likely affect the population dynamics of all oviparous species. Thus, a thorough understanding of the dynamics of the egg stage is critical to an assessment of changes in lizard abundance (this study; Overall this volume) and perhaps also the presence or absence of a species in an area (Muth 1980). This is not to say that the relationship between variation in the physical environment and variation in population density will always be as tightly linked as we have demonstrated for A. limifrons. With annual population turnover and the potential for two generations per breeding season, the number of lizards at the end of the breeding season largely reflects the success of recruitment during that breeding season. This particular life cycle may lead to instability, as r, the intrinsic rate of increase, is directly related to the amplitude of population fluctuations (May 1981).

Acknowledgments

We would like to thank A. S. Rand for conducting the late wet-season censuses at Lutz site in 1990, 1991, and 1992 and also for his comments on the manuscript, and Milton Garcia for conducting the majority of the lizard censuses. Funding was provided by the Environmental Sciences Program of the Smithsonian Tropical Research Institute, Balboa, Panama.

CHAPTER 13
SPATIAL AND TEMPORAL VARIATION IN STRUCTURE OF A DIVERSE LIZARD ASSEMBLAGE IN ARID AUSTRALIA

Craig D. James

Understanding the origin and structure of communities of species has been a central theme of ecology. A driving question behind the study of animal and plant communities is: What forces, if any, determine which species are found together, and what influences the abundance of component species (Andrewartha and Birch 1954; Hutchinson 1959)? Such questions have served to focus thought on community structure. Naturally enough, opinions are as diverse as the biological communities people study, and a debate currently ensues over this general question. As with most debates, views are polarized into two main schools of thought–those who argue for equilibrial or deterministic community structure, and those who support nonequilibrial, chaotic, or stochastic community structure (Wiens, 1984).

Equilibrial communities are those in which species composition and relative abundance of species are in some way predictable. Equilibrial communities can be stable through time, track fluctuating resources, or follow successional trajectories. They are resistant to perturbation (have inertia) and are resilient following perturbation (return to normal). These characteristics led Grossman (1982) to define such communities as "deterministic," because future structure could be predicted from knowledge of present conditions. In contrast, nonequilibrial communities are those in which species composition, and relative abundance of species, changes frequently and unpredictably, and individual species respond independently to environmental changes. Grossman (1982) defined these communities as "stochastic" because they are strongly influenced by local stochastic events. Knowing the present structure and condition does not allow a confident prediction of the future structure of a stochastic community.

Forces creating structure in each type of community are also postulated to be different. Biotic interactions (Cody 1974; Paine 1974; Pianka et al. 1979; Brown and Munger 1985), and habitat partitioning (Pianka 1972; Brown and Lieberman 1973; Schoener 1974) are generally thought to control structure in deterministic communities, whereas density-independent responses to unpredictable environmental variation are more significant in stochastic communities (Andrewartha and Birch 1954; Connell 1978; Sousa 1979; Sale 1984; Wiens 1989a). However, communities may lie at various points along a continuum of organization by stochastic or deterministic forces, and may "slide" along this continuum through time (Wiens 1984). For example, communities may experience crunch periods when biotic interactions have a significant and long-lasting deterministic effect on structure, but at other times these forces may be weak, resulting in unpredictable community organization (Wiens 1977). Finally Yant et al. (1984) distinguish "... deterministic processes with stochastic environmental parameters from purely stochastic processes," which they say "... discourages the investigation of mechanisms leading to apparent stochasticity." In other words, communities may be variable in response to stochastic environmental factors, but still be driven by deterministic processes in the longer term. Observing apparent stochastic variation in community structure should not preclude deeper investigation of processes and an assessment of the relative importance of deterministic versus stochastic forces.

Communities structured by deterministic or stochastic processes should display different characteristics of temporal change. Two features commonly attributed to nonsuccessional deterministic communities are persistence of species from one time to the next, and concordance of species' rank abundances through time. The reverse of these should be true for stochastic communities. Empirical data to test such hypotheses have mostly come from fish (marine and freshwater; Grossman 1982; Grossman et al. 1982; Sale 1984; Moyle and Vondracek 1985), terrestrial invertebrate (Boecklen and Price 1991; Root and Cappuccino 1992), bird (James and Boecklen 1984; Wiens 1989a), and mammal assemblages (Brown and Heske 1990; Brown and Zeng 1989). The purpose of this paper is to illustrate patterns of spatial and temporal change in structure of a lizard assemblage. I restrict this analysis to lizards rather than examining a trophic guild or some other functional grouping that reflects energy flows (Drake 1990), and hence, I prefer to use the word "assemblage" rather than "community." The word "structure" is used to describe the dual attributes of species composition and relative abundance of a group of coexisting species.

This study is based on a lizard assemblage in the spinifex-grassland deserts of Australia. These grasslands are renowned for enormously rich assemblages of lizard species compared with other continental deserts (Pianka 1969a, 1986). High species richness of lizards in spinifex grasslands is apparent at both regional scales (Cogger 1984) and local scales (Morton and James 1988). A typical 10 ha patch of spinifex grassland habitat supports 30 to 40 species of lizards, two to three times the number of species on similar-sized patches of habitat in southern African or North American deserts (Pianka 1967, 1969a, 1971, 1973). Explanations for why species richness of lizards in spinifex grasslands differs from desert habitats in the Americas or Africa have been advanced by a number of researchers (Pianka 1972, 1989; Morton and James 1988). These explanations invoke evolutionary pathways, resource partitioning, and biogeographic patterns, and these issues will not be addressed further here. I restrict this analysis to an exploration of patterns of coexistence and abundance of species at one site.

Pianka devoted much effort to studying lizard assemblages in spinifex grasslands at microhabitat scales (Pianka 1969b, 1986). These studies demonstrated that lizard species partitioned spinifex habitat into nocturnal and diurnal niches; arboreal, terrestrial, and fossorial niches; and microhabitat niches around spinifex hummocks (e.g., Pianka 1986, table 10.5, p. 130). Pianka viewed lizard assemblages in spinifex grasslands as stable groups of species, modulated by biotic interactions and resource partitioning. Pianka's results gave a snap-shot of organization of assemblages at a scale of a few m² but did not indicate how patterns vary through time and space. Populations of lizards in arid Australia are extremely volatile in response to rainfall (James 1991a). Therefore, the structure of a lizard assemblage observed at one time may differ at some time in the future. For example, are all species recorded on a 10 ha grassland site syntopic, and how does the assemblage structure of syntopic species vary through time? Such questions are fundamental to understanding whether assemblages are orderly and structured, or randomly aggregated groups of species (Sale 1984).

I examine the assemblage structure of lizards in a spinifex grassland by looking at two spatial scales: the species assemblage on a 50 ha area, and assemblages on 12 trapping subsites throughout the 50 ha, each of 0.2 ha. At these two scales, lizard species richness was recorded across the range of habitats in the spinifex landscape ($\approx$ β-diversity), and assemblages of coexisting species on discrete subsites ($\approx$ α-diversity). Even at the smallest scale of this analysis, microhabitat features that were the focus of Pianka's observations are obscured because habitat characteristics are averaged over a large area relative to daily activities of lizards. Also, I only analyze adult distribution and abundance data. Results for juveniles will be presented elsewhere.

Two hypotheses that characterize temporal variation in the structure of deterministic communities are tested: (1) Species' relative abundances are autocorrelated through time and (2) Species composition on a site is consistent through time. To address these two hypotheses, analyses are organized around four questions: (1) Were there identifiable assemblages of species on small-scale trapping sites, and how were these assemblages related to habitat features? (2) How did species abundance respond to changing environmental conditions? (3) How similar were assemblages of species on subsites through time? (4) How similar through time was the assemblage on the entire study site? The first three questions relate primarily to the smaller spatial scale (subsites), and the fourth question to temporal variation in species composition across the entire study site.

Methods and Materials

Study site and habitat

The study site was 50 ha of spinifex grassland on a dune-swale system, at Ewaninga, 40 km south of Alice Springs in central Australia (133° 54' E, 24° 00' S). Topography consists of irregular longitudinal sand dunes up to 7 m above interdune swales. Sand on dune slopes and crests was looser with higher infiltration rates than the harder, clay-sand soils of the swales (Buckley 1979).

Ground-cover vegetation was dominated by spinifex, grasses of the genus *Triodia* endemic to Australia (Jacobs 1984). The hummock-shaped growth-form of spinifex presents pungent, outwardly projecting leaves forming a dense, although mostly sparse, ground cover making visual observation of lizards difficult. Two species of spinifex were present on the study site: *Triodia pungens* (soft spinifex) and *T. basedowii* (hard spinifex). *Triodia pungens* produces hummocks up to 1.5 m diameter and 60 cm high, whereas *T. basedowii* hummocks are typically smaller and more compact (50 cm diameter and 30 cm high). Hummock size and percentage cover of spinifex varied across the study site due to differences in species of spinifex present, topography, soil hardness, and regeneration time since fire (Fig. 13.1). Spinifex existed as a virtual monoculture over large areas of the study site or with a scattered shrub overstory. Shrubs and mallee eucalypts from 2 m–5 m in height were common but widely scattered on the site. The most common shrubs were species in the genera *Acacia*, *Eremophila*, *Grevillea*, and *Cassia*, which produced projected ground covers of about 5% of the total area.

Fire is a ubiquitous feature of spinifex grasslands, creating a mosaic of spinifex patches of different ages and affecting shrub cover (Griffin 1984). Spinifex burned less than ten years previously formed a groundcover of small hummocks of high density and low percentage cover. Middle-aged spinifex cover (20–30 yrs since fire) consisted of mature hummocks with a structure described above. Spinifex burned more than 40 years ago formed a low percentage cover and low density of large rings of senescent spinifex with a growing fringe. In one area of the study site, naturally sheltered from fire for approximately 50 years, spinifex of this character could be found with an overstory cover of mallee (*Eucalyptus gamophylla*) up to 30% projected ground cover.

Quantification of habitat

Nine habitat variables were measured at each trapping subsite (Fig. 13.1) using on-ground assessment, and from vertical aerial photography at a scale of 1:5,000. This scale allowed resolution of spinifex hummocks down to 10 cm diameter. Habitat variables were only measured once (i.e., autumn

Figure 13.1. Map of Ewaninga study site showing spinifex cover and shrub cover from aerial photographs. White areas are bare ground, and darker areas indicate increasing percentage cover of spinifex and shrubs. The vegetation is overlain with access roads (thick solid lines), sand-dune crests (thin dotted lines), and permanent pit-trapping subsites (crosses enclosed in dashed circles) consisting of four 25 m drift-fence arms with four pit traps on each arm and one pit trap at the center. Habitat variables were measured over an area enclosed in the dashed line around each subsite.

1986) because the perennial vegetation changed very little over three years of study. Variables measured from aerial photographs were assessed for a circular area centered on the subsite with a radius of approximately 25 m. Hence, features for each subsite were measured over an area of approximately 2,000 m² encompassing all pit traps (Fig. 13.1). Variables calculated from aerial photographs were: mean area of spinifex clumps (m²), percentage cover of spinifex, density of spinifex (no./m²), percentage of dead spinifex, percentage shrub cover, and density of shrubs (no./m²). Species of spinifex, topographic position (i.e., relative height above swales in m), and soil hardness (g/cm³) were measured on each subsite. Percentage or proportional variables were arcsine transformed. All nine variables were standardized to mean zero and unit variance and reduced to composite variables using principal component analysis with varimax transformation and no data rotation. Principal component loadings were used to recalculate composite habitat variables using a linear equation of the form:

$$Z_i = a_1 X_1 + a_2 X_2 + a_3 X_3 \dots a_j X_j$$

where Z_i is the principal component value for subsite i, a_j is the coefficient (loading) for the jth variable, and X_j is the value of the jth habitat variable for a subsite. Composite habitat variables were used in correlational analyses against species abundance.

Climatic conditions

Climate around Alice Springs is arid with an average annual rainfall of 268 mm. Rainfall in central Australia is highly variable and unpredictable in timing and intensity compared with other continental deserts (Slatyer 1962; Stafford Smith and Morton 1990). During the study (October 1985–December 1987), a range of environmental conditions was experienced (James 1991b). Lizard censuses were conducted during three spring seasons and two autumn seasons and each census consisted of 30–50 days of continuous pit trapping. The study began during a dry period, and there was little rain during the first season. The second season (November 1986–March 1987) was wet resulting in a flush of ephemeral plant growth, and high invertebrate abundance. The third season was again dry. These changes in hydric conditions profoundly influenced lizard population dynamics and reproductive activity (James 1991a,b).

Trapping design and handling methods for lizards

Twelve permanent pit-trapping subsites were established across the study site (Fig. 13.1). Each subsite consisted of 17 pit traps and 100 m of drift-fence laid out in a cross shape (e.g., four pit traps and 25 m of drift-fence on each arm of the cross with one pit trap in the center). Pit traps were 20 L plastic buckets, 45 cm deep and 33 cm wide at the mouth. Subsites were chosen to be around 100–150 m apart, covering the range of topographic and vegetational variation typical at Ewaninga. A fully orthogonal replication of each habitat feature (topographic position, different species of spinifex, different percentage cover of shrubs and spinifex) was not possible, but major habitat features were represented on more than one subsite.

Lizards captured in pit traps were measured, weighed, and released at the point of capture. Individuals of 26 species were toe-clipped for future identification, but this was not possible for all species. From recapture records, movements and population densities could be calculated, but because these data were not available for all species, only abundance data are presented here (see below). Without prior knowledge of the home ranges of any of the species, subsites were spaced with an expectation that most species would rarely travel between subsites. Thus, subsites would be essentially independent. This was mostly true for those species for which individual recognition was possible. Table 13.1 shows toe-clipped species that moved between subsites. A few individuals of most species moved between subsites, and most movements were by large-bodied species low in abundance. Frequencies of movements between subsites were generally low compared with total number of captures. The extent to which species that were not individually identifiable moved between subsites is unknown, but most such species were small bodied and thus unlikely to move large distances. Due to the low incidence of exchange among subsites, they were considered to be independent for statistical analyses.

Pit traps do not catch all species with the same efficiency. Some species are underrepresented in pit-trap samples because their behavior (e.g., slow moving) or microhabitat preferences (e.g., arboreal species) bias against them being pit trapped. Unequal catchability is not a problem for these analyses because trapping techniques were identical throughout the study.

Individuals dying in pit traps (less than 5% of total captures) were preserved and used to determine size at sexual maturity by dissection and to distinguish adults from juveniles. Sexual maturity was judged by size and condition of testes or ovaries: testes were small and without obvious convolutions of the vas deferens in immature males; immature females had small follicles (< 0.5 mm diameter) and thin, strap-like oviducts.

Analyses of lizard data

Abundance data and transformation

Total number of lizards captured on each subsite during each census was used as an indication of abundance. These data could not be converted into density estimates because not all individuals were marked for recognition when recaptured, and sample sizes for some species were low. Number of captures per subsite per census was standardized to account for unequal

Table 13.1. Incidence of movement of toe-clipped species from one subsite to another. Only those species that were recorded moving between subsites are shown; another nine species that were toe-clipped (shown in Table 13.3) were not recorded moving between subsites and are not shown. "Percentage of recaptured individuals" is the number of individuals that moved divided by the total number of individuals that were recaptured. "Percentage of all captures" is the frequency of between-subsite movements as a proportion of all captures.

Species	No. of individuals moving	% of recaptured individuals	% of all captures
Diplodactylus ciliaris	3	17.3	1.9
Diplodactylus conspicillatus	16	21.1	4.6
Diplodactylus stenodactylus	2	8.3	0.9
Nephrurus levis	2	6.9	1.1
Rhynchoedura ornata	2	9.5	1.0
Ctenophorus isolepis	2	20.0	2.1
Diporiphora winneckei	2	10.5	2.6
Ctenotus helenae	2	7.7	1.3
Ctenotus pantherinus	3	4.8	0.8
Ctenotus piankai	2	6.5	1.4
Ctenotus quattuordecimlineatus	2	2.1	0.3
Eremiascincus fasciolatus	2	16.7	2.7
Varanus gouldii	2	50.0	14.3

durations of trapping periods. These standardized abundance data represent both population size as well as relative activity of a species. That is, number of individuals of a species captured per unit time reflects three variables: (1) density of animals in the vicinity of pit traps; (2) proportion of the population active at that time; and (3) distance individuals move when surface-active. Density (variable 1) is confounded by the other two variables. Both proportionate activity and movement activity are related to environmental conditions. Pit traps are passive devices, relying on animals moving into contact with drift-fences. When animals move more, a greater number will be captured in pit traps for a given population density.

In this analysis, I cannot separate the influence of these three variables on overall numbers of lizards captured for reasons given above. However, in detailed studies on the genus *Ctenotus*, I was able to estimate density and movement (James 1991a), allowing me to approximate the magnitude of changes in activity versus density that may have occurred in other species (see "Discussion"). I use three terms when referring to number of lizards captured: "population density" is the number of animals per unit area which I am unable to quantify for all species; "activity" is the combination of the second and third variables described above—changes in the proportion of the population that are surface-active and movements of individuals. Finally, I use "abundance" to refer to catch/pit trap/day data analyzed in this paper.

Three rare species were not sampled effectively by pit traps and therefore were omitted from most analyses (*Varanus tristis, Varanus gilleni*, and *Tiliqua multifasciata*). Thus, most analyses were performed on 36 species. Descriptions of the species can be found in Cogger (1992) and Storr et al. (1981, 1983, 1990).

Correlational relationships of distribution and abundance of lizards were calculated using average abundance of adults over five censuses, and standardized abundance during each census against principal components for subsites. Significance of multiple nonindependent correlations was assessed with a critical P value calculated using the sequential Bonferroni method (Rice 1989). In this case, correlations of abundance of 36 species against each PC were calculated. P values were ranked from smallest to largest ($P_1...P_i$) and the smallest value was judged significant if $P_1 \leq 0.05/36 = 0.00138$. Thereafter P_2 is significant if $\leq 0.05/35$, until $P_i > 0.05/(1 + k - i)$. This method gives a conservative test of significant correlations.

Correspondence of species assemblages

Correspondence analysis (CA) of average adult abundance was used to examine the correspondence of species assemblages on subsites with gradients of habitat characteristics. Thirty-six lizard species were used in the analyses, after removing three species for which fewer than three individuals were captured.

Variation in population size

Coefficients of variation of adult population size over five censuses were compared using analyses of variance (ANOVA). In the first analysis, subsites were replicates and species were grouped into treatments based on time of activity: diurnal, crepuscular, and nocturnal. This analysis examines the hypothesis that time of activity has an impact on temporal variability in population size. The second analysis treats species as replicates: subsites are grouped as swale or dune treatments to test for significant differences in the variability of adult abundance with respect to topographic position.

Assemblage similarity

Similarity of assemblage structure among censuses was calculated using the proportional similarity index:

$$C_{ij} = 1 - 1/2 \sum_{k=1}^{s} \left| P_{ik} - P_{jk} \right|$$

where C_{ij} is percentage similarity between censuses i and j, and P_{ik} and P_{jk} are the proportions of the fauna in each census that comprise species k (Sale

1984). C_{ij} varies from 0.0 (no overlap) to 1.0 (total overlap). Similarities for each subsite were calculated in two ways: (1) similarity of census 1 with all future censuses and (2) similarity of each census compared with an overall mean matrix for all censuses. Method (1) demonstrates changes in assemblage structure through time relative to the first census, whereas method (2) shows how each census deviated from a hypothetically "equilibrial" assemblage.

Time and subsite variation

Multidimensional scaling (MDS) was used to investigate patterns in variation of structure of lizard assemblages on subsites and through time. Abundance data for MDS was further transformed such that each variable had mean zero and unit variance. The analysis was performed on data for each species on each subsite at each sampling time (a matrix of 12 subsites × 5 censuses × 36 species). A similarity matrix was generated using the Bray-Curtis measure, and hybrid MDS was run using the SSH (Semi-Strong Hybrid) module of PATN (Belbin 1992). A threshold stress value of less than 0.15 was used to determine the adequacy of multidimensional linear axes for reducing variance in the original data, as suggested by Clarke (1993).

Computing tools

Statistical analyses were performed on a PC and on a Macintosh Powerbook 170 with the software packages PATN (Belbin 1992), SYSTAT v5.2.1 (SYSTAT 1992), and SuperANOVA v1.1 (Abacus Concepts 1989). Statistical analyses were guided by software manuals and Snedecor and Cochran (1956), Digby and Kempton (1987), and Steel and Torrie (1980).

Results

Habitat characteristics of subsites

Principal components analysis resolved three components from nine habitat variables accounting for 83% of the variation in the original data (Table 13.2). PC I accounted for 45.2% of the variance and was dominated by five variables describing spinifex structure and one variable describing shrub density. Large positive values for this PC describe subsites with low density and high percentage cover of large spinifex hummocks (i.e., *T. pungens*) and no shrub cover (e.g., Subsites 3 and 10–Fig. 13.2). A large negative value describes a subsite with a high density and low percentage cover of small spinifex hummocks (*T. basedowii*) and shrub cover (e.g., Subsite 8). PC II accounts for a further 25.8% of the variance and is primarily influenced by percentage cover of spinifex and shrubs, and soil hardness (Table 13.2). A subsite with a high negative score for this PC has a low percentage cover

Table 13.2. Loadings from principal component analysis (PCA) on nine microhabitat variables. This solution is for an unrotated matrix with varimax transformation. Values in bold face indicate variables that are the main contributors to a given factor. For example, PC I is dominated by spinifex characteristics and shrub density.

Variable	PC I	PC II	PC III
Species of spinifex	**−0.666**	0.303	−0.087
Mean area of spinifex hummocks	**0.767**	−0.508	−0.032
Percentage cover of spinifex	**0.697**	**0.661**	0.125
Density of spinifex	**−0.886**	0.179	−0.261
Percentage dead spinifex	**0.852**	−0.405	−0.158
Percentage shrub cover	−0.132	**−0.943**	0.055
Density of shrubs	**−0.776**	−0.199	−0.219
Topography	−0.324	0.01	**0.923**
Soil hardness	0.56	**0.641**	−0.143
Variance explained	45.2%	25.8%	11.2%

of spinifex, high percentage cover of shrubs and soft soil. Subsite 1 stands out from other subsites based on this PC because of the very high percentage canopy cover produced by mallee eucalypts (*E. gamophylla*). This habitat probably results from a long history of protection from fire, also reflected in the spinifex cover of that subsite (large, senescent hummocks). PC III accounts for a further 11.2% of the variance, and is strongly influenced by subsite topography. A high positive value for this PC indicates subsites on dune crests (e.g., Subsites 2, 5, and 12), and a negative value indicates swale subsites. PC loadings in Table 13.2 were used to recalculate habitat variables that were then used as regressor variables in analyses below.

Were there identifiable assemblages of species, and how were these assemblages related to habitat features?

Over three years, 3,926 adult lizards were captured or recaptured. Thirty-nine species were captured in pit traps (Table 13.3), and another species (*Varanus giganteus*), too large to fall into pit traps, was regularly sighted. Species ranged in adult body size (snout-vent length [SVL]) from 35 mm to over 1000 mm, and in body mass from 3 g to 3 kg.

Although 39 species were captured, the maximum number of species recorded for each subsite over three years varied from 23 to 32 (Table 13.4). Not all species present on the entire study site were present on each subsite. Can this variation in species richness of subsites be explained by habitat characteristics? This question was examined by correlating lizard abundance data with habitat PCs. Correlations between average adult abundance and

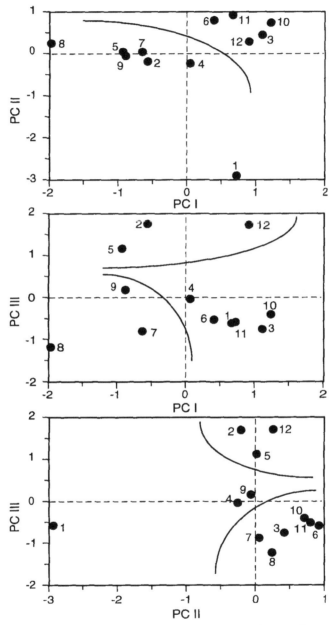

Figure 13.2. Plots of principal components scores for 12 subsites. Subsites are clustered according to their habitat characteristics and there is good discrimination among subsites. Note the unusual characteristics of subsite 1 relative to others.

PCs with P ≤ 0.05 are shown in Table 13.5. Correlations were significant at the traditional 0.05 level for sixteen species, but only five species were considered significant after sequential Bonferroni adjustment of the critical P value. *Egernia inornata* was most common on subsites with a sparse and low percentage cover of spinifex. Significant correlations of *Pogona vitticeps* and *Rhynchoedura ornata* with PC II reflect the effect of Subsite 1, with its high score for shrub cover (Fig. 13.2). When this subsite is removed, these correlations are not significant. *Ctenotus brooksi* and *Nephrurus levis* were more abundant on dune crests than in swales and were positively correlated with PC III.

The relationship between abundance and PCs of habitat structure was also examined for each species on each census to assess variation of correlations through time. As with correlations against average abundance, very few significant results were found after sequential Bonferroni correction for table-wide significance. Significant correlations tended to be for the same species identified above; however, in all but one case a species' distribution and abundance were never consistently correlated with habitat PC (*Pogona vitticeps* was significantly correlated with PC II during every spring census). Three other points about results in Table 13.6 are important: (1) Most of the significant correlations are with PC II. These results reflect the effect of unusual habitat characteristics of subsite 1 that deviate from the main group of subsites (Fig. 13.2) and "drive" the significance of the correlations. (2) Most significant correlations (and most correlations with $P < 0.05$) were found during spring 1986. At this time, lizard activity and abundance were high. Variation in activity of populations of a species seems to obscure patterns of distribution and abundance related to habitat structure. (3) Very few species show consistent correlations through time, highlighting variation in distribution and abundance of species.

These correlations with habitat features indicate that distribution and abundance of only a few species are clearly associated with general habitat architecture of a subsite. The next section analyzes patterns of association of entire assemblages of species with habitat architecture.

Correspondence analysis on average adult abundance of 36 species identified an association between shrub cover and assemblage structure. This result appears to have been driven primarily by arboreal species (*Pogona vitticeps*) and species associated with leaf-litter under shrubs (*Lerista*). A correspondence analysis using only terrestrial, diurnal species (18 species), shows structuring based on topographic variation (Fig. 13.3): *Ctenotus leonhardii* and *Ctenotus piankai* dominated assemblages in swale subsites; *Ctenotus pantherinus*, *Carlia triacantha*, and *Ctenophorus isolepis* dominate in subsites of intermediate height; and *Ctenotus quattuordecimlineatus* dominated on dune crests.

Table 13.3. Species of lizards captured on the Ewaninga study site from October 1985–December 1987. Asterisks indicate species that were individually marked for future identification. Number of lizards is the total number of adults captured (including recaptures).

Family Species	Number of captures	Mean adult SVL (mm)	N for SVL
Gekkonidae (8 species)			
Diplodactylus ciliaris°	176	68.0	147
Diplodactylus conspicillatus°	349	53.6	258
Diplodactylus stenodactylus°	208	50.4	172
Gehyra purpurascens°	4	51.5	4
Gehyra variegata°	21	50.8	9
Heteronotia binoei°	28	42.7	10
Nephrurus levis°	180	73.9	143
Rhynchoedura ornata°	198	43.6	160
Pygopodidae (5 species)			
Delma australis	4	68.0	2
Delma nasuta	33	80.2	23
Delma tincta	7	71.6	5
Lialis burtonis	19	184.4	14
Pygopus nigriceps	36	154.5	31
Chamaeleonidae (5 species)			
Ctenophorus isolepis°	94	55.7	82
Ctenophorus nuchalis°	9	100.5	2
Diporiphora winneckei°	78	55.5	66
Moloch horridus°	4	73.0	4
Pogona vitticeps°	15	160.1	8
Scincidae (17 species)			
Carlia triacantha	203	38.7	18
Cryptoblepharus plagiocephalus	12	41.0	3
Ctenotus brooksi°	10	46.1	10
Ctenotus helenae°	157	80.2	214
Ctenotus leonhardii°	267	65.8	204
Ctenotus pantherinus ocellifer°	349	85.7	317
Ctenotus piankai°	142	50.1	105
Ctenotus quattuordecimlineatus°	599	61.5	515
Ctenotus schomburgkii°	10	42.0	6
Egernia inornata°	102	72.2	73
Eremiascincus fasciolatus°	75	75.5	52

Table 13.3 (cont.)

Family Species	Number of captures	Mean adult SVL (mm)	N for SVL
Lerista desertorum	34	72.5	2
Lerista labialis	211	54.7	9
Lerista xanthura	34	41.4	5
Menetia greyii	5	35	1
Morethia ruficauda	111	60.4	14
Tiliqua multifasciata ∞	2	195.5	2
Varanidae (5 species)			
Varanus brevicauda°	116	91.1	114
Varanus giganteus †	0	—	—
Varanus gilleni°∞	3	135	1
Varanus gouldii°	14	230.5	11
Varanus tristis° ∞●	1	—	—

∞ Omitted from analyses because of rarity.
† Sighted but not captured in pit traps.
● Only one juvenile was captured in pit traps.

How did abundance of species vary in response to changing environmental conditions?

Species abundance varied greatly during the study in response to both seasonal and moisture patterns. A variety of patterns of response is illustrated in Fig. 13.4. In general, species were more abundant in spring than in autumn, and more abundant during wet periods (spring 1986). However, it is difficult to disentangle lizard responses to seasonal changes in temperature from responses to changes in rainfall. Some species were more abundant during spring than autumn (e.g., *Ctenophorus isolepis, Diplodactylus conspicillatus*), while other species were more abundant during autumn than spring (e.g., *Rhynchoedura ornata*); one species had a relatively constant abundance (*Ctenotus pantherinus*); still other species fluctuated widely and were not trapped at certain times (*Nephrurus levis*).

All captures for 12 subsites were summed for each trip, and the ratio of maximum to minimum abundance for each species through time was calculated. This ratio ranged from 1 to 26 (Fig. 13.5), indicating that for some species abundance changed enormously, whereas for other species, changed little. Half of the lizard species had greater than a two-fold range in abundance over the five censuses.

Variation in abundance of each species was examined by ANOVAs of coefficients of variation of abundance over all five censuses. No significant

Table 13.4. Number of lizard species recorded from each subsite at different times, and trapping effort for each census period.

Census date	\multicolumn{12}{c}{Subsite code}												Trapping effort days/pit trap days
	1	2	3	4	5	6	7	8	9	10	11	12	
Spring 85	24	23	18	24	22	21	20	17	21	20	20	20	57/11,628
Autumn 86	4	8	5	9	8	8	7	7	10	8	8	10	40/8,160
Spring 86	14	17	14	16	15	14	12	14	19	13	10	22	38/6,222
Autumn 87	5	10	11	8	12	9	7	8	8	7	8	11	40/8,160
Spring 87	9	22	19	21	9	18	15	15	17	16	18	19	63/12,852
All trips	28	27	27	32	27	27	24	23	27	25	24	32	238/47,022

Table 13.5. Correlation coefficients and probability levels for all correlations between average abundance of adults and principal components describing habitat structure in which $P \leq 0.05$. Correlations were only considered significant if the P value was below that determined by the sequential Bonferroni adjustment of table-wide significance. Significant correlations are asterisked. Correlations in which $P > 0.05$ have been omitted to simplify the table.

Species	PC I		PC II		PC III	
	r^2	P	r^2	P	r^2	P
Diplodactylus ciliaris	–	–	–	–	0.383	0.0320
D. conspicillatus	–	–	0.339	0.0470	–	–
D. stenodactylus	0.500	0.0103	–	–	–	–
Heteronotia binoei	–	–	–	–	0.138	0.0243
Nephrurus levis	–	–	–	–	0.671	0.0011*
Rhynchoedura ornata	–	–	0.705	0.0006*	–	–
Pygopus nigriceps	–	–	0.377	0.0337	–	–
Ctenophorus isolepis	0.434	0.0198	–	–	–	–
Moloch horridus	0.341	0.0460	–	–	–	–
Pogona vitticeps	–	–	0.868	0.0001*	–	–
Ctenotus brooksi	–	–	–	–	0.850	0.0001*
C. piankai	–	–	0.351	0.0425	–	–
C. quattuordecimlineatus	–	–	–	–	0.385	0.0314
Egernia inornata	0.686	0.0009*	–	–	–	–
Lerista labialis	–	–	0.356	0.0407	0.399	0.0276
L. xanthura	–	–	0.422	0.0221	–	–

Key to Species, from bottom to top

■ *Ctenotus leonhardii*   ■ *Varanus brevicauda*
■ *Ctenotus piankai*   ▨ *Morethia ruficauda*
▨ *Ctenotus pantherinus*   ▨ *Ctenotus quattuordecimlineatus*
▥ *Ctenotus helenae*   ▨ *Diporiphora winneckei*
▨ *Carlia triacantha*   ▨ *Ctenotus brooksi*
■ *Ctenophorus isolepis*

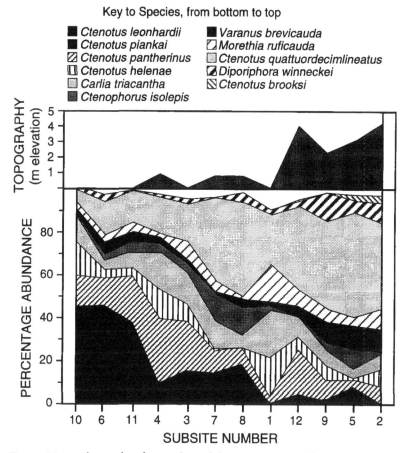

Figure 13.3. Relative abundance of 11 of the most common diurnal, terrestrial species plotted in association with a topographic profile for each subsite, identified by correspondence analysis.

differences in CVs of abundances of species grouped by activity time (diurnal vs. crepuscular vs. nocturnal) occurred, but abundances of species on dune subsites were significantly more variable than abundances of species in swales ($\bar{x}$ CV of abundance on dunes = 0.51 ± 0.30; $\bar{x}$ CV of abundance in swales = 0.43 ± 0.29; $F_{1,197}$ = 4.12, P = 0.044).

How similar through time were assemblages on subsites?

Assemblages of species on each subsite varied through time. Similarity of structure at a census compared with overall average structure varied from 90% to 25% (Fig. 13.6). Overall, average similarity was only 58% indicating large changes in species composition and relative abundance of species.

Table 13.6. Correlation coefficients (r^2) and probability levels in parentheses for correlations between abundance of adults at one census and principal components describing habitat structure in which $P \le 0.05$. Criteria for assessing significant correlations (asterisked) are described in the text and Table 13.5. Correlations in which $P > 0.05$ have been omitted to simplify the table.

Species		Census Date				
		Sp 85	Aut 86	Sp 86	Aut 87	Sp 87
Diplodactylus ciliaris	PC III	–	–	0.375 (0.034)	–	–
Diplodactylus stenodactylus	PC I	0.394 (0.029)	–	0.552 (0.006)	–	–
Nephrurus levis	PC III	–	–	–	–	0.429 (0.021)
Rhynchoedura ornata	PC II	–	–	0.686 (0.0009)°	0.582 (0.004)	
Delma australis	PC I	–	–	–	–	0.391 (0.03)
Pygopus nigriceps	PC II	–	–	0.593 (0.003)	–	–
Lialis burtonis	PC II	–	–	0.391 (0.03)	–	–
Ctenophorus isolepis	PC I	–	–	0.557 (0.005)	–	–
Ctenophorus nuchalis	PC II	–	–	–	0.604 (0.003)	–
Pogona vitticeps	PC II	0.724 (0.0004)°	–	0.762 (0.0002)°	–	0.853 (0.0001)°
Carlia triacantha	PC II	–	–	–	0.394 (0.029)	–
Cryptoblepharus plagiocephalus	PC II	0.487 (0.012)	–	–	–	–
Ctenotus brooksi	PC III	0.450 (0.017)	–	0.735 (0.0003)°	–	–
Ctenotus helenae	PC I	–	0.350 (0.043)	–	0.356 (0.045)	0.436 (0.02)
Ctenotus pantherinus	PC I	–	0.421 (0.023)	–	–	–
Ctenotus piankai	PC I	0.371 (0.035)	–	–	–	–
	PC II	–	–	0.376 (0.034)	–	–
Ctenotus quattuor-decimlineatus	PC III	0.347 (0.044)	–	0.392 (0.03)	–	–
Egernia inornata	PC I	0.556 (0.005)	–	0.333 (0.049)	–	0.689 (0.0009)°

Table 13.6 (cont.)

Species		Sp 85	Aut 86	Sp 86	Aut 87	Sp 87
				Census date		
Eremiascincus fasciolatus	PC II	–	–	–	0.536 (0.007)	–
Lerista desertorum	PC III	0.354 (0.041)	–	–	–	–
Lerista labialis	PC II	0.621 (0.002)	0.440 (0.019)	0.585 (0.004)	–	–
	PC III	–	–	–	–	0.519 (0.008)
Lerista xanthura	PC II	0.404 (0.026)	0.338 (0.048)	0.701 (0.0006)°	–	–
Morethia ruficauda	PC I	0.353 (0.042)	–	–	–	–
Varanus brevicauda	PC I	–	0.390 (0.03)	–	–	–
	PC II	–	–	0.338 (0.048)	–	–
	PC III	–	–	–	0.416 (0.024)	0.530 (0.01)

However, some change patterns repeated from time to time. Many subsites had a saw-tooth pattern with higher similarities for spring censuses than autumn censuses, indicating seasonal differences in activity. The remainder of the subsites had a relatively flat or steadily declining profile.

A time-series examination of similarities is even more variable. When every census is compared with assemblage structure at the first census, similarities range from 9%–65%, and the largest range for an individual subsite is 9%–48% (Fig. 13.7). Species composition and abundance of species changed dramatically on subsites through time. The next analysis attempts to classify and resolve patterns in structural changes through space and time.

Numbers of species captured on each subsite during each season are shown in Table 13.3. More species were captured during spring than during autumn (mean species richness in spring = 17.4 ± 4.0; mean species richness in autumn = 8.2 ± 1.9; $F_{1,58}$ = 109.9, $P < 0.0001$). Number of subsites a species was captured on from one census to the next also varies. Across 39 species, adult lizards were trapped on significantly more subsites during spring than autumn (mean number of subsites occupied by adult lizards during spring = 5.4 ± 4.2; mean number of subsites occupied by adult lizards during autumn = 2.7 ± 3.2; $F_{1,193}$ = 17.5, $P < 0.0001$). Also, abundance and species richness of lizards on subsites was higher during the first spring than any subsequent census, suggesting a possible effect of trapping on behavior.

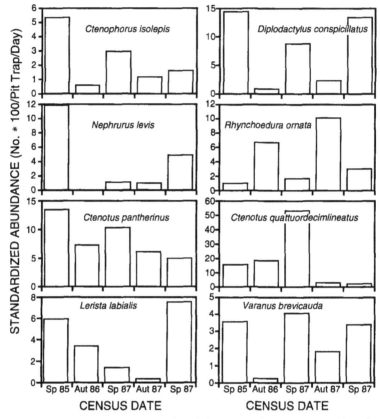

Figure 13.4. Representative examples of changes in the abundance of lizards over five census periods. Note the variety of patterns and lack of correlation in population changes among species.

Multidimensional scaling of the subsite-time*species data matrix revealed three-dimensional structure in the data (stress value = 0.14). The axes of this multidimensional space seem to be season (x-axis), adult activity related to rainfall and reproduction (z-axis), and topography and habitat (y-axis). A fourth dimension that relates to fire successional stage is also apparent in the solution but is not easily displayed. Patterns are illustrated by three views of the three-dimensional plot, rotated to different perspectives (Figs. 13.8A–C).

Two nonoverlapping planes of data represent seasonal dimensions (Fig. 13.8A). One plane contains most spring samples, while the other plane contains most autumn samples. Outliers from these seasonal planes are all subsite 1, and are discussed further below. Seasonal- and adult-activity axes (x-z plane) interact to produce discrete clusterings of points for each different sampling time (Fig. 13.8B). Thus, although all spring samples differ

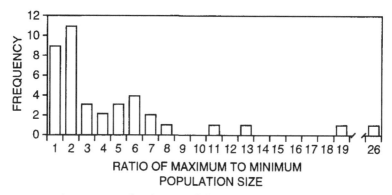

Figure 13.5. Frequency distribution of the ratio of the maximum:minimum abundance of adult lizards over five census periods. Abundances varied up to 26-fold among sampling seasons for a single species.

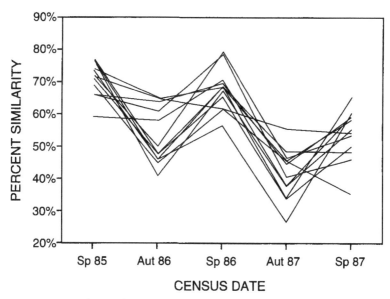

Figure 13.6. Similarity of the lizard assemblage on each subsite compared with an average assemblage for the subsite. The overall mean similarity value is 58%.

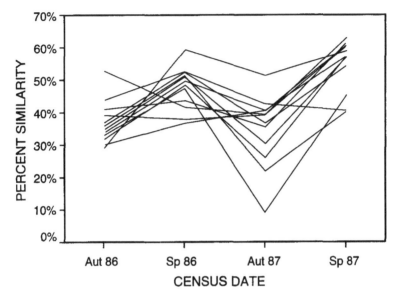

Figure 13.7. A time series of similarity measures for lizard assemblages on each subsite by comparing each census with the first census (spring 1985).

from all autumn samples, virtually all subsites at one census date cluster as a unit and do not overlap with subsites from other census dates (except subsite 1). Clustering reinforces the above result that structure of lizard assemblages on subsites is identifiable more by the specific *time* of sampling than by either habitat or topographic features. Temporal separation is driven by rainfall and *season*: at one extreme, samples from spring 1986 were taken when adults were very mobile, and reproductive activity was high, compared with autumn 1987 when adults were inactive, nonreproductive, and abundance was low.

The topographic/habitat dimension of differentiation (y-axis) reflects habitat separation among subsites within a census. Points at one extreme are dune crest subsites, and at the other extreme are swale subsites (Fig. 13.8C). This result concurs with the correspondence analysis that identified assemblages of species related to topographic variation.

Finally, samples from subsite 1 consistently deviate from the cluster of points for the other 11 subsites (Fig. 13.8B). The axis of this deviation does not follow any of the three axes identified above. I interpret this deviation to reflect the unusual habitat characteristics of subsite 1 (Fig. 13.2), probably the result of a long period of shelter from fire compared to other subsites.

How similar through time was the assemblage on the entire study site?

All previous analyses have focussed on patterns evident on and among subsites. The final component of these analyses is to examine patterns of

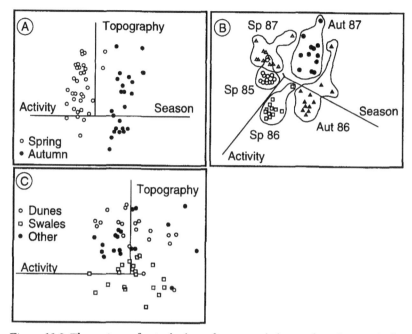

Figure 13.8. Three views of a single three-dimensional plot resulting from multidimensional scaling of abundance of adult lizards on subsites at different times. (A) shows strong seasonal separation of the lizard assemblages. (B) shows data in the x–z plane and illustrates discrete clustering of each census—lizard assemblages on all subsites were more similar to each other at one time than were subsites with similar habitat features through time. (C) shows the distribution of points in the y-plane and illustrates separation of lizard assemblages on subsites by topographic features.

species composition across the entire site. Figure 13.9 shows a plot of the day of the study on which each species was first recorded. From this figure, nearly all species present were recorded within the first 60 days of pit trapping (i.e., during the first spring census). Similarity of species composition (presence/absence data) for the entire study site from one censuses to the next averaged 78.4%. This value reflects the high similarity of each spring census and absence of some species during autumn censuses. Similarity values of species composition for spring censuses only were 91% for spring 1985 to 1986, and 86% for the comparison of spring 1986 to 1987.

Discussion

The main results of these analyses can be summarized as follows: (1) Individual species of lizards vary in the range of habitats they occupy throughout the spinifex landscape. The distribution and abundance of a species was rarely explained by linear correlations with habitat variables measured over the area of subsites trapped (0.2 ha). (2) Species vary in their response to changing environmental conditions. Adult abundances were usually higher in spring than autumn, and following rain, but reverse patterns were also observed for some species. Abundances fluctuated widely over time. (3) The similarity of assemblages on subsites from one time to the next averaged around 50%, and ranged from 25%–90%. Thus, species composition and abundance of species on subsites changed substantially through time. (4) Composition and abundance of species that co-occur on a subsite at any one time are strongly affected by current season, recent rainfall, and fire succession stage (three temporal dimensions in MDS) and by position in the landscape (topography). (5) Characteristics of spinifex cover were not important in determining the structure of an assemblage at a subsite.

Analyses reveal a large amount of variability in relative abundance of species, species composition on a subsite, and hence variability in assemblage structure through time. However, species composition was more consistent across the entire study site. These results suggest important effects of spatial and temporal scales on the organization and structure of lizard assemblages.

Variation in abundance and assemblage structure on subsites

Although species abundance varied across the 12 subsites, very few significant correlations existed between abundance and habitat structure (as principal components). In contrast, a large proportion of studies that have examined the distribution and abundance of lizards has found significant correlations of distribution with habitat characteristics (Schoener 1974; Pianka 1969a,b). The lack of significant correlations in this study is not surprising because each subsite trapped contained hundreds to thousands of spinifex hummocks and dozens of shrubs, providing numerous different microenvironments. If different species were recognizing microhabitat features on scales of a few meters (as suggested by Pianka's results) each subsite potentially contained many different niches, and therefore many species could be found on a subsite. In other words, averaged habitat measurements for a subsite could have obscured fine-scale heterogeneity, and hence PCs may not reflect vegetation structure at a scale relevant to lizards.

Variation in abundance seemed to be at the individual species level. Cocorrelated seasonal patterns of response were observed among species, but the magnitudes of changes in relative abundance were not cocorrelated. Many species disappeared from individual subsites for one or two concurrent censuses, and some species disappeared from the entire subsite on some censuses. Changes in distribution and abundance of species are reflected in changes in number of species present on a subsite through time and changes in the number of subsites a species occupied at different times. Variation in the number or types of subsites (or habitats) on which a species was captured may reflect expansion and contraction of the distribution of species on meso-scales, from source habitat to marginal habitat (Brown and Kurzius 1989). Populations of a species on subsites of marginal quality (for that species) are likely to be more variable in activity than populations on subsites of high quality. Individuals are also more likely to become inactive earlier on subsites of marginal quality compared to subsites of high quality. Alternatively, the intrinsic variability of primary production and invertebrate populations may be higher in some habitats (e.g., dune crests) than others (e.g., swales).

Fluctuations in abundance of adult lizards could be due to emigration and mortality, immigration and recruitment, and changes in activity. Of these, immigration and emigration of adults were probably minor factors for most species: movements between subsites made up a relatively low proportion of recaptures, although some species are more mobile than others. However, at least one species (*Ctenotus leonhardii*) may have increased at two sites because of immigration of adults from a nearby (≈ 75 m) mulga (*Acacia aneura*) habitat.

Mortality and recruitment were potentially significant sources of change in adult abundance, but in only one year (1986/87) was recruitment to adult populations significant. After a wet winter and spring, reproductive activity was high (e.g., data for *Ctenotus* in James 1991b). Juveniles were abundant during autumn but only a small proportion of these juveniles reached adult size by the following spring and so do not feature in the present analyses.

Mortality rates for most species were not available either because of low number of captures or lack of individual recognition. Annual mortality rates for adult *Ctenotus* were estimated to vary from 60%–70%, but mortality rates varied greatly from year-to-year (James 1991a,b). In any case, recruitment or mortality alone cannot explain several-fold changes in abundance of species from one year to the next.

Large changes in abundance of adult lizards, and the disappearance of some species at some times, primarily reflect changes in lizard activity. Unlike many other vertebrates, lizards can become inactive underground for long periods. Most species are inactive (brummate) over winter because of low temperatures. But even during seasons when temperatures are high enough for activity, lizards may aestivate and only be active for a proportion of the time (Whitford and Creusere 1977; Rose 1981). Cycles of inactivity are exaggerated in arid environments where low rainfall limits primary productivity and prey abundance. Many individuals in the adult population may be inactive for a large proportion of the time when conditions are dry, then be highly active following rain. But the increase in abundance is not totally attributable to changes in the proportion of adults active. Changes in

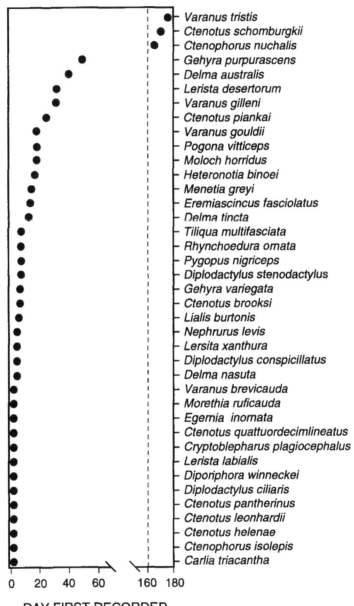

Figure 13.9. Dates on which a species was first recorded on the study site. The first census in spring 1985 lasted for 61 days, and most species were recorded during that first census. Three remaining species were first recorded in autumn 1987, after over 120 days of trapping, and were rare.

small-scale movements of lizards resulted in more animals being captured in pit traps. For *Ctenotus*, average movements of adults varied over a three-fold range among censuses (James 1991a).

In studies of the structure of other vertebrate assemblages, changes in species composition or abundance are interpreted as a gain or loss of individuals in the adult population. In mammal assemblages, changes in adult abundance can be due to recruitment or mortality, and also due to periods of inactivity (Brown and Kurzius 1989; Brown and Zeng 1989). In coral-reef fish communities, species settle from the plankton onto isolated patch reefs, and species turnover is due to adult mortality and recruitment of young (Sale and Dybdahl 1978; Sale 1977; Williams and Sale 1981; Sale et al. 1984). The functional result from changes in lizard activity (or mammals) is equivalent to that of changes in abundance seen in other taxonomic assemblages. That is, inactive individuals are functionally similar to temporary local extinction in reef-fish assemblages because an inactive individual is not consuming resources nor interacting with other lizard species (or other species in the same trophic guild). However, one important difference between an assemblage with large numbers of inactive individuals or inactive species and one in which individuals are not present is that structure of a lizard assemblage can respond rapidly to environmental change.

Spatial and temporal scales

At Ewaninga, 39 lizard species were recorded but only 23–32 of these species were recorded on the same subsite over three years. During a given season, the number of species that occurred on a subsite ranged only from 4–24. Not all species at Ewaninga actually overlap spatially and temporally. Temporal variation in activity (both diurnal and seasonal) and habitat partitioning reduce the number of species that are syntopic. Hence, larger numbers of sympatric species recorded by Pianka ($N = 42$; 1969a, 1986) may represent the accumulation of species over a larger area encompassing a range of habitats (β-diversity), as well as an accumulation over time.

Stochastic patterns at small spatial scales may eventually sum to deterministic patterns at larger scales (Anderson et al. 1981; Ogden and Ebersole 1981). So an important question is: at what scale is an assemblage defined? In the Introduction, I opted for the term "assemblage" in preference to "community" because I was restricting this analysis to a taxonomically defined group and because I was interested in the patterns at one spinifex site. (I do not claim that this study represents the patterns of the lizard communities of the spinifex grasslands of all of arid Australia, which represents some 35% of the continent's surface area.) A relevant scale to examine an assemblage is one in which individuals of a species interact in their normal daily activities. Patterns of species richness and similarity are therefore relevant at subsite-scales.

Connell and Sousa (1983) suggested appropriate spatial and temporal scales for assessing stability in natural communities. They defined spatial scale as the smallest area required by a species that provides conditions necessary to replace existing adults. An appropriate minimum time scale is for one complete turnover of individuals. Using these guidelines, an appropriate scale for larger and more mobile lizard species in spinifex grasslands may be tens of hectares, studied over a decade or more. During this study, viewing species composition at the largest spatial scale (50 ha) resulted in a more deterministic perception of species composition of the assemblage than examination at the subsite scale. However, study over a still longer time may reveal more variation than I found in the species composition for the study site as species become locally extinct or establish new populations.

At small spatial scales, structure of patch-reef fish assemblages is highly variable due to patchy recruitment. At larger spatial scales in reef ecosystems, relative abundance of recruits is concordant. In lizard assemblages, variation in species composition on a subsite (an analogue of patch reefs?) appears to be due to individualistic changes in species activity. Yet at the scale of the entire study site, concordant patterns of variation in abundance and species composition occur over time. The spatial and temporal scale at which lizard assemblages of spinifex grasslands are examined is critical with a potentially profound impact on observed results.

Deterministic or stochastic assemblages?

Evidence presented here suggests that assemblage structure of lizards on a small area is variable. The stochastic nature of the lizard assemblage at Ewaninga is similar to that reported for lizards on a site censused 10 years apart in the Great Victoria Desert, W.A. (chapter 10 in Pianka 1986). In Pianka's study, relative abundances of species changed by an average of 2.7 times, one species disappeared, and five new species were reported. The similarity of the assemblages at two times was 61% (my calculation from table 10.3 in Pianka 1986 [p. 128]). Pianka concluded that fluctuations in abundance of species were not tracking prey availability, were not autocorrelated, and that "stochasticity in this system is considerable."

Two hypotheses for deterministic community organization posed in the Introduction appear to be rejected by results from individual subsites. First, species abundances were not autocorrelated through time. Concordant responses of populations in fish assemblages have been demonstrated using Kendall's rank abundance correlation (Grossman 1982; Grossman et al. 1982; Moyle and Vondracek 1985; Ebeling et al. 1990). However, this technique has been criticized because it is not a reliable indicator of regulation of local species populations (Ebeling et al. 1990; Yant et al. 1984). More importantly, significant rank correlation may not indicate deterministic structuring forces (Schlosser 1982; Strong 1983). Nonetheless, this study (without attempting rank correlations) shows that species' responses to changing seasons and environmental conditions are clearly asynchronous. The second hypothesis, that species composition is consistent through time, is also not supported for individual subsites.

Assemblages following successional trajectories are also considered to be deterministic because species composition, although changing, is predictable at some future point given present conditions. The spinifex environment frequently burns, and postfire successional pathways may explain some variation in assemblage structure across subsites. I am unable to quantify postfire regeneration time for each of the subsites accurately. However, differences between subsite 1, which had not been burned for a long period, and other subsites suggest that structure of lizard assemblages changes over the long run as habitats regenerate from fire. Detecting successional patterns in lizard assemblages may prove difficult because of enormous seasonal variation on small scales.

The crucial aim of community investigations such as this is to understand why species' complement and abundances are as they are. Are local assemblages the result of density-dependent habitat selection, competitive and predatory interactions that result in a set of species evolutionarily coadapted to coexist? Or, do assemblages result from historical and stochastic events that affect which species in a regional pool coexist on a site? In addressing these questions, I turn to organizational forces in lizard assemblages.

Assemblages are affected by a range of forces classified as "deterministic" or "stochastic," but these forces may vary in intensity and effect through time (Wiens 1984), and might operate at different spatial scales (Wiens 1989b). Rather than trying to pigeon-hole an assemblage as stochastically or deterministically structured, a more fruitful avenue may be to consider a conceptual continuum of organization from deterministic to stochastic, (sensu Wiens 1984). It would be useful to try to assess the significance of the major structuring forces, and to determine which forces dominate over others, and when. Wiens (1984, p. 451) suggested a set of attributes that could be used to assess where an assemblage lies on such a continuum. Boecklen and Price (1991) used this framework to assess the organization of assemblages of sawflies on arroyo willows. Their approach was laudable, but I cannot repeat it here because I do not yet have sufficient data to address in detail some of the attributes identified by Wiens (i.e., biotic coupling, competitive interactions, saturation with species). However, some discussion of these attributes is warranted.

As suggested above, forces structuring lizard assemblages in spinifex grasslands shift in importance through time. In hot, arid environments around the world, rainfall events are a major driving climatic variable.

Extreme variation in size and intensity of rainfall events in central Australia make it an environment where opportunistic activity and opportunistic reproduction are crucial to survival. The most frequent situation is one of xeric conditions (from a mammalian perspective). During prolonged periods of normal conditions, density-independent processes probably dominate. Loose patterns of association with habitat structure, lack of cocorrelation among species' distributions, and species assemblages that are varying subsets of the total species pool are characteristics demonstrated by these analyses for such times. Opportunistic use of the habitat and mortality due to environmental harshness (e.g., lack of prey) may also be key characteristics at such times. Under these conditions, it is hard to believe that competitive and predatory interactions had a strong influence on adult lizard assemblages at Ewaninga. This assessment would place lizard assemblages further toward the nonequilibrial end of Wiens' continuum.

During rare mesic periods, the story seems to be different. Subsite-scale assemblages have more species than during dry periods, have greater absolute abundance of individuals, and are therefore likely to experience stronger competitive interactions. Although community interactions under mesic conditions could be assessed for only one year out of three, some examples from that period illustrate the potential importance such periods may have on long-term structure. Juvenile *Ctenotus* dominated the lizard community at Ewaninga during autumn 1986 because wet conditions during the past spring led to above-average reproductive activity and hatching success (James 1991b). The overwhelming abundance of juvenile *Ctenotus* almost certainly increased competitive interactions among species (either direct or diffuse). Rates of juvenile mortality varied among subsites over the following 12 months, and were highly correlated with patterns of distribution and abundance of adults measured before recruitment (James unpubl.). Hence, distribution and abundance of adults may be explained by patterns of juvenile mortality. Even though short-term patterns of structure in these lizard assemblages appear to be dominated by stochastic events, deterministic processes may occur infrequently and still have an appreciable impact on structure (Yant et al. 1984).

Even if deterministic processes are fleeting but have long-lingering impacts, stochastic events that dominate the temporal variability of the lizard assemblages also drive the intensity and timing of deterministic processes. Attributes of assemblages such as species saturation and density dependence are temporally variable and are driven by stochastic events. Clearly, more data are required to evaluate dynamics and processes of organization in lizard assemblages of spinifex grasslands. Results of this study suggest that on small spatial and temporal scales (i.e., hectares and seasons) structure of assemblages is driven primarily by stochastic processes, resulting in nonequilibrial states, whereas at larger spatial and temporal scales (km^2 and years), the structure of assemblages may be more deterministic.

Acknowledgments

I am deeply grateful to Mac Beavis for helping with field work when things were getting out of hand. Statistical analyses were very much improved by suggestions and help from Lee Belbin, Fiona Gell, and Jeff Wood. I am pleased to acknowledge very useful comments from Greg Skilleter, Julian Caley, Steve Morton, Chris Dickman, Jon Losos, and an anonymous reviewer. This work was financially supported by grants from the Australian Museum Postgraduate Research Award Scheme; a University College of the Northern Territory Postgraduate Research Award; the North Australian Research Unit; a Joyce W. Vickery Scientific Research Award from the Linnean Society of NSW; and the Australian Geographic Society Scientific Research Awards. Logistic support was provided by the CSIRO Arid Zone Research Institute, Alice Springs.

CHAPTER 14
HISTORICAL CONTINGENCY AND LIZARD COMMUNITY ECOLOGY

Jonathan B. Losos

Community ecologists usually focus on the following questions: (1) What processes are operating within a community? (2) What processes led to the currently observed structure of a community? (3) What accounts for the differences and/or similarities among communities?

Although the first question concerns what is happening within present-day communities, the latter two questions inquire about the processes responsible for patterns observed in extant communities. These latter questions directly address the historical genesis of community patterns: through what route and guided by what processes have communities attained their current state? Such questions are critical to investigation of present-day community patterns for several reasons. On one hand, observed patterns often could be the result of a number of different processes (Case and Sidell 1983). Conversely, the same process can lead to different end-states whether starting conditions are identical or not (Drake 1990, 1991; Drake et al. 1993). Consequently, often it is not possible to draw inferences about the processes that shaped a community solely from inspection of the current structure of that community; what is needed is information on how a community came to its current state (Ricklefs 1987; Brooks and McLennan 1991; Losos 1992; Gorman 1993).

Recent years have seen widespread acceptance of the idea that historical phenomena must be studied in an explicitly historical context (Lauder 1982; Cracraft 1990; Harvey and Pagel 1991; Brooks and McLennan 1991). Ecologists have long been aware that historical contingencies may be responsible for differences among communities, but analyses of historical ecology have generally looked to the history of the environment for explanation (Ricklefs 1987). Examples include enhanced speciation in forest refugia as a cause of tropical diversity (e.g., Haffer 1969) and oceanic islands having lower species richness than continental islands because the former were never connected with the mainland (e.g., Case 1975; Wilcox 1978).

A complementary approach that incorporates historical information into ecological analyses considers the history of the taxa that make up a community (e.g., Duellman and Pianka 1990). In some cases, the fossil record is sufficient to permit inferences about paleo-community structure and processes (e.g., Russell 1991; Warheit 1992), but a more generally applicable method is to examine the phylogeny of lineages present in a community.

A phylogenetic perspective can permit insight about the diversification and evolutionary changes that have occurred and resulted in the currently observed community. Further, comparison of lineages in multiple communities can provide insight into the causes of differences and similarities among communities (e.g., Brooks and McLennan 1991; Cadle and Greene 1993).

Lizards have played a key role in the development of community ecology theory. To name just a few examples, studies of lizard communities have been important in the formulation of ideas concerning species diversity (Vanzolini and Williams 1970; Pianka 1972), island biogeography (MacArthur and Williams 1967; Schoener 1970; Case 1975, 1983), resource partitioning (Schoener 1968b; Pianka 1969b), niche complementarity (Pianka 1973; Schoener 1974), competition (Dunham 1980; Pacala and Roughgarden 1985), and predation (Schall and Pianka 1980). Here, using lizard communities as an example, I discuss and review how phylogenetic information can be integrated into studies of community ecology.

The Role of History in Determining Community Structure

Species diversity

Ecologists have long been interested in why some communities have more species than others. Species diversity is a function of two factors: the pool of available species and the number of species that the community can contain. Most theories emphasize the latter aspect and investigate proximate ecological factors as an explanation for the diversity of a community. However, if communities are not saturated, then species diversity could also be a function of the number of species potentially available to join the community—the larger the regional pool of species, the greater the diversity of local communities (MacArthur 1965; Ricklefs 1987). The size of the pool of available species, which includes all species within a community plus all other species physically capable of immigrating into the community, is ultimately determined by rates of extinction and speciation. Because lineages differ in their propensity to speciate or perish, the pool of available species may differ as a result of among-region differences in which lineages are present. These differences, in turn, may result in differences in local diversity.

Variation in local diversity also may result from historical contingency. Hypotheses that look to proximate ecological conditions as an explanation for differences in diversity assume that the ecological "types" of species available in the species pools of different communities are comparable. But if certain "types" are not present in a pool for whatever reason (e.g., random extinction, constraint on the evolution of appropriate phenotype in lineages present in the species pool), then the "niche" normally utilized by that type may not be filled in a given community (e.g., Pianka 1989, p. 354).

Examination of the lineages present in different communities can allow an assessment of how important historical factors may be in contributing to observed patterns. As an example, I will consider differences in species richness of desert lizard communities based on the data of Pianka (1986). Pianka and colleagues have demonstrated that the number of lizard species inhabiting desert communities varies remarkably among continents, from as few as four in North American deserts to as many as 42 in Australian deserts (Pianka 1986; see Table 14.1).

How can this difference in diversity be explained? A variety of proximate explanations have been advanced: lizards have replaced snakes and mammalian carnivores in Australia; fewer nonlizard competitors and predators are present in the more diverse communities; resource levels (e.g., low fertility of soils makes deserts inhospitable to endotherms) or habitat structure differs among continents (reviewed and discussed in Pianka 1986, 1989; Morton and James 1988). Alternatively, we must entertain the possibility that the history of the deserts themselves is responsible for differences in diversity. Perhaps climatic conditions have promoted speciation in Australian deserts but not in North American deserts (discussed in Cogger 1984; Pianka 1986), or perhaps Australian deserts are considerably older than North American deserts (Pianka 1986).

An alternative hypothesis recognizes that different lizard lineages occur in the deserts of North America and Australia and suggests that the disparity in species richness results from interlineage variation in the propensity to survive and speciate. Consider the lizard families present in Australian and North American desert communities (Table 14.1). Five families are present in each, but only one, the Gekkonidae, is found in both (although no skinks occurred at Pianka's North American study sites, two skink species extend their ranges to include relictual mesic and/or rocky habitats in some North American deserts [Stebbins 1985; Greene pers. comm.; Vitt pers. comm.]). Further, among the Gekkonidae, all Australian geckos are members of the diplodactyline, gekkonine, and pygopodine lineages, whereas North American taxa belong to the Eublepharinae (Kluge 1987), with the exception of one species that occurs in extreme southwestern North America (Stebbins 1985). Consequently, an alternative hypothesis is that the lineages in Australia are intrinsically more prone to speciate and/or coexist than those in North America, perhaps for reasons unrelated to proximate differences in the deserts of the two continents.

Is this a reasonable alternative hypothesis? Comparison of the lineages present on the two continents indicates that the lineages present for Australia's great diversity are absent in North America, whereas lineages present in

Table 14.1. Comparison of the lizard fauna of Australian and North American deserts (numbers below are the range, with the mean, when available, in parentheses; data from Pianka 1986).

	North America	Australia
Total Species Number	4–11 (7.4)	18–42 (29.8)
Species Number by Family		
Agamidae°	0	2–8
Gekkonidae	1	5–9
Helodermatidae	1	0
Iguanidae†	3–8	0
Pygopodidae	0	1–2
Scincidae	0	6–18
Teiidae	1	0
Varanidae	0	1–5
Xantusiidae	1	0
Species Number by Habits		
Nocturnal	0–2 (1.0)	8–13 (10.2)
Arboreal	0–3 (0.9)	1–9 (5.4)
Fossorial	0 (0)	1–2 (1.2)

° In the taxonomy of Frost and Etheridge (1989), Agamidae is now considered a subfamily (Agaminae) of the Chamaeleonidae.
† In the taxonomy of Frost and Etheridge (1989), the families represented would be Crotaphytidae, Iguanidae, and Phrynosomatidae.

North America but not Australia tend not to radiate. The three families responsible for Australia's heightened diversity are the Varanidae, Scincidae, and Gekkonidae (including pygopodids). The Varanidae are currently restricted to the Old World, the one gekkonid lineage in North American deserts is not speciose anywhere in its range (Grismer 1988), and skinks barely occur in North American deserts (see above). Thus, one would not have expected these lineages to have contributed substantially to North American desert diversity. By contrast, two families found in North America and not in Australia, the Helodermatidae and Xantusiidae, are both depauperate in species (2 and 19 species respectively) and fail to compensate for the lineages absent from North American deserts. A third family that occurs in North American deserts is the Teiidae. Although teiids have diversified in the tropics of Central and South America, their diversity in deserts of North and South American deserts is relatively low (see Peters and Donoso-Barros 1970; Stebbins 1985). Finally, North American iguanians and Australian iguanians (referred to as Agamidae here, but considered a subfamily [Agaminae] of the Chamaeleonidae by Frost and Etheridge 1989), which may be considered broadly as ecological analogues, are approximately equal in diversity in deserts on the two continents (note that the Iguanidae has been split into eight families by Frost and Etheridge 1989, three of which are found in North American deserts).

One could thus interpret differences in diversity in two ways. On one hand, one might argue that lizard families do not intrinsically differ in speciation or extinction rates and only ecological factors regulate how many species can coexist. On the other hand, differences in diversity may be a function of intrinsic properties of the lineages present on the two continents and unrelated to proximate environmental effects. Perhaps some lineages are more prone to speciate or less susceptible to extinction than others (perhaps as a result of interlineage differences in population density or substructuring, levels of gene flow, or type of species-recognition signal). Alternatively, some lineages intrinsically may be more capable of partitioning resources more finely than others, as Pianka (1972) suggested for Australian skinks.

To make these possibilities more concrete, consider what postulates are implicit in the view that differences in desert-lizard diversity result from proximate ecological causes: (1) New world skinks and/or teiids are the ecological analogues of Australian skinks and would have diversified to an equal extent if they had originally occurred in Australian deserts, even though their diversity is low in North American deserts. Equivalently, one would contend that had Australian skink lineages occurred in North American deserts, they would not have radiated. (2) Eublepharine geckos, which are nowhere speciose (only 5 genera and 22 species worldwide; Grismer 1988), would have radiated widely and into arboreal niches in Australia; similarly, arboreal Australian geckos would not have diversified in North America. (3) Varanids would not be able to survive or proliferate in North America. (4) Helodermatids and xantusiids would possibly have radiated in Australia.

These points suggest a reappraisal of the conclusion that differences in diversity result from environmental differences between deserts. Certainly, some of the differences between Australian and North American deserts are due to ecological differences among the sites. For example, one large difference is the number of arboreal species per community ($\bar{x}$ = 5.4 in Australia, 0.9 in North America; Pianka 1986), which has been attributed to the greater availability of trees in Australian deserts (Pianka 1986, 1989; Morton and James 1988). Occupation of the arboreal niche has occurred in four Australian families (Agamidae, Gekkonidae, Scincidae, Varanidae). Consequently, it is clearly an adaptive shift undertaken repeatedly in response to environmental conditions, rather than a result of the inherent tendencies of a single lineage.

By contrast, other aspects of the difference in diversity could have resulted from historical contingency. Much of the difference in species richness between Australia and North America results because Australia has considerably more nocturnal, carnivorous (sensu stricto), fossorial, and skink-like species. I suggest the possibility that this is a function of the lineages present in each continent rather than a consequence of environmental differences among continents.

The disparity in the number of nocturnal species ($\bar{x}$ = 10.2 in Australia, 1.0 in North America; Pianka 1986) is almost completely the result of the occurrence of the Gekkonoidea (sensu Kluge 1987) in Australia and the Eublepharinae in North America. Gekkonoids are both speciose and ecologically diverse throughout their range; eublepharines are neither. Ecological explanations of differences in the number of nocturnal species that are particular to differences in deserts on these continents thus appear too narrowly focused (see also Cogger 1984).

Similarly, why is there a paucity of carnivorous and fossorial lizards in North America? Ecological explanations (e.g., lack of resources, competition from other taxa) are certainly possibly correct. But one also must entertain the possibility that lineages present in North America are not capable of producing such forms. In South America, several genera (*Callopistes, Tupinambis*) within the Teiioidea (sensu Estes et al. 1988) have evolved that are superficially similar to varanids; other Teiioidea (e.g., *Bachia*) have moved toward fossoriality by becoming elongate with reduced limbs. Thus, one could argue that the Teiioidea, as represented in North America by *Cnemidophorus*, has the potential to attain morphologies and ecologies similar to those exhibited in Australia. On the other hand, the closest relatives of *Cnemidophorus* (the *Ameiva* species group: *Ameiva, Teius, Kentropyx*, and *Dicrodon*; Gorman 1970; Presch 1983) vary little morphologically despite occurring in a wide range of habitats (e.g., Vitt and Carvalho 1992), which might suggest that North American *Cnemidophorus* do not have the evolutionary potential to fill carnivorous or fossorial niches (however, the *Ameiva* group of macroteiids does vary in dental morphology and diet; Presch 1974). Iguanians, which are considerably more diverse than teiids in North America, show no tendency anywhere toward fossoriality and only a limited trend toward an analogue to varanids (*Crotaphytus, Gambelia*). Thus, it is not clear whether the lack of carnivorous and fossorial lizard taxa in North America is due to lack of ecological opportunity or to internal constraints preventing the production of such forms in North American iguanian and teiid lineages.

Finally, why has no radiation comparable to the Australian skinks occurred in North America? The closest ecological analogue to Australian skinks is again *Cnemidophorus*. In some areas, not included in Pianka's study sites, as many as five species of *Cnemidophorus* co-occur (Wright and Vitt 1993), but nowhere does this genus display the ecological or morphological diversity of Australian skinks, even when nonskink-like genera (e.g., *Lerista*) are excluded.

The bottom line is that lack of adaptive radiation in teiids, as represented by *Cnemidophorus*, and geckos, represented by *Coleonyx*, is responsible for the low diversity in North America. Certainly, it is conceivable that these taxa have the capability to speciate and radiate, and had they been in Australia, they would have produced a radiation comparable to that displayed by Australian taxa. But the other possibility is that, due to particular quirks of these lineages, they have not radiated, for reasons possibly completely unrelated to the proximate conditions of their surroundings. Perhaps their population biology is such that speciation rarely occurs? Perhaps they are so highly specialized and constrained that major morphological change (e.g., limb loss, evolution of toe-pads; eublepharines exhibit few subdigital specializations; Grismer 1988) is not permissible?

These possibilities are difficult to investigate. However, posing the questions can lead to an important reorientation of the focus of research. Questions of speciation and morphological constraint require population genetic and evolutionary approaches. Only by integrating these approaches with ecological comparisons of community structure can we address questions such as those posed above. More broadly, the point is that there often will be plausible explanations for differences in diversity among communities. By ignoring historical phenomenon, one implicitly makes a variety of assumptions, some of which are reasonable, others less so. Almost certainly, both historical and ecological factors are always important in determining diversity (Duellman and Pianka 1990).

Ironically, although large scale intercontinental comparisons allow us to examine independent radiations and allow a comparative approach, they also come with a drawback. If all deserts had equal numbers of species, we would accept this as evidence for ecological regulation of species diversity. But when differences arise, there will always be the problem of confounding variables, the presence of different lineages in different communities. Thus, in this respect, intracontinental comparisons may be more compelling because the same lineages are present in each community.

Community structure

The current structure of a community is the result not only of processes currently operating, but also of those that shaped the community during its genesis. Because the same pattern could be produced by several, often quite distinct, processes, inferring process from pattern can be problematic (Case and Sidell 1983; Drake 1990).

Phylogenetic analyses can be useful in distinguishing possible explanations for observed patterns. An example is the cause of nonrandom size-structuring in communities. Many authors have suggested that species in a community differ more in body size (a presumed indicator of resource use) than expected by chance (Schoener 1986b; Dayan et al. 1990; Taper and Case 1992; and references therein). Although numerous processes could produce such a pattern, they can be divided into two categories: ecological and coevolutionary theories. Ecological theories assume that there is a potential source pool of species varying in size, but the only species that can colonize a community are those that are sufficiently different in size from species already present. Coevolutionary explanations suggest that size evolution occurred subsequent to sympatry. Coevolution could take the form of divergence (= character displacement), convergence, or parallel directional evolution (e.g., Schoener 1970; Roughgarden and Pacala 1989; Abrams 1990; and references therein). Processes responsible for either ecological sorting by size or coevolution (size assortment and size adjustment; sensu Case and Sidell 1983) include interspecific competition, predation, and interspecific hybridization.

Case and Sidell (1983; see also Case 1983) proposed the first test to distinguish between the two processes. They suggested comparing all possible sets of sympatric species to those sets that actually occur (e.g., in an archipelago, compare all possible 3-species communities to those 3-species communities that actually exist). If size assortment has occurred, then species in real communities should be more different in size than would be expected from the set of all possible communities. By contrast, tests for coevolution (termed "size adjustment" by Case and Sidell 1983) compare the difference in size between sympatric species against all possible combinations of populations of these species. The expectation is that populations of generally similar-sized species will be under much greater selective pressure to diverge in size when sympatric than will populations of species that are not generally similar in size. Thus, the more similar two species are in size, as judged by all possible combinations of populations of the two species, the more greatly divergent the two sympatric populations should be. Case (1983) used this protocol to investigate patterns of size distributions in *Cnemidophorus* assemblages.

Although this method is implicitly phylogenetic, by comparing populations to their close relatives, a more directly phylogenetic approach might be preferable when possible. I suggested such an approach to analyze size-structured communities in Lesser Antillean *Anolis* lizards (Losos 1990a,b). In the northern Lesser Antilles, five of six two-species islands contain a large and a small species, and 10 of 11 one-species islands contain an intermediate-sized species (Schoener 1970). I predicted that if character displacement were responsible for this pattern, then phylogenetic character reconstruction should indicate that size change occurred coincidentally with the attainment of sympatry of two previously similar-sized taxa. Phylogenetic analysis indicated exactly this pattern, but also suggested that character displacement may have occurred only once (Fig. 14.1). Consequently, character displacement is probably responsible for the evolution of different-sized species which are capable of coexisting, but the occurrence of pairs of dissimilar-sized species on five islands must be the result of size assortment. Thus, the phylogenetic analysis not only indicates that both processes probably have been operating, but also gives some indication of their relative importance.

In a similar vein, Arnold (1981, 1990, 1993) used a phylogenetic approach to understand differences in habitat use among various lizard clades. A number of lacertid and gekkonid clades display a trend in which more derived taxa progress along a continuum, using and adapting to increasingly more environmentally challenging habitats (often xeric habitats; Fig. 14.2). Arnold suggested that this pattern results from competitive pressures, which force newly arisen taxa to utilize marginal and previously nonutilized niches. Speciation in these taxa, in turn, leads to taxa that must shift into even more marginal niches to avoid competitive exclusion.

A second important role for phylogenetic analyses is in statistical comparative methods. Comparative analyses that ignore phylogenetic relationships are invalid because they assume that the character value of each species is independent of the value of all other species (e.g., Felsenstein 1985; Harvey and Pagel 1991). A variety of statistical methods has been proposed to incorporate phylogenetic information into statistical analyses (reviewed in Harvey and Pagel 1991; Losos and Miles 1994; see also chapters by Garland and Martins this volume).

Reconstruction of community evolution

Phylogenetic methods allow one to trace a community's historical development. Such reconstructions can reveal patterns and suggest hypotheses not apparent from consideration of the structure of extant communities.

One reasonably well-understood example involves communities of *Anolis* lizards in the Greater Antilles, which are very similar in composition. This convergence has resulted from the evolution of the same set of "ecomorphs" on each island (Williams 1983). Phylogenetic analysis of the anole radiations on Puerto Rico and Jamaica revealed that not only are extant communities convergent in structure, but they attained their current state by passing through essentially the same set of intermediate community structures (Losos 1992; Fig. 14.3). As discussed above, many processes can produce similar patterns in present-day communities; consequently, only an historical analysis could present evidence for an hypothesis of parallel community evolution.

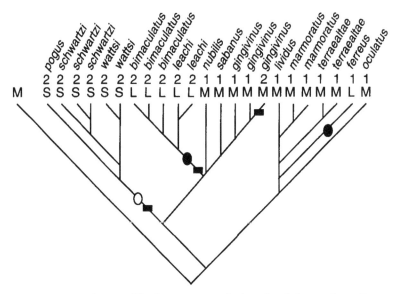

Figure 14.1. Evolution of body size in *Anolis* lizards of the northern Lesser Antilles (based on Roughgarden and Pacala 1989 and Losos 1990a,c). Numbers indicate the number of *Anolis* species on the island occupied by each taxon. Letters indicate body size (small, medium, or large). Circles represent major evolutionary changes in body size (solid = increase; open = decrease); bars represent the transition from an ancestor on a one-species island to a descendant on a two-species island. The statistical analysis in Losos (1990a,c) used actual values rather than categorical variables. Figure redrawn from Losos (1992) with permission.

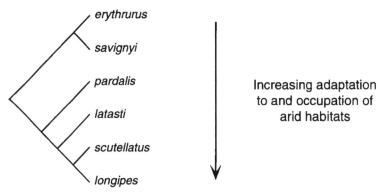

Figure 14.2. Evolution of *Acanthodactylus* in northern Africa near the Sahara. Modified from Arnold (1981).

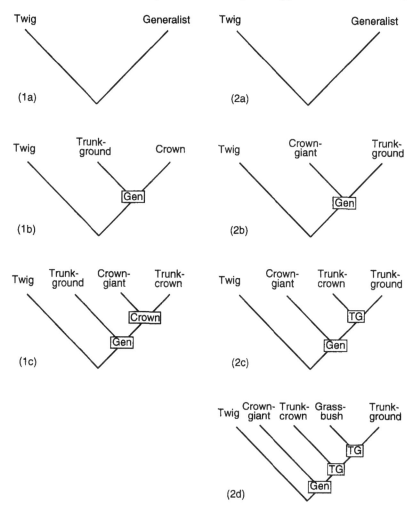

Figure 14.3. The evolution of *Anolis* community structure in (1) Jamaica and (2) Puerto Rico. Evolution of habitat use, as predicted by morphology, was reconstructed using parsimony methods (see Losos 1992 for methodological details). Not only are anole communities on the two islands very similar today (the only difference being the presence of grass-bush anoles in Puerto Rico), but the intermediate stages in community evolution on both islands were also quite similar. The names of the ecomorphs refer to the habitat they normally utilize. Redrawn with permission from Losos (1992).

Phylogenetic analysis provides an important additional perspective on anole community structure. Puerto Rico has five ecomorph types, whereas Jamaica only has four; the missing type in Jamaica is the "grass-bush" ecomorph. One might reasonably ask why the grass-bush niche is vacant in Jamaica, particularly because grassy areas exist in Jamaica and no other taxa obviously has usurped that ecological role. Phylogenetic analysis suggests that the wrong question is being addressed (Williams 1972; Losos 1992). The grass-bush ecomorph is the fifth, and last, type to have evolved in Puerto Rico. Thus, at the most proximate level, the reason the grass-bush ecomorph is absent in Jamaica is because it is the fifth type to evolve in the ecomorph radiation sequence, and Jamaica has only progressed to the four-ecomorph stage. Thus, the more appropriate question is not: Why do grass-bush anoles not occur on Jamaica? but, rather: Why has Jamaica only made it to the four-ecomorph stage? The answer may have nothing to do with grass-bush anoles and their habitat, but may, instead, pertain to species-packing, rates of speciation, or age of the Jamaican radiation.

Discussion

I have argued that studies of community structure are incomplete if phylogenetic information is not considered. This view is part of a broader perspective that historical contingencies often play a large role in determining the composition of communities (Ricklefs 1987; Duellman and Pianka 1990; Cadle and Greene 1993). As is currently being recognized in all fields of organismal biology, historical analyses are an important complement to the study of function and structure of extant entities, be they organisms, taxa, or communities.

The application of phylogenetic principles to questions of community ecology will not always be straightforward, however, because community composition is a result of speciation, extinction, immigration, and in situ evolutionary change. Reconstructing the historical sequence of events in communities containing multiple lineages may prove particularly difficult because phylogenetic reconstructions within a lineage only provide information on relative order of branching. Thus, without supplemental information on the absolute timing of events (as might be provided by fossil material or molecular clocks, for instance), it may prove difficult to determine the relative ordering of either speciation events occurring in several lineages or of immigration and speciation events. In this respect, communities composed of monophyletic radiations, as often occurs on islands, may prove more tractable. On the other hand, in comparative community studies, it may be most beneficial to go to the other extreme and examine communities that contain multiple lineages present in each of the communities (e.g., Brooks and

McLennan 1991; Gorman 1993). While making it more difficult to reconstruct the precise sequence of events, this latter approach will avoid the problem of confounding differences in community structure with differences in lineage characteristics (Losos and Miles 1994).

Importance of monophyly

Phylogenetic systematists stress the importance of only considering monophyletic groups (Brooks and McLennan 1991; Wiley et al. 1991). This rule makes sense for systematic and evolutionary studies, in which the study of paraphyletic groups (i.e., taxa that do not include all of the descendants of a common ancestor) can lead to mistaken inferences. However, as an absolute rule, this prescription is inappropriate for ecological studies. Most studies of community ecology, for example, consider a subset of the species present in a community that might be expected to interact with each other or the environment in a similar way (often termed a guild or assemblage; Terborgh and Robinson 1986). Guilds are not monophyletic units, but, rather, are often composed of members of a number of lineages. Further, monophyletic lineages often contain members of several guilds, particularly when some taxa have evolved substantially and have become ecologically distinct. For example, studies of mammalian communities often exclude bats because they have diverged to the extent that they interact with the environment in a completely different way than earthbound mammals. Thus, in determining which species to include in studies of community ecology, taxa should be categorized in an ecologically relevant fashion; such classifications will often not be completely concordant with phylogenetic classifications.

Reliance on current Linnean classification, which is based on a mixture of phylogenetic and morphological criteria, may be the worst choice, however. For example, most phylogenetic analyses agree that snakes are most closely related to scleroglossan lizards; lizards, thus, are a paraphyletic group (Estes et al. 1988; Fig. 14.4). Many studies of lizard community ecology do not include snakes because they are morphologically, ecologically, and behaviorally quite distinct from "standard" lizards. However, excluding snakes is only legitimate if other legless and/or fossorial squamates are also excluded. The loss of limbs has evolved a minimum of 12 times in squamates, usually in association with cluttered, narrow, or fossorial habitats (Gans 1975; Gans et al. 1978; Edwards 1985; Shine 1986; Fig. 14.4). Thus, there is little justification, either ecologically or phylogenetically, for including most legless squamate lineages in studies of lizard community ecology because they are called lizards, but excluding one (snakes) because it has a different name. If ecological criteria (e.g., foraging mode, way-of-life, adaptive zone) are used to delimit the boundaries of an ecological study, they must be applied to all taxa and not just certain lineages.

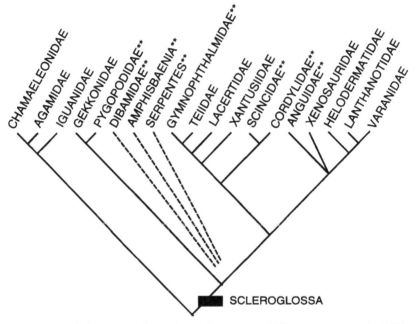

Figure 14.4. Phylogenetic relationships of squamates (following Estes et al. 1988). Dibamids, amphisbaenians, and snakes all lie within the Scleroglossa, but their position is unclear. Asterisks refer to taxa in which limb reduction has occurred in one or more lineages to the extent that limbs are functionally useless for locomotion.

Testing phylogenetically inspired hypotheses

In recent years, ecologists have emphasized the importance of experimentally testing hypotheses. Phylogenetic analyses pertain to the historical development of a community. These analyses describe historical patterns, from which inferences about processes can be drawn. Because such inferences are by necessity retrospective, they are not amenable to direct experimental testing. However, hypotheses derived from phylogenetic analyses often can be tested on extant taxa (Brooks and McLennan 1991; Losos and Miles 1994).

For example, analyses might suggest that competitive pressures led to character divergence. One could test this by experimentally placing two species together and investigating whether (1) competition occurs and, if so, whether (2) microevolutionary divergence occurs. Natural selection often can lead to rapid microevolutionary change (reviewed in Mayr 1963; Williams 1992; Losos et al. in press) so that such hypotheses could be tested in relatively short (in evolutionary terms) experiments. Similarly, hypotheses about differences in community structure could be examined by introducing lineages from one community into another. For example, introduction of varanid lizards into North American deserts would permit one to test whether ecological or biogeographical factors are responsible for the absence of varanid-like forms there. Of course, such experiments can only ethically be performed in circumstances in which no long-term effects on the biota will occur. However, such experiments have already inadvertently been established in many areas through unintentional (at least from a scientific viewpoint) human-assisted introductions (e.g., Wilson and Porras 1983; Losos et al. 1993).

Phylogenetic information can be used to test hypotheses generated from ecological studies, as well as vice-versa. For example, based on studies of extant taxa, it could be concluded that a particular process, such as competition, is important. If evolutionary outcomes that this process should produce can be identified, then predictions testable by reference to a phylogeny can be generated (e.g., Arnold 1981; Gorman 1993).

Only rarely will ecological forces be so strong as to completely erase the vestiges of history. Thus, to fully understand the structure of extant communities will require a synthetic approach that conducts studies of ongoing processes in the context of historical patterns of community development and evolution.

Acknowledgments

I thank T. Case, H. Greene, C. James, L. Vitt, and an anonymous reviewer for providing very useful comments on previous versions of this chapter.

REFERENCES

Abacus Concepts. 1989. *SuperANOVA: Accessible general linear modelling.* Abacus Concepts, Inc., Berkeley.

Abrams, P. A. 1990. Ecological versus evolutionary consequences of competition. *Oikos* 57:147–151.

Abts, M. L. 1987. Environment and variation in life history traits of the chuckwalla, *Sauromalus obesus. Ecol. Monogr.* 57:215–232.

Ackerman, R. A., R. C. Seagrave, R. Dmi'el, and A. Ar. 1985. Water and heat exchange between parchment-shelled reptile eggs and their surroundings. *Copeia* 1985:703–711.

Adolph, S. C. 1990a. Influence of behavioral thermoregulation on microhabitat use by two *Sceloporus* lizards. *Ecology* 71:315–327.

———. 1990b. Perch height selection by juvenile *Sceloporus* lizards: Interspecific differences and relationship to habitat use. *J. Herpetol.* 24:69–75.

Adolph, S., and W. P. Porter. 1993. Temperature, activity, and lizard life histories. *Amer. Nat.* 142:273–295.

Alberts, A. C. 1991. Phylogenetic and adaptive variation in lizard femoral gland secretions. *Copeia* 1991:69–79.

———. 1992. Pheromonal self-recognition in desert iguanas. *Copeia* 1992:229–232.

Altman, S. A. 1987. The impact of locomotor energetics on mammalian foraging. *J. Zool. London* 211:215–225.

Anderson, G.R.V., A. H. Ehrlich, P. R. Ehrlich, J. D. Roughgarden, B. C. Russell, and F. H. Talbot. 1981. The community structure of coral reef fishes. *Amer. Nat.* 117:476–495.

Anderson, P. K. 1989. *Dispersal in rodents: A resident fitness hypothesis.* Amer. Soc. Mammal. Spec. Publ. 9. Provo, Utah.

Anderson, R. A. 1993. Analysis of foraging in a lizard, *Cnemidophorus tigris*: Salient features and environmental effects. In *Biology of whiptail lizards (genus* Cnemidophorus), edited by J. W. Wright and L. J. Vitt, 83–116. Oklahoma Mus. Nat. Hist., Norman.

Andrewartha, H. G., and L. C. Birch. 1954. *The distribution and abundance of animals.* University of Chicago Press, Chicago.

Andrews, R. M. 1971. Structural habitat and time budget of a tropical *Anolis* lizard. *Ecology* 52:262–270.

———. 1976. Growth rate in island and mainland anoline lizards. *Copeia* 1976:477–482.

———. 1979a. Reproductive effort of female *Anolis limifrons* (Sauria: Iguanidae). *Copeia* 1979:620–626.

————. 1979b. Evolution of life histories: A comparison of *Anolis* from matched island and mainland habitats. *Breviora* 454:1–51.

————. 1982a. Patterns of growth in reptiles. In *Biology of the Reptilia. Vol. 13. Physiology D. Physiological ecology*, edited by C. Gans and H. Pough, 273–320. Academic Press, New York

————. 1982b. Spatial variation in egg mortality of the lizard *Anolis limifrons. Herpetologica* 38:165–171.

————. 1988. Demographic correlates of variable egg survival for a tropical lizard. *Oecologia* 76:376–382.

————. 1991. Population stability of a tropical lizard. *Ecology* 72: 1204–1217.

Andrews, R. M., and T. Asato. 1977. Energy utilization of a tropical lizard. *Comp. Biochem. Physiol.* 58A:57–62.

Andrews, R. M., A. S. Rand, and S. Guerrero. 1983. Seasonal and spatial variation in the annual cycle of a tropical lizard. In *Advances in herpetology and evolutionary biology: Essays in honor of Ernest E. Williams*, edited by A. G. Rhodin and K. Miyata, 441–454. Harvard University Press, Cambridge, Mass.

Andrews, R. M., and A. S. Rand. 1974. Reproductive effort in anoline lizards. *Ecology* 55:1317–1327.

————. 1982. Seasonal breeding and long-term population fluctuations in the lizard *Anolis limifrons*. In *The ecology of a tropical forest: Seasonal rhythms and long-term changes*, edited by E. G. Leigh, Jr., A. S. Rand, and D. M. Windsor, 405–412. Smithsonian Institution Press, Washington, D.C.

————. 1983. Limited dispersal of juvenile *Anolis limifrons. Copeia* 1983:429–434.

————. 1990. Reproducción estacional y fluctuaciones poblacionales a largo plazo de la lagartija *Anolis limifrons*. Adición: nuevas percepciones derivadas de la continuación de un estudio a largo plazo de la lagartija *Anolis limifrons*. In *Ecologiá de un bosque tropical: Ciclos estacionales y cambios a largo plazo*, edited by E. G. Leigh, Jr., A. S. Rand, and D. M. Windsor, 469–479. Smithsonian Institution Press, Washington, D.C.

Andrews, R. M., and J. D. Nichols. 1990. Temporal and spatial variation in survival rates of the tropical lizard *Anolis limifrons. Oikos* 57:215–221.

Andrews, R. M., and O. J. Sexton. 1981. Water relations of the eggs of *Anolis auratus* and *Anolis limifrons. Ecology* 62:556–562.

Antonovics, J., and R. B. Primack. 1982. Experimental ecological genetics in *Plantago* IV: The demography of seedling transplants of *P. lanceolata. J. Ecology* 70:55–75.

Arcese, P. 1989. Intrasexual competition, mating system and natal dispersal in song sparrows. *Anim. Behav.* 38:958–979.

Arnold, E. N. 1981. Competition, evolutionary change and montane distributions. In *The evolving biosphere,* edited by P. L. Forey, 217–228. Cambridge University Press, Cambridge, UK.

———. 1984. Ecology of lowland lizards in the eastern United Arab Emirates. *J. Zool. London* 204:329–354.

———. 1990. Why do morphological phylogenies vary in quality? An investigation based on the comparative history of lizard clades. *Proc. R. Soc. Lond.* B 240:135–172.

———. 1993. Historical changes in the ecology and behaviour of semaphore geckos (*Pristurus*, Gekkonidae) and their relatives. *J. Zool. London* 229:353–384.

Arnold, S. J. 1981. Behavioral variation in natural populations. I. Phenotypic, genetic, and environmental correlations between chemoreceptive responses to prey in the garter snake, *Thamnophis elegans. Evolution* 35:489–509.

———. 1983. Morphology, performance and fitness. *Amer. Zool.* 23: 347–361.

Arnold, S. J., and A. F. Bennett. 1988. Behavioural variation in natural populations. V. Morphological correlates of locomotion in the garter snake (*Thamnophis radix*). *Biol. J. Linn. Soc.* 34:175–190.

Arnold, S. J., and M. J. Wade. 1984. On the measurement of natural and sexual selection: theory. *Evolution* 38:709–719.

Auffenberg, W. 1981. *The behavioral ecology of the Komodo Monitor.* University Presses of Florida, Gainesville.

———. 1984. Notes on the feeding behaviour of *Varanus bengalensis* (Sauria: Varanidae). *J. Bombay Nat. Hist. Soc.* 80:286–302.

Autumn, K., R. B. Weinstein, and R. J. Full. 1994. Low cost of locomotion increases performance at low temperature in a nocturnal lizard. *Physiol. Zool.* (in press).

Avery, R. A. 1975. Clutch size and reproductive effort in the lizard *Lacerta vivipara* Jacquin. *Oecologia* 19:165–170.

———. 1982. Field studies of body temperatures and thermoregulation. In *Biology of the Reptilia. Vol. 12. Physiology C,* edited by C. Gans and F. H. Pough, 93–166. Academic Press, New York.

Avery, R. A., C. F. Mueller, S. M. Jones, J. A. Smith, and D. J. Bond. 1987. Speeds and movement patterns of European lacertid lizards: A comparative study. *J. Herpetol.* 21:324–329.

Avise, J. C. 1989. Gene trees and organismal histories: A phylogenetic approach to population biology. *Evolution* 43:1192–1208.

Avise, J. C., J. Arnold, R. M. Ball, E. Bermingham, T. Lamb, J. E. Neigel, C. A. Reeb, and N. C. Saunders. 1987. Intraspecific phylogeography: The mitochondrial DNA bridge between population genetics and systematics. *Ann. Rev. Ecol. Syst.* 18:489–522.

Ballinger, R. E. 1973a. Experimental evidence of the tail as a balancing organ in the lizard *Anolis carolinensis*. *Herpetologica* 29:65–66.

———. 1973b. Comparative demography of two viviparous iguanid lizards (*Sceloporus jarrovi* and *Sceloporus poinsetti*). *Ecology* 54:269–283.

———. 1977. Reproductive strategies: Food availability as a source of proximal variation in a lizard. *Ecology* 58:628–635.

———. 1979. Intraspecific variation in demography and life history of the lizard, *Sceloporus jarrovi*, along an altitudinal gradient in southeastern Arizona. *Ecology* 60:901–909.

———. 1983. Life-history variations. In *Lizard ecology: Studies of a model organism*, edited by R. B. Huey, E. R. Pianka, and T. W. Schoener, 241–260. Harvard University Press, Cambridge, Mass.

Ballinger, R. E., and J. D. Congdon. 1980. Food resource limitation of body growth rates in *Sceloporus scalaris* (Sauria: Iguanidae). *Copeia* 1980:921–923.

———. 1981. Population ecology and the life history strategy of a montane lizard (*Sceloporus scalaris*) in southeastern Arizona. *J. Nat. Hist.* 15:213–222.

Ballinger, R. E., D. L. Droge, and S. M. Jones. 1981. Reproduction in a Nebraska sandhills population of the Northern Prairie Lizard *Sceloporus undulatus garmani*. *Amer. Midl. Nat.* 106:157–164.

Barbault, R. 1975. Dynamique des populations de lézards. *Bull. Ecol.* 6:1–22.

———. 1976. Population dynamics and reproductive patterns of three African skinks. *Copeia* 1976:483–490.

———. 1988. Body size, ecological constraints, and the evolution of life-history strategies. *Evol. Biol.* 22:261–286.

Barbault, R., and Y. P. Mou. 1988. Population dynamics of the common wall lizard, *Podarcis muralis*, in southwestern France. *Herpetologica* 44:38–47.

Barton, N. H. 1992. The genetic consequences of dispersal. In *Animal dispersal: Small mammals as a model*, edited by N. C. Stenseth and W. Z. Lidicker, Jr., 37–59. Chapman and Hall, London.

Baum, D. A., and A. Larson. 1991. Adaptation reviewed: A phylogenetic methodology for studying character macroevolution. *Syst. Zool.* 40:1–18.

Bauwens, D., and C. Thoen. 1980. An enclosure design allowing quantification of dispersal in lizard population studies. *British J. Herpetol.* 6:165–168.

Bauwens, D., and C. Thoen. 1981. Escape tactics and vulnerability to predation associated with reproduction in the lizard *Lacerta vivipara*. *J. Anim. Ecol.* 50:733–743.

Bauwens, D., B. Heulin, and T. Pilorge. 1987. Variations spatio-temporelles des caractéristiques démographiques dans et entre populations du lézard *Lacerta vivipara*. In *Colloque national CNRS, biologie des populations, Lyon, 4–6 Septembre 1986*, 531–536. Université Claude Bernard—Lyon I, Lyon.

Bauwens, D., T. Garland, Jr., A. M. Castilla, and R. Van Damme. 1994. Evolution of sprint speed in lacertid lizards: Morphological, physiological, and behavioral coadaptation. *Evolution* (in press).

Baverstock, P. R., and M. Adams. 1987. Comparative rates of molecular, chromosomal and morphological evolution in some Australian vertebrates. In *Rates of evolution*, edited by K.S.W. Campbell and M. F. Day, 175–188. Allen and Unwin Press, London.

Beaupre, S. J. 1993. An ecological study of oxygen consumption in the mottled rock rattlesnake, *Crotalus lepidus lepidus*, and the black-tailed rattlesnake, *Crotalus molossus molossus*, from two populations. *Physiol. Zool.* 66:348–365.

Beaupre, S. J., A. E. Dunham, and K. L. Overall. 1993a. Metabolism of a desert lizard: The effects of mass, sex, population of origin, temperature, time of day, and feeding on oxygen consumption of *Sceloporus merriami*. *Physiol. Zool.* 66:128–147.

———. 1993b. The effects of consumption rate and temperature on apparent digestibility coefficient, urate production, metabolizable energy coefficient and passage time in canyon lizards (*Sceloporus merriami*) from two populations. *Funct. Ecol.* 7:273–280.

Belbin, L. 1992. *PATN: Pattern analysis package technical reference*. CSIRO, Canberra, Australia.

Bell, G. 1980. The costs of reproduction and their consequences. *Amer. Nat.* 116:45–76.

Bell, G., and V. Koufopanou. 1986. The cost of reproduction. In *Oxford Surveys in Evolutionary Biology*, edited by R. Dawkins and M. Ridley, 83–131. Oxford University Press, Oxford.

Bellairs, A. 1970. *The life of reptiles*. Universe Books, New York.

Bennett, A. F. 1980. The thermal dependence of lizard behaviour. *Anim. Behav.* 28:752–762.

———. 1983. Ecological consequences of activity metabolism. In *Lizard ecology: Studies of a model organism*, edited by R. B. Huey, E. R. Pianka, and T. W. Schoener, 11–23. Harvard University Press, Cambridge, Mass.

———. 1987a. Interindividual variability: An underutilized resource. In *New directions in ecological physiology*, edited by M. E. Feder, A. F. Bennett, W. W. Burggren, and R. B. Huey, 147–169. Cambridge University Press, Cambridge, UK.

———. 1987b. Evolution of the control of body temperature: Is warmer better? In *Comparative physiology: Life in water and on land,* edited by P. Dejours, L. Bolis, C. R. Taylor, and E. R. Weibel, 421–431. Liviana Press, Padova.

———. 1990. The thermal dependence of locomotor capacity. *Amer. J. Physiol. 259 (Reg. Integ. Comp. Physiol. 28)*:R253–R258.

———. 1991. The evolution of activity capacity. *J. Exp. Biol.* 160:1–23.

Bennett, A. F., and R. B. Huey. 1990. Studying the evolution of physiological performance. In *Oxford surveys in evolutionary biology. Vol. 7,* edited by D. J. Futuyma and J. Antonovics, 251–284. Oxford University Press, Oxford.

Bennett, A. F., and T. T. Gleeson. 1979. Metabolic expenditure and the cost of foraging in the lizard *Cnemidophorus murinus. Copeia* 1979:573–577.

Bernardo, J. 1991. Manipulating egg size to study maternal effects on off-spring traits. *Trends Ecol. Evol.* 6:1–2.

———. 1993. Experimental analysis of allocation in two divergent, natural salamander populations. *Amer. Nat.* (in press).

Berven, K. A. 1982. The genetic basis of altitudinal variation in the wood frog *Rana sylvatica.* I. An experimental analysis of life history traits. *Evolution* 36:962–983.

Beuchat, C. A. 1982. Physiological and ecological consequences of viviparity in a lizard. Ph. D. Diss. Cornell University, Ithaca, New York.

———. 1988. Temperature effects during gestation in a viviparous lizard. *J. Therm. Biol.* 13:135–142.

Beuchat, C. A., and D. Vleck. 1990. Metabolic consequences of viviparity in a lizard, *Sceloporus jarrovi. Physiol. Zool.* 63:555–570.

Bissinger, B. E., and C. E. Simon. 1979. Comparison of tongue extrusion in representatives of six families of lizards. *J. Herpetol.* 13:133–139.

Blackburn, D., and H. E. Evans. 1986. Why are there no viviparous birds? *Amer. Nat.* 128:165–190.

Blair, W. F. 1960. *The rusty lizard: A population study.* University of Texas Press, Austin.

Blondel, J., P. Perret, M. Maistre, and P. Dias. 1992. Do harlequin Mediterranean environments function as source-sink for blue tits (*Parus caeruleus* L.)? *Landscape Ecol.* 6:213–219.

Bock, W. J. 1959. Preadaptation and multiple evolutionary pathways. *Evolution* 13:194–211.

———. 1980. The definition and recognition of biological adaptation. *Amer. Zool.* 20:217–227.

Bock, W. J., and G. von Wahlert 1965. Adaptation and the form-function complex. *Evolution* 19:269–299.

Boecklen, W. J., and P. W. Price. 1991. Nonequilibrial community structure of sawflies on arroyo willow. *Oecologia* 85:483–491.

Bradshaw, A. D. 1965. Evolutionary significance of phenotypic plasticity in plants. *Adv. Genetics* 13:115–155.

Braña, F. A., A. Bea, and M. J. Arrayago. 1991. Egg retention in lacertid lizards: Relationships with reproductive ecology and the evolution of viviparity. *Herpetologica* 47:218–226.

Branch, B. 1988. *Field guide to the snakes and other reptiles of southern Africa.* Ralph Curtis, Sanibel Island, Florida.

Brandt, C. A. 1992. Social factors in immigration and emigration. In *Animal dispersal: Small mammals as a model*, edited by N. C. Stenseth and W. Z. Lidicker, Jr., 96–141. Chapman and Hall, London.

Brinkman, D. 1980. Structural correlates of tarsal and metatarsal functioning in *Iguana* (Lacertilia: Iguanidae) and other lizards. *Can. J. Zool.* 58:277–289.

———. 1981. The hind limb step cycle of *Iguana* and primitive reptiles. *J. Zool. London* 181:91–103.

Brockelman, W. Y. 1975. Competition, the fitness of offspring, and optimal clutch size. *Amer. Nat.* 109:677–699.

Brodie, E. D., III. 1992. Correlational selection for color pattern and anti-predator behavior in the garter snake *Thamnophis ordinoides*. Evolution 46:1284–1298.

Brodie, E. D., III, and T. Garland, Jr. 1993. Quantitative genetics of snake populations. In *Snakes: Ecology and behavior,* edited by R. A. Seigel and J. T. Collins, 315–362. McGraw-Hill, New York.

Brooks, D. R., and D. A. McLennan. 1991. *Phylogeny, ecology, and behavior: A research program in comparative biology.* University of Chicago Press, Chicago.

Brown, J. H., and E. J. Heske. 1990. Temporal changes in a Chihuahuan Desert rodent community. *Oikos* 59:290–302.

Brown, J. H., and G. A. Lieberman. 1973. Resource utilization and coexistence of seed eating desert rodents in a sand-dune habitat. *Ecology* 54:788–797.

Brown, J. H., and J. C. Munger. 1985. Experimental manipulation of a desert rodent community: Food addition and species removal. *Ecology* 66:1545–1563.

Brown, J. H., and M. A. Kurzius. 1989. Spatial and temporal variation in guilds of North American desert rodents. In *Patterns in the structure of mammalian communities,* edited by D. W. Morris, Z. Abramsky, B. J. Fox, and M. R. Willig, 71–90. Texas Tech University, Lubbock.

Brown, J. H., and Z. Zeng. 1989. Comparative population ecology of eleven species of rodents in the Chihuahuan desert. *Ecology* 70:1507–1525.

Brown, J. L., and G. H. Orians. 1970. Spacing patterns in mobile animals. *Ann. Rev. Ecol. Syst.* 1:239–269.

Buckley, R. C. 1979. *Soils and vegetation on central Australian sandridges: A review.* CSIRO Technical Memorandum 79/19.

Bull, C. M. 1978. Dispersal of the Australian reptile tick *Aponomma hydrosauri* by host movement. *Austral. J. Zool.* 26:689–697.

———. 1987. A population study of the viviparous Australian lizard *Trachydosaurus rugosus* (Scincidae). *Copeia* 1987:749–757.

———. 1988. Mate fidelity in an Australian lizard *Trachydosaurus rugosus*. *Behav. Ecol. Sociobiol.* 23:45–49.

———. 1990. Comparisons of displaced and retained partners in a monogamous lizard. *Austral. Wildl. Res.* 17:135–140.

Bull, C. M., A. McNally, and G. Dubas. 1991. Asynchronous seasonal activity of male and female sleepy lizards. *J. Herpetol.* 25:436–441.

Bull, C. M., G. S. Bedford, and B. A. Schulz. 1993. How do sleepy lizards find each other? *Herpetologica* 49:290–296.

Bull, J. J., and R. Shine. 1979. Iteroparous animals that skip opportunities for reproduction. *Amer. Nat.* 114:296–316.

Bulova, S. J. In press. Ecological correlates of population and individual variation in antipredator behavior of two species of desert lizards. *Copeia.*

Bunce, R.G.H., and D. C. Howard, eds. 1990. *Species dispersal in agricultural habitats.* Belhaven Press, London.

Burghardt, G. M. 1970a. Chemical perception in reptiles. In *Advances in chemoreception, vol. I. Communication by chemical signals*, edited by J. W. Johnston, D. G. Moulton, and A. Turk, 241–308. Appleton-Century-Crofts, New York.

———. 1970b. Intraspecific geographical variation in chemical food cue preferences of newborn garter snakes (*Thamnophis sirtalis*). *Behaviour* 36:246–257.

———. 1973. Chemical release of prey attack: Extension to naive newly hatched lizards, *Eumeces laticeps*. *Copeia* 1973:178–181.

Burghardt, G. M., B. A. Allen, and H. Frank. 1986. Exploratory tongue-flicking by green iguanas in laboratory and field. In *Chemical signals in vertebrates 4: Ecology, evolution, and comparative biology*, edited by D. Duvall, D. Muller-Schwarze, and R. M. Silverstein, 305–321. Plenum Press, New York.

Burkholder, C. L., and W. W. Tanner. 1974. Life history and ecology of the Great Basin sagebrush swift, *Sceloporus graciosus graciosus* Baird and Girard 1852. *Brigham Young Univ. Sci. Bull. Ser.* 19, 1044.

Cadle, J. E., and H. W. Greene. 1993. Phylogenetic patterns, biogeography, and the ecological structure of neotropical snake assemblages. In

Historical and geographical determinants of community diversity, edited by R. E. Ricklefs and D. Schluter. University of Chicago Press, Chicago (in press).

Cagle, F. R. 1946. The growth of the slider turtle. *Amer. Midl. Nat.* 36:685–729.

Calow, P., and A. S. Woollhead. 1977. The relationship between ration, reproductive effort and age-specific mortality in the evolution of life-history strategies: Some observations on freshwater triclads. *J. Anim. Ecol.* 46:765–781.

Camp, C. L. 1923. Classification of the lizards. *Bull. Amer. Mus. Nat. Hist.* 48:289–481.

Campbell, E. C., G. S. Campbell, and W. K. Barlow. 1973. A dewpoint hygrometer for water potential measurement. *Agric. Meteorol.* 12:113–121.

Campbell, G. S., and W. H. Gardner. 1971. Psychometric measurement of soil water potential: Temperature and bulk density effects. *Soil Sci. Soc. Amer. Proc.* 35:8–12.

Cannatella, D. C., and K. de Queiroz. 1989. Phylogenetic systematics of the anoles: Is a new taxonomy warranted? *Syst. Zool.* 38:57–69.

Cannon, J. G., and M. J. Kluger. 1985. Altered thermoregulation in the iguana *Dipsosaurus dorsalis* following exercise. *J. Therm. Biol.* 10:41–45.

Carothers, J. H. 1986. An experimental confirmation of morphological adaptation: Toe fringes in the sand-dwelling lizard *Uma scoparia. Evolution* 40:871–874.

Carpenter, C. C. 1967. Aggression and social structure in iguanid lizards. In *Lizard ecology: A symposium,* edited by W. W. Milstead, 87–105. University of Missouri Press, Columbia.

Carter, D. L., M. M. Mortland, and W. D. Kemper. 1986. Physical and mineralogical methods. In *Methods of soil analysis. Part I., second edition,* edited by A. Klute, 413–423. Amer. Soc. Agron., Inc., Soil Sci. Soc. Amer., Inc., Madison, Wis.

Case, T. J. 1975. Species numbers, density compensation, and colonizing ability of lizards on islands in the Gulf of California. *Ecology* 56:3–18.

———. 1983. Sympatry and size similarity in *Cnemidophorus.* In *Lizard ecology: Studies of a model organism,* edited by R. B. Huey, E. R. Pianka, and T. W. Schoener, 297–325, Harvard University Press, Cambridge, Mass.

Case, T. J., and R. Sidell. 1983. Pattern and chance in the structure of model and natural communities. *Evolution* 37:832–849.

Caswell, H. 1989. *Matrix population models.* Sinauer Associates, Sunderland, Mass.

Chepko-Sade, B. D., and Z. T. Halpin, eds. 1987. *Mammalian dispersal pattern: The effect of social structure on population genetics*. The University of Chicago Press, Chicago.

Cheverud, J. M., and M. M. Dow. 1985. An autocorrelation analysis of genetic variation due to lineal fission in social groups of Rhesus Macaques. *Amer. J. Physiol. Anthro.* 67:113–121.

Cheverud, J. M., M. M. Dow, W. Leutenegger. 1985. The quantitative assessment of phylogenetic constraints in comparative analyses: Sexual dimorphism in body weight among primates. *Evolution* 39:1335–1351.

Chiszar, D., and K. M. Scudder. 1980. Chemosensory searching by rattlesnakes during predatory episodes. In *Chemical signals in vertebrates and aquatic invertebrates*, edited by D. Muller-Schwarze and R. M. Silverstein, 125–139. Plenum Press, New York.

Chiszar, D., K. M. Scudder, R. Boyd, A. Radcliffe, H. Yun, H. M. Smith, T. Boyer, B. Atkins, and F. Feiler. 1986a. Trailing behavior in cottonmouths (*Agkistrodon piscivorus*). *J. Herpetol.* 20:269–272.

Chiszar, D., K. M. Scudder, and F. Feiler. 1986b. Trailing behavior in banded rock rattlesnakes (*Crotalus lepidus klauberi*) and prairie rattlesnakes (*C. viridis viridis*). *J. Comp. Psychol.* 100:368–371.

Christian, K. A., and C. R. Tracy. 1981. The effect of the thermal environment on the ability of hatchling Galapagos land iguanas to avoid predation during dispersal. *Oecologia* 49:218–223.

Christian, K. A., and S. Waldschmidt. 1984. The relationship between lizard home range and body size: A reanalysis of the data. *Herpetologica* 40:68–75.

Clarke, K. R. 1993. Non-parametric multivariate analyses of changes in community structure. *Austr. J. Ecol.* 18:117–143.

Clobert, J., C. M. Perrins, R. H. McCleery, and A. G. Gosler. 1988. Survival rate in the great tit *Parus major* in relation to sex, age, and immigration status. *J. Anim. Ecol.* 57:287–306.

Clutton-Brock, T. H. 1984. Reproductive effort and terminal investment in iteroparous animals. *Amer. Nat.* 123:212–229.

———. 1988. Reproductive success. In *Reproductive success: Studies of individual variation in contrasting breeding systems*, edited by T. H. Clutton-Brock, 472–485. University of Chicago Press, Chicago.

Cochran, R. A., and J. L. Rives. 1985. *Soil survey of Big Bend National Park*. U.S. Dept. Agric. Soil Cons. Serv.

Cockburn, A. 1992. The process of dispersal. In *Animal dispersal: Small mammals as a model*, edited by N. C. Stenseth and W. Z. Lidicker, Jr., 65–95. Chapman and Hall, London.

Coddington, J. A. 1988. Cladistic tests of adaptational hypotheses. *Cladistics* 4:3–22.

Cody, M. L. 1974. *Competition and structure of bird communities*. Princeton University Press, Princeton, New Jersey.

Cogger, H. G. 1984. Reptiles in the Australian arid zone. In *Arid Australia*, edited by H. G. Cogger and E. E. Cameron, 235–252. Australian Museum, Sydney.

———. 1992. *Reptiles and amphibians of Australia*. Reed Books, Sydney, Australia.

Cole, L. C. 1954. The population consequences of life history phenomenon. *Quart. Rev. Biol.* 29:103–137.

Coleman, R. M., and R. G. Gross. 1991. Parental investment theory: The role of past investment. *Trends Ecol. Evol.* 6:404–406.

Collette, B. B. 1961. Correlations between ecology and morphology in anoline lizards from Havana, Cuba and southern Florida. *Bull. Mus. Comp. Zool.* 125:137–162.

Congdon, J. D., A. E. Dunham, and R. C. van Loben Sels. 1993a. Demographics of common snapping turtles (*Chelydra serpentina*): Implications for conservation and management of long-lived organisms. *Amer. Zool.* (in press).

———. 1993b. Delayed sexual maturity and demographics of Blanding's Turtles (*Emydoidea blandingi*): Implications for conservation and management of long-lived organisms. *Cons. Biol.* (in press).

Congdon, J. D., and J. W. Gibbons. 1987. Morphological constraint on egg size: A challenge to optimal egg size theory? *Proc. Natl. Acad. Sci. USA* 84:4145–4147.

Connell, J. H. 1978. Diversity in tropical rainforests and coral reefs. *Science* 199:1302–1310.

Connell, J. H., and W. P. Sousa. 1983. On the evidence needed to judge ecological stability or persistence. *Amer. Nat.* 121:789–824.

Cooper, W. E., Jr. 1989a. Absence of prey odor discrimination by iguanid and agamid lizards in applicator tests. *Copeia* 1989:472–478.

———. 1989b. Prey odor discrimination in the varanoid lizards *Heloderma suspectum* and *Varanus exanthematicus*. *Ethology* 81:250–258.

———. 1989c. Strike-induced chemosensory searching occurs in lizards. *J. Chem. Ecol.* 15:1311–1320.

———. 1990a. Prey odour discrimination by lizards and snakes. In *Chemical signals in vertebrates 5*, edited by D. W. MacDonald, D. Muller-Schwarze, and S. E. Natynczuk, 533–538. Oxford University Press, Oxford.

———. 1990b. Prey odor detection by teiid and lacertid lizards and the relationship of prey odor detection to foraging mode in lizard families. *Copeia* 1990:237–242.

———. 1990c. Prey odor discrimination by anguid lizards. *Herpetologica* 46:183–190.

————. 1991a. Responses to prey chemicals by a lacertid lizard, *Podarcis muralis*: Prey chemical discrimination and poststrike elevation in tongue-flick rate. *J. Chem. Ecol.* 17:849–863.

————. 1991b. Discrimination of integumentary prey chemicals and strike-induced chemosensory searching in the ball python, *Python regius*. *J. Ethol.* 9:9–23.

————. 1992. Prey odor discrimination and post-strike elevation in tongue flicking by a cordylid lizard, *Gerrhosaurus nigrolineatus*. *Copeia* 1992:146–154.

————. 1993. Duration of poststrike elevation in tongue-flicking rate in the savannah monitor lizard. *Ethol. Ecol. Evol.* 5:1–18.

————. In press a. Lingual and searching responses to prey chemicals by three iguanian lizard species. *Anim. Behav.*

————. In press b. Prey chemical discrimination and foraging mode in gekkonoid lizards. *Herp. Monogr.*

————. In press c. Chemical discrimination by tongue-flicking in lizards: A review with hypotheses on its origin and its ecological and phylogenetic relationships. *J. Chem. Ecol.*

Cooper, W. E., Jr., and A. C. Alberts. 1990. Responses to chemical food stimuli by an herbivorous actively foraging lizard, *Dipsosaurus dorsalis*. *Herpetologica* 46:259–266.

————. 1991. Tongue-flicking and biting in response to chemical food stim-uli by an iguanid lizard (*Dipsosaurus dorsalis*) having sealed vomeronasal ducts: Vomerolfaction may mediate these behavioral responses. *J. Chem. Ecol.* 17:135–146.

Cooper, W. E., Jr., and G. M. Burghardt. 1990. Vomerolfaction and vomo-dor. *J. Chem. Ecol.* 16:103–105.

Cooper, W. E., Jr., and L. J. Vitt. 1989. Prey odor discrimination by the broad-headed skink (*Eumeces laticeps*). *J. Exp. Zool.* 249:11–16.

Cooper, W. E., Jr., and N. Greenberg. 1992. Reptilian coloration and behavior. In *Biology of the Reptilia. Vol. 18. Brain, hormones, and behavior*, edited by C. Gans and D. Crews, 298–422. University of Chicago, Chicago.

Cooper, W. E., Jr., L. J. Vitt, and J. P. Caldwell. 1994. Movement and sub-strate tongue-flicks in phrynosomatid lizards. *Copeia* 1994:234–237.

Cooper, W. E., Jr., L. J. Vitt, R. Hedges, and R. B. Huey. 1990. Locomotor impairment and defense in gravid lizards (*Eumeces laticeps*): Behavioural shift in activity may offset costs of reproduction in an active forager. *Behav. Ecol. Sociobiol.* 27:153–157.

Cracraft, J. 1990. The origin of evolutionary novelties: Pattern and process at different hierarchical levels. In *Evolutionary Innovations*, edited by M. H. Nitecki, 21–44. University of Chicago Press, Chicago.

Crenshaw, J. W., Jr. 1955. The life history of the southern spiny lizard, *Sceloporus undulatus undulatus* Latreille. *Amer. Midl. Nat.* 54:257–298.

Crews, D. 1975. Psychobiology of reptilian reproduction. *Science* 189:1059–1065.

Crowley, S. R., and R. D. Pietruszka. 1983. Aggressiveness and vocalization in the leopard lizard (*Gambelia wislizenii*): The influence of temperature. *Anim. Behav.* 31:1055–1060.

Cunningham, B., and E. Huene. 1938. Further studies on water absorption by reptile eggs. *Amer. Nat.* 70:590–595.

Curio, E. 1976. *The ethology of predation.* Springer-Verlag, New York.

Curio, V. E., and H. Möbius. 1978. Versuche zum Nachweis eines Riechvermögens von *Anolis l. lineatopus* (Rept., Iguanidae). *Z. Tierpsychol.* 47:281–292.

Darwin, C. 1859. *The origin of species by means of natural selection.* Harvard University Press, Cambridge, Mass.

Davis, G. J., and R. W. Howe. 1992. Juvenile dispersal, limited breeding sites, and the dynamics of metapopulations. *Theor. Pop. Biol.* 41:184–207.

Davis, J., and R. G. Ford. 1983. Home range in the western fence lizard (*Sceloporus occidentalis occidentalis*). *Copeia* 1983:933–940.

Davis, J. I., and K. C. Nixon. 1992. Populations, genetic variation, and the delimitation of phylogenetic speies. *Syst. Biol.* 41:421–435.

Dayan, T., D. Simberloff, E. Tchernov, and Y. Yom-Tov. 1990. Feline canines: Community-wide character displacement among the small cats of Israel. *Amer. Nat.* 136:39–60.

de Queiroz, K. 1992. Phylogenetic relationships and rates of allozyme evolution among lineages of sceloporine sand lizards. *Biol. J. Linn. Soc.* 45:333–362.

DeLury, D. B. 1958. The estimation of population size by a marking and recapture procedure. *J. Fish. Res. Bd. Canada* 15:19–25.

DeMarco, V. G. 1989. Annual variation in the seasonal shift in egg size and clutch size in *Sceloporus woodi. Oecologia* 80:525–532.

———. 1993. Metabolic rates of female viviparous lizards (*Sceloporus jarrovi*) throughout the reproductive cycle: Do pregnant females adhere to standard allometry? *Physiol. Zool.* 66:166–180.

DeMarco, V. G., and L. J. Guillette, Jr. 1992. Physiological cost of pregnancy in a viviparous lizard (*Sceloporus jarrovi*). *J. Exp. Zool.* 262:383–390.

Derickson, W. K. 1976. Ecological and physiological aspects of reproductive strategies in two lizards. *Ecology* 57:445–458.

Deslippe, R. J., and R. T. M'Closkey. 1991. An experimental test of mate defense in an iguanid lizard (*Sceloporus graciosus*). *Ecology* 42:1218–1224.

Dessauer, H. C., and C. J. Cole. 1989. Diversity between and within nominal forms of unisexual teiid lizards. In *Evolution and ecology of unisexual vertebrates*, edited by R. M. Dawley and J. P. Bogert, 49–71. New York State Museum, Albany.

Dhondt, A. A., A. Adriansen, E. Matthysen, and B. Kempenaers. 1990. Non-adaptive clutch sizes in tits. *Nature* 348:723–725.

Dhondt, A. A., and R. Eykerman. 1980. Competition and the regulation of numbers in great and blue tits. *Ardea* 68:121–132.

Dial, B. E. 1978. Aspects of the behavioral ecology of two Chihuahuan desert geckos (Reptilia, Lacertilia, Gekkonidae). *J. Herpetol.* 12: 209–216.

———. 1990. Predator-prey signals: Chemosensory identification of snake predators by eublepharid lizards and its ecological consequences. In *Chemical signals in vertebrates 5*, edited by D. W. MacDonald, D. Muller-Schwarze, and S. E. Natynczuk, 555–565. Oxford University Press, Oxford.

Dial, B. E., P. J. Weldon, and B. Curtis. 1989. Chemosensory identification of snake predators (*Phyllorhynchus decurtatus*) by banded geckos (*Coleonyx variegatus*). *J. Herpetol.* 23:224–229.

Dial, B. E., R. E. Gatten, Jr., and S. Kamel. 1987. Energetics of concertina locomotion in *Bipes biporus* (Reptilia: Amphisbaenia). *Copeia* 1987:470–477.

Digby, P.G.N., and R. A. Kempton. 1987. *Multivariate analysis of ecological communities*. Chapman and Hall, London.

Dillon, W. R., and M. Goldstein. 1984. *Multivariate analysis: Methods and applications*. J. Wiley and Sons, New York.

Djawdan, M. 1993. Locomotor performance of bipedal and quadrupedal heteromyid rodents. *Funct. Ecol.* 7:195–202.

Djawdan, M., and T. Garland, Jr. 1988. Maximal running speeds of bipedal and quadrupedal rodents. *J. Mammal.* 69:765–772.

Dobson, F. S., and W. T. Jones. 1985. Multiple causes of dispersal. *Amer. Nat.* 126:855–858.

Douglas, M. E. 1987. An ecomorphological analysis of niche packing and niche dispersion in stream fish clades. In *Community and evolutionary ecology of North American stream fishes*, edited by W. J. Matthews and D. C. Heins, 144–149. University of Oklahoma Press, Norman.

Drake, J. A. 1990. Communities as assembled structures: Do rules govern pattern? *Trends Ecol. Evol.* 5:159–164.

———. 1991. Community-assembly mechanics and the structure of an experimental species ensemble. *Amer. Nat.* 137:1–26.

Drake, J. A., T. E. Flum, G. J. Witteman, T. Voskuil, A. M. Hoylman, C. Creson, D. A. Kenny, G. R. Huxel, C. S. Larue, and J. R. Duncan. 1993. The construction and assembly of an ecological landscape. *J. Anim. Ecol.* 62:117–130.

Drent, R. H., and S. Daan. 1980. The prudent parent: Energetic adjustments in avian breeding. In *The integrated study of bird populations*, edited by H. Klomp and J. W. Woldendorp, 225–252. North-Holland, Amsterdam.

Dubas, G. 1987. Biotic determinants of the home range size of the scincid lizard *Trachydosaurus rugosus* (Gray). Ph. D. Diss., Flinders University of South Australia, Adelaide.

Dubas, G., and C. M. Bull. 1991. Diet choice and food availability in the omnivorous lizard, *Trachydosaurus rugosus*. *Wildl. Res.* 18:147–155.

———. 1992. Food addition and home range size of the lizard, *Tiliqua rugosa*. *Herpetologica* 48:301–306.

Duellman, W. E., and E. R. Pianka. 1990. Biogeography of nocturnal insectivores: Historical events and ecological filters. *Ann. Rev. Ecol. Syst.* 21:57–68.

Dufaure, J. P., and J. Hubert. 1961. Table de development du lézard vivipare: *Lacerta (Zootica) vivipara* Jacquin. *Arch. Anat. Micr. Morph. Exp.* 50:309–328.

Dunham, A. E. 1978. Food availability as a proximate factor influencing individual growth rates in the iguanid lizard *Sceloporus merriami*. *Ecology* 59:770–778.

———. 1980. An experimental study of interspecific competition between the iguanid lizards *Sceloporus merriami* and *Urosaurus ornatus*. *Ecol. Monogr.* 50:309–330.

———. 1981. Populations in a fluctuating environment: The comparative population ecology of the iguanid lizards *Sceloporus merriami* and *Urosaurus ornatus*. *Misc. Publ. Univ. Michigan Mus. Zool.* No. 158:1–62.

———. 1982. Demography and life history variation among populations of the iguanid lizard *Urosaurus ornatus*: Implications for the study of life history phenomena in lizards. *Herpetologica* 38:201–221.

———. 1983. Realized niche overlap, resource abundance, and the intensity of interspecific competition. In *Lizard ecology: Studies of a model organism*, edited by R. B. Huey, E. R. Pianka, and T. W. Schoener, 261–280. Harvard University Press, Cambridge, Mass.

———. 1993. Population responses to global change: Physiologically structured models, operative environments, and population dynamics. In *Evolutionary, population, and community responses to global change*, edited by P. Kariva, J. Kingsolver, and R. Huey, 95–119. Sinauer Associates, Sunderland, Mass.

Dunham, A. E., and D. B. Miles. 1985. Patterns of covariation in life-history traits of squamate reptiles: The effects of size and phylogeny reconsidered. *Amer. Nat.* 126:231–257.

Dunham, A. E., B. W. Grant, and K. L. Overall. 1989. Interfaces between biophysical and physiological ecology and the population ecology of terrestrial vertebrate ectotherms. *Physiol. Zool.* 62:335–355.

Dunham, A. E., D. B. Miles, and D. N. Reznick. 1988a. Life history patterns in squamate reptiles. In *Biology of the Reptilia. Vol. 16. Ecology B. Defense and life history*, edited by C. Gans and R. B. Huey, 441–522. A. R. Liss, New York.

Dunham, A. E., P. J. Morin, and H. M. Wilbur. 1988b. Methods for the study of reptile populations. In *Biology of the Reptilia. Vol. 16. Ecology B. Defense and life history*, edited by C. Gans and R. B. Huey, 331–386. A. R. Liss, New York.

Duvall, D., D., Chiszar, W. K., Hayes, J. K., Leonhardt, and M. J. Goode. 1990. Chemical and behavioral ecology of foraging in prairie rattlesnakes (*Crotalus viridis viridis*). *J. Chem. Ecol.* 16:87–101.

Ebeling, A. W., S. J. Holbrook, and R. J. Schmitt. 1990. Temporally concordant structure of a fish assemblage: Bound or determined? *Amer. Nat.* 135:63–73.

Edwards, J. L. 1985. Terrestrial locomotion without appendages. In *Functional vertebrate morphology*, edited by M. Hildebrand, D. M. Bramble, K. F. Liem, and D. B. Wake, 159–172. Belknap Press, Cambridge, Mass.

Efron, B. 1982. *The jackknife, the bootstrap and other resampling plans.* SIAM, Philadelphia.

Egan, A. J. 1984. Reproduction in the sleepy lizard, *Trachydosaurus rugosus*. Honours thesis, University of Adelaide.

Ehmann, H., and H. Cogger. 1985. Australia's endangered herpetofauna—A review of criteria and policies. In *Biology of Australasian frogs and reptiles*, edited by G. Grigg, R. Shine, and H. Ehmann, 435–437. Surrey Beatty and Sons, Chipping Norton, NSW.

Endler, J. A. 1986. *Natural selection in the wild.* Princeton University Press, Princeton, New Jersey.

Estes, R., and G. Pregill, eds. 1988. *Phylogenetic relationships of the lizard families: Essays commemorating Charles L. Camp.* Stanford University Press. Stanford.

Estes, R., K. de Queiroz, and J. Gauthier. 1988. Phylogenetic relationships within Squamates. In *Phylogenetic relationships of the lizard families: Essays commemorating Charles L. Camp*, edited by R. Estes and G. Pregill, 119–281. Stanford University Press, Stanford.

Etheridge, R. 1964. The skeletal morphology and systematic relationships of sceloporine lizards. *Copeia* 1964:610–631.

Etheridge, R., and K. de Queiroz. 1988. A phylogeny of Iguanidae. In *Phylogenetic relationships of the lizard families: Essays commemorating Charles L. Camp*, edited by R. Estes and G. Pregill, 283–367. Stanford University Press, Stanford.

Evans, L. T. 1961. Structure as related to behavior in the organization of populations of reptiles. In *Vertebrate speciation*, edited by W. F. Blair, 148–178. University of Texas Press, Houston.

Everitt, B. 1974. *Cluster analysis*. Heinemann Educational Books, London.

Fahrig, L. 1990. Interacting effects of disturbance and dispersal on individual selection and population stability. *Comm. Theor. Biol.* 1:275–299.

Fahrig, L., and J. Paloheimo. 1988. Effect of spatial arrangement of habitat patches on local population size. *Ecology* 69:468–475.

Falconer, D. S. 1989. *Introduction to quantitative genetics*. Longman, London.

Fellers, G. M., and C. A. Drost. 1991. Ecology of the island night lizard, *Xantusia riversiana*, on Santa Barbara Island. *Herp. Monogr.* 5:28–78.

Felsenstein, J. 1985. Phylogenies and the comparative method. *Amer. Nat.* 125:1–15.

Ferguson, G. W., and C. H. Bohlen. 1972. *The regulation of prairie swift (lizard) populations—A progress report*. Third Midwest Prairie Conference Proceedings, Kansas State University, Manhattan.

Ferguson, G. W., and H. L. Snell. 1986. Endogenous control of seasonal change of egg, hatchling, and clutch size of the lizard *Sceloporus undulatus garmani*. *Herpetologica* 42:185–191.

Ferguson, G. W., and L. G. Talent. 1993. Life-history traits of the lizard *Sceloporus undulatus* from two populations raised in a common laboratory environment. *Oecologia* 93:88–94.

Ferguson, G. W., and S. F. Fox. 1984. Annual variation of survival advantage of large juvenile side-blotched lizards, *Uta stansburiana*: Its causes and evolutionary significance. *Evolution* 38:342–349.

Ferguson, G. W., and T. Brockman. 1980. Geographic differences of growth rate of *Sceloporus* lizards (Sauria: Iguanidae). *Copeia* 1980:259–264.

Ferguson, G. W., C. H. Bohlen, and H. P. Wooley. 1980. *Sceloporus undulatus*: Comparative life history and regulation of a Kansas population. *Ecology* 61:313–322.

Ferguson, G. W., J. L. Hughes, and K. L. Brown. 1983. Food availability and territorial establishment of juvenile *Sceloporus undulatus*. In *Lizard ecology: Studies of a model organism*, edited by R. B. Huey, E. R. Pianka, and T. W. Schoener, 134–148. Harvard University Press, Cambridge, Mass.

Ferguson, G. W., K. L. Brown, and V. G. DeMarco. 1982. Selective basis for the evolution of variable egg and hatchling size in some iguanid lizards. *Herpetologica* 38:178–188.

Ferner, J. W. 1973. An ecological study of *Sceloporus undulatus erythrocheilus* (Reptilia, Iguanidae) in Colorado. *Diss. Abstr.* 33:4035–B.

———. 1993. Ontogeny of dispersal distances in young Spanish imperial eagles. *Behav. Ecol. Sociobiol.* 32:259–263.

Fisher, R. A. 1930. *The genetical theory of natural selection*. Oxford University Press, Oxford.

Fitch, H. S. 1935. Natural history of the alligator lizards. *Trans. Acad. Sci. St. Louis* 29:1–38.

———. 1973. A field study of Costa Rican lizards. *Univ. Kans. Sci. Bull.* 50:39–126.

———. 1978. Sexual size differences in the genus *Sceloporus*. *Univ. Kansas Sci. Bull.* 51:441–461.

———. 1989. A field study of the slender glass lizard, *Ophisaurus attenuatus*, in northeastern Kansas. *Occas. Pap. Mus. Nat. Hist. Univ. Kansas.* 125:1–50.

Fitch, H. S., and A. V. Fitch. 1967. Preliminary experiments on physical tolerances of the eggs of lizards and snakes. *Ecology* 48:160–165.

Fitch, W. M. 1971. Toward defining the course of evolution: Minimum change for a specific tree topology. *Syst. Zool.* 20:406–416.

Fletcher, D.J.C., and C. D. Michener, eds. 1987. *Kin recognition in animals*. John Wiley and Sons, Chichester, UK.

Ford, N. B., and R. A. Seigel. 1989. Phenotypic plasticity in reproductive traits: Evidence from a viviparous snake. *Ecology* 70:1768–1774.

Foster, R. B., and N.V.L. Brokaw. 1982. Structure and history of the vegetation of Barro Colorado Island. In *The ecology of a tropical forest: Seasonal rhythms and long-term changes,* edited by E. G. Leigh, Jr., A. S. Rand, and D. M. Windsor, 67–81. Smithsonian Institution Press, Washington, D.C.

Fox, S. F. 1975. Natural selection on morphological phenotypes of the lizard *Uta stansburiana*. *Evolution* 29:95–107.

———. 1978. Natural selection on behavioral phenotypes of the lizard *Uta stansburiana*. *Ecology* 59:834–847.

———. 1983. Fitness, home-range quality, and aggression in *Uta stansburiana*. In *Lizard ecology: Studies of a model organism,* edited by R. B. Huey, E. R. Pianka, and T. W. Schoener, 134–148. Harvard University Press, Cambridge, Mass.

Fox, W. 1963. Special tubules for sperm storage in female lizards. *Nature* 198:500–501.

Frost, D. R., and R. Etheridge. 1989. A phylogenetic analysis and taxonomy of iguanian lizards (Reptilia: Squamata). *Misc. Publ. Univ. Kansas Mus. Nat. Hist.* 81:1–65.

Full, R. J. 1991. The concepts of efficiency and economy in land locomotion. In *Efficiency and economy in animal physiology,* edited by R. W. Blake, 97–131. Cambridge University Press, Cambridge, UK.

Gadgil, M., and W. H. Bossert. 1970. Life historical consequences of natural selection. *Amer. Nat.* 104:1–24.

Gans, C. 1975. Tetrapod limblessness: Evolution and functional corollaries. *Amer. Zool.* 15:455–467.

Gans, C., H. C. Dessauer, and D. Baic. 1978. Axial differences in the musculature of uropeltid snakes: The freight-train approach to burrowing. *Science* 199:189–192.

Garland, T., Jr. 1983. Scaling the ecological cost of transport to body mass in terrestrial mammals. *Amer. Nat.* 121:571–587.

———. 1984. Physiological correlates of locomotory performance in a lizard: An allometric approach. *Amer. J. Physiol. (Reg. Integ. Comp. Physiol. 16)* 247:R806–R815.

———. 1985. Ontogenetic and individual variation in size, shape and speed in the Australian agamid lizard *Amphibolurus nuchalis*. *J. Zool. London* 207:425–439.

———. 1988. Genetic basis of activity metabolism. I. Inheritance of speed, stamina, and antipredator displays in the garter snake *Thamnophis sirtalis*. *Evolution* 42:335–350.

———. 1992. Rate tests for phenotypic evolution using phylogenetically independent contrasts. *Amer. Nat.* 140:509–519.

———. 1993. Locomotor performance and activity metabolism of *Cnemidophorus tigris* in relation to natural behaviors. In *Biology of whiptail lizards (genus Cnemidophorus)*, edited by J. W. Wright and L. J. Vitt, 163–210. Oklahoma Mus. Nat. Hist., Norman.

Garland, T., Jr., A. F. Bennett, and C. B. Daniels. 1990a. Heritability of locomotor performance and its correlates in a natural population. *Experientia* 46:530–533.

Garland, T., Jr., A. W. Dickerman, C. M. Janis, and J. A. Jones. 1993. Phylogenetic analysis of covariance by computer simulation. *Syst. Biol.* 42:265–292.

Garland, T., Jr., and C. M. Janis. 1993. Does metatarsal/femur ratio predict maximal running speed in cursorial mammals? *J. Zool. London* 229:133–151.

Garland, T., Jr., and J. B. Losos. 1994. Ecological morphology of locomotor performance in squamate reptiles. In *Ecological morphology: Integrative organismal biology*, edited by P. C. Wainwright and S. M. Reilly. University of Chicago Press, Chicago (in press).

Garland, T., Jr., and P. L. Else. 1987. Seasonal, sexual, and individual variation in endurance and activity metabolism in lizards. *Amer. J. Physiol. (Reg. Integ. Comp. Physiol. 21)* 252:R439–R449.

Garland, T., Jr., and S. C. Adolph. 1991. Physiological differentiation of vertebrate populations. *Ann. Rev. Ecol. Syst.* 22:193–228.

Garland, T., Jr., E. Hankins, and R. B. Huey. 1990b. Locomotor capacity and social dominance in male lizards. *Funct. Ecol.* 4:243–250.

Garland, T., Jr., F. Geiser, and R. V. Baudinette. 1988. Comparative locomotor performance of marsupial and placental mammals. *J. Zool. London* 215:505–522.

Garland, T., Jr., P. H. Harvey, and A. R. Ives. 1992. Procedures for the analysis of comparative data using phylogenetically independent contrasts. *Syst. Biol.* 41:18–32.

Gatten, R. E., Jr., K. Miller, and R. Full. 1992. Energetics of amphibians at rest and during locomotion. In *Environmental physiology of the Amphibia*, edited by M. E. Feder and W. W. Burggren, 314–377. University of Chicago Press, Chicago.

Gauthier, J., R. Estes, and K. de Queiroz. 1988. A phylogenetic analysis of Lepidosauromorpha. In *Phylogenetic relationships of the lizard families: Essays commemorating Charles L. Camp*, edited by R. Estes and G. Pregill, 15–98. Stanford University Press, Stanford.

Gavaud, J. 1986. Vitellogenesis in the lizard *Lacerta vivipara* Jacquin. II. Vitellogenin synthesis during the reproductive cycle and its control by ovarian steroids. *Gen. Comp. Endocrinol.* 63:11–23.

Gibbons, J. W. 1967. Variation in growth rate in three populations of the painted turtle, *Chrysemys picta. Herpetologica* 23:296–303.

Gillis, R., and R. E. Ballinger. 1992. Reproductive ecology of red-chinned lizards (*Sceloporus undulatus erythrocheilus*) in southcentral Colorado: Comparisons with other populations of a wide-ranging species. *Oecologia* 89:236–243.

Gingerich, P. D. 1983. Rates of evolution: Effects of time and temporal scaling. *Science* 222:159–161.

Gittleman, J. L., and M. Kot. 1990. Adaptation: Statistics and a null model for estimating phylogenetic effects. *Syst. Zool.* 39:227–241.

Gittleman, J. L., and H. -K. Luh. 1992. On comparing comparative methods. *Ann. Rev. Ecol. Syst.* 23:383–404.

Gleeson, T. T. 1979. Foraging and transport costs in the Galapagos marine iguana, *Amblyrhynchus cristatus. Physiol. Zool.* 52:549–557.

———. 1980. Metabolic recovery from exhaustive activity by a large lizard. *J. Appl. Physiol.: Respirat. Environ. Exercise Physiol.* 48:689–694.

———. 1981. Preferred body temperature, aerobic scope, and activity capacity in the monitor lizard, *Varanus salvator. Physiol. Zool.* 54:423–429.

———. 1985. Glycogen synthesis from lactate in skeletal muscle of the lizard *Dipsosaurus dorsalis. J. Comp. Physiol.* B 156:277–283.

———. 1991. Patterns of metabolic recovery from exercise in amphibians and reptiles. *J. Exp. Biol.* 160:187–207.

Gleeson, T. T., and A. F. Bennett. 1982. Acid-base imbalance in lizards during activity and recovery. *J. Exp. Biol.* 98:439–453.

Glinski, T. H., and C. O. Krekorian. 1985. Individual recognition in free-living adult male desert iguanas, *Dipsosaurus dorsalis. J. Herpetol.* 19:541–544.

Gonzalez, A., G. Alvarado-Dufree, and C. T. Diaz. 1975. *Canal zone water quality study, final report.* Water and Laboratories Branch, Panama Canal Company, Canal Zone, Panama.

Good, D. A. 1988a. Phylogenetic relationships among gerrhonotine lizards (Squamata: Anguidae): An analysis of external morphology. *Univ. California Publ. Zool.* 121:1–139.

———. 1988b. The phylogenetic position of fossils assigned to the gerrhonotinae (Squamata: Anguidae). *J. Vert. Paleo.* 8:188–195.

Gordon, R. E. 1960. The influence of moisture on variation in the eggs and hatchlings of *Anolis carolinensis* Voigt. *Nat. Hist. Misc., No. 173.* Chicago Acad. Sci., Chicago.

Gorman, G. C. 1970. Chromosomes and the systematics of the family Teiidae (Sauria, Reptilia). *Copeia* 1970:230–245.

Gorman, O. T. 1993. Evolutionary ecology and historical ecology: Assembly, structure, and organization of stream fish communities. In *Systematics, historical ecology, and North American freshwater fishes*, edited by R. L. Mayden, 659–688. Stanford University Press, Stanford.

Goszczynski, J. 1986. Locomotor activity of terrestrial predators and its consequences. *Acta Theriologica* 31:79–95.

Gould, S. J., and E. S. Vrba. 1982. Exaptation—A missing term in the science of form. *Paleobiology* 3:4–15.

Gould, S. J., and R. C. Lewontin. 1979. The spandrels of San Marco and the Panglossian paradigm: A critique of the adaptationist program. *Proc. Royal Soc. London.* B 205:581–598.

Grafen, A. 1989. The phylogenetic regression. *Phil. Trans. Royal Soc. London,* Series B 326:119–157.

Grant, B. W. 1990. Trade-offs in activity time and physiological performance for thermoregulating desert lizards, *Sceloporus merriami. Ecology* 71:2323–2333.

Grant, B. W., and A. E. Dunham. 1988. Thermally imposed time constraints on the activity of the desert lizard *Sceloporus merriami. Ecology* 69:167–176.

Grant, B. W., and A. E. Dunham. 1990. Elevational covariation in environmental constraints and life histories of the desert lizard *Sceloporus merriami. Ecology* 71:1765–1776.

Grant, B. W., and W. P. Porter. 1992. Modeling global macroclimatic constraints on ectotherm energy budgets. *Amer. Zool.* 32:154–178.

Grant, P. R. 1983. Conclusion: Lizard ecology, viewed at a short distance. In *Lizard ecology: Studies of a model organism*, edited by R. B. Huey, E. R. Pianka, and T. W. Schoener, 411–417. Harvard University Press, Cambridge, Mass.

Graves, B. M., and M. Halpern. 1989. Chemical access to the vomeronasal organs of the lizard *Chalcides ocellatus. J. Exp. Zool.* 249:150–157.

———. 1991. Discrimination of self from conspecific chemical cues in *Tiliqua scincoides* (Sauria: Scincidae). *J. Herpetol.* 25:125–126.

Greene, H. W. 1986. Diet and arboreality in the emerald monitor, *Varanus prasinus*, with comments on the study of adaptation. *Fieldiana Zool. New Ser.* 31:1–12.

———. 1988. Antipredator mechanisms in reptiles. In *Biology of the Reptilia. Vol. 16. Ecology B. Defense and life history*, edited by C. Gans and R. B. Huey, 1–152. Alan R. Liss, New York.

Greenwood, P. J. 1980. Mating systems, philopatry, and dispersal in birds and mammals. *Anim. Behav.* 28:1140–1162.

Greer, A. E. 1989. *The biology and evolution of Australian lizards.* Surrey Beatty and Sons, Chipping Norton, NSW.

Gregory, P. T. 1982. Reptilian hibernation. In *Biology of the Reptilia. Vol. 13. Physiology D. Physiological ecology*, edited by C. Gans and F. H. Pough, 53–140. Academic Press, New York.

Griffin, G. F. 1984. Vegetation patterns. In *Anticipating the inevitable: A patch-burn strategy for fire management at Uluru (Ayers Rock-Mt. Olga) National Park*, edited by E. C. Saxon, 25–37. CSIRO, Melbourne, Australia.

Grismer, L. L. 1988. Phylogeny, taxonomy, classification, and biogeography of eublepharid geckos. In *Phylogenetic relationships of the lizard families: Essays commemorating Charles L. Camp*, edited by R. Estes and G. Pregill, 369–469. Stanford University Press, Stanford.

Grossberg, R. K. 1988. Life-history variation within a population of the colonial ascidian *Botryllus schlosseri* I. The genetic and environmental control of seasonal variation. *Evolution* 42:900–920.

Grossman, G. D. 1982. Dynamics and organization of a rocky intertidal fish assemblage: The persistence and resilience of taxocene structure. *Amer. Nat.* 119:611–637.

Grossman, G. D., P. B. Moyle, and J. O. Whitaker. 1982. Stochasticity in structural characteristics of an Indiana stream fish assemblage: A test of community theory. *Amer. Nat.* 120:423–454.

Guillette, L. J. 1982. The evolution of viviparity and placentation in the high elevation, Mexican lizard *Sceloporus aeneus. Herpetologica* 38:94–103.

Gutzke, W.H.N., and G. C. Packard. 1985. Hatching success in relation to egg size in painted turtles (*Chrysemys picta*). *Can J. Zool.* 63:67–70.

Gutzke, W.H.N., and G. C. Packard. 1987. Influence of the hydric and thermal environments on eggs and hatchlings of bull snakes *Pituophis melanoleucus. Physiol. Zool.* 60:9–17.

Guyer, C. 1988a. Food supplementation in a tropical mainland anole, *Norops humilis*: Demographic effects. *Ecology* 69:350–361.

———. 1988b. Food supplementation in a tropical mainland anole, *Norops humilis*: Effects on individuals. *Ecology* 69:362–369.

Guyer, C., and J. M. Savage. 1986. Cladistic relationships among anoles (Sauria: Iguanidae). *Syst. Zool.* 35:509–531.

———. 1992. Anole systematics revisited. *Syst. Biol.* 41:89–110.

Haffer, J. 1969. Speciation in Amazonian forest birds. *Science* 165:131–137.

Hailey, A., C. A. Rose, and E. Pulford. 1987. Food consumption, thermoregulation and ecology of the skink *Chalcides bedriagai*. *Herpet. J.* 1:144–153.

Haines, R. W. 1942. The tetrapod knee joint. *J. Anat.* 76:270–301.

Haldane, J.B.S. 1949. Suggestions as to quantitative measurement of rates of evolution. *Evolution* 3:51–56.

Halpern, M. 1992. Nasal chemical senses in reptiles: Structure and function. In *Biology of the Reptilia. Vol. 18. Brain, hormones, and behavior*, edited by C. Gans and D. Crews, 423–523. University of Chicago Press, Chicago.

Hanski, I. 1991a. Metapopulation dynamics: Brief history and conceptual domain. *Biol. J. Linn. Soc.* 42:3–16.

———. 1991b. Single-species metapopulation dynamics: Concepts, models and observations. *Biol. J. Linn. Soc.* 42:17–38.

Hanski, I., A. Peltonen, and L. Kaski. 1991. Natal dispersal and social dominance in the common shrew *Sorex araneus*. *Oikos* 62:48–58.

Harvey, P. H., and M. D. Pagel. 1991. *The comparative method in evolutionary biology*. Oxford University Press, Oxford.

Hasegawa, M. 1990. Demography of an island population of the lizard, *Eumeces okadae*, on Miyake-Jima, Izu islands. *Res. Pop. Ecol.* 32:119–133.

Hasting, A. 1992. Age dependent dispersal is not a simple process: Density dependence, stability, and chaos. *Theor. Pop. Biol.* 41:388–400.

Henle, K. 1990. Population ecology and life history of three terrestrial geckos in arid Australia. *Copeia* 1990:759–781.

———. 1991. Life history patterns in lizards of the arid and semiarid zone of Australia. *Oecologia* 88:347–358.

Herrenkohl, L. R. 1979. Prenatal stress reduces fertility and fecundity in female offspring. *Science* 206:1097–1099.

Hertz, P. E., R. B. Huey, and T. Garland, Jr. 1988. Time budgets, thermoregulation, and maximal locomotor performance: Are ectotherms olympians or boy scouts? *Amer. Zool.* 28:927–938.

Heulin, B. 1984. Contribution l'étude de la biologie des populations de *Lacerta vivipara*: Stratégie démographique et utilisation de l'espace dans une population du massif forestier de Paimpont. Ph.D. Thesis, University of Rennes I, Rennes.

Hews, D. K. 1993. Food resources affect female distribution and male mating opportunities in the iguanian lizards *Uta palmeri*. *Anim. Behav.* 46:279–291.

Hildebrand, M. 1985. Walking and running. In *Functional vertebrate morphology*, edited by M. Hildebrand, D. M. Bramble, K. F. Liem, and D. B. Wake, 38–57. Harvard University Press, Cambridge, Mass.

Hillel, D. 1971. *Soil and water: Physical principles and processes*. Academic Press, New York.

Ho, S., S. Kleis, R. McPherson, G. J. Heisermann, and I. P. Callard. 1982. Regulation of vitellogenesis in reptiles. *Herpetologica* 38:40–50.

Houston, A. I., and J. M. McNamara. 1992. Phenotypic plasticity as a state-dependent life-history decision. *Evol. Ecol.* 6:243–253.

Howe, R. W., G. J. Davis, and V. Mosca. 1991. The demographic significance of "sink" populations. *Biol. Cons.* 57:239–255.

Huey, R. B. 1982. Phylogenetic and ontogenetic determinants of sprint performance in some diurnal Kalahari lizards. *Koedoe* 25: 43–48.

Huey, R. B., A. E. Dunham, K. L. Overall, and R. A. Newman. 1990. Variation in locomotor performance in demographically known populations of the lizard *Sceloporus merriami*. *Physiol. Zool.* 63:845–872.

Huey, R. B., A. F. Bennett, H. John-Alder, and K. A. Nagy. 1984. Locomotor capacity and foraging behavior of Kalahari Lacertid lizards. *Anim. Behav.* 32: 41–50.

Huey, R. B., and A. E. Dunham. 1987. Repeatability of locomotor performance in natural populations of the lizard *Sceloporus merriami*. *Evolution* 41:1116–1120.

Huey, R. B., and A. F. Bennett. 1986. A comparative approach to field and laboratory studies in evolutionary biology. In *Predator-prey relationships: Perspectives and approaches from the study of lower vertebrates*, edited by M. E. Feder and G. V. Lauder, 82–98. University of Chicago Press, Chicago.

Huey, R. B., and A. F. Bennett. 1987. Phylogenetic studies of co-adaptation: Preferred temperatures versus optimal performance temperatures of lizards. *Evolution* 41:1098–1115.

Huey, R. B., and E. R. Pianka. 1981. Ecological consequences of foraging mode. *Ecology* 62:991–999.

Huey, R. B., and P. E. Hertz. 1982. Effects of body size and slope on sprint speed of a lizard *Stellio (Agama) stellio*. *J. Exp. Biol.* 97:401–409.

———. 1984. Effects of body size and slope on acceleration of a lizard (*Stellio stellio*). *J. Exp. Biol.* 110:113–123.

Huey, R. B., and R. D. Stevenson. 1979. Integrating thermal physiology and ecology of ectotherms: A discussion of approaches. *Amer. Zool.* 19:357–366.

Huey, R. B., E. R. Pianka, and T. W. Schoener, eds. 1983. *Lizard ecology: Studies of a model organism*. Harvard University Press, Cambridge, Mass.

Hutchinson, G. E. 1959. Homage to Santa Rosalia, or why are there so many kinds of animals? *Amer. Nat.* 93:145–159.

Ims, R. A. 1990. Determinants of natal dispersal and space use in grey-sided voles, *Clethrionomys rufocanus*: A combined field and laboratory experiment. *Oikos* 57:106–113.

Jackson, J. F., and S. R. Telford, Jr. 1974. Reproductive ecology of the Florida scrub lizard, *Sceloporus woodi*. *Copeia* 1974:689–694.

Jacobs, S.W.L. 1984. Spinifex. In *Arid Australia*, edited by H. G. Cogger and E. E. Cameron, 131–142. Australian Museum, Sydney.

Jaksic, F. M., and H. Núñez. 1979. Escaping behavior and morphological correlates in two *Liolaemus* species of central Chile (Lacertilia: Iguanidae). *Oecologia* 42:119–122.

Jaksic, F. M., H. Núñez, and F. P. Ojeda. 1980. Body proportions, microhabitat selection, and adaptive radiation of *Liolaemus* lizards in central Chile. *Oecologia* 45:178–181.

James, C. D. 1991a. Population dynamics, demography and life history of sympatric scincid lizards (*Ctenotus*) in central Australia. *Herpetologica* 47:194–210.

———. 1991b. Annual variation in reproductive cycles of scincid lizards (*Ctenotus*) in central Australia. *Copeia* 1991:742–758.

James, C. D., and R. Shine. 1988. Life-history strategies of Australian lizards: Comparison between tropics and the temperate zone. *Oecologia* 75:307–316.

James, F. C., and W. J. Boecklen. 1984. Interspecific morphological relationships and the densities of birds. In *Ecological communities: Conceptual issues and the evidence*, edited by D.R.J. Strong, D. Simberloff, L. G. Abele, and A. B. Thistle, 458–477. Princeton University Press, Princeton, New Jersey.

Janson, C. H. 1992. Measuring evolutionary constraints: A Markov model for phylogenetic transitions among seed dispersal syndromes. *Evolution* 46:136–158.

Jayne, B. C., and A. F. Bennett. 1990. Selection on locomotor performance capacity in a natural population of garter snakes. *Evolution* 44:1204–1229.

John-Alder, H. B. 1983. Effects of thyroxin supplementation on metabolic rate and aerobic capacity in a lizard. *Amer. J. Physiol. (Reg. Integ. Comp. Physiol. 13)* 244:R659–R666.

———. 1984a. Reduced aerobic capacity and locomotory endurance in thyroid-deficient lizards. *J. Exp. Biol.* 109:175–189.

———. 1984b. Seasonal variations in activity, aerobic energetic capacities, and plasma thyroid hormones (T3 and T4) in an iguanid lizard. *J. Comp. Physiol.* B 154:409–419.

John-Alder, H. B., and A. F. Bennett. 1981. Thermal dependence of endurance and locomotory energetics in a lizard. *Amer. J. Physiol. (Reg. Integ. Comp. Physiol. 10)* 241:R342–R349.

John-Alder, H. B., C. H. Lowe, and A. F. Bennett. 1983. Thermal dependence of locomotory energetics and aerobic capacity of the Gila monster (*Heloderma suspectum*). *J. Comp. Physiol.* 151:119–126.

John-Alder, H. B., T. Garland, Jr., and A. F. Bennett. 1986. Locomotory capacities, oxygen consumption, and the cost of locomotion of the shingle-back lizard (*Trachydosaurus rugosus*). *Physiol. Zool.* 59:523–531.

Johnson, M. L., and M. S. Gaines. 1990. Evolution of dispersal: Theoretical models and empirical tests using birds and mammals. *Ann. Rev. Ecol. Syst.* 21:449–480.

Jones, R. E. 1978. Control of follicular selection. In *The vertebrate ovary*, edited by R. E. Jones, 763–788. Plenum Press, New York.

Jones, R. E., R. R. Tokarz, and F. T. LaGreek. 1975. Endocrine control of clutch size in reptiles V. Patterns of FSH-induced follicular formation and growth in immature ovaries of *Anolis carolinensis*. *Gen. Comp. Endocrinol.* 26:354–367.

Jones, R. E., R. R. Tokarz, F. T. LaGreek, and K. T. Fitzgerald. 1976. Endocrine control of clutch size in reptiles VI. Patterns of FSH-induced ovarian stimulation in adult *Anolis carolinensis*. *Gen. Comp. Endocrinol.* 30:101–116.

Jones, S. M., and D. L. Droge. 1980. Home range size and spatial distributions of two sympatric lizard species (*Sceloporus undulatus, Holbrookia maculata*) in the sand hills of Nebraska. *Herpetologica* 36:127–132.

Jones, S. M., and R. E. Ballinger. 1987. Comparative life histories of *Holbrookia maculata* and *Sceloporus undulatus* in western Nebraska. *Ecology* 68:1828–1838.

Jones, S. M., R. E. Ballinger, and W. P. Porter. 1987a. Physiological and environmental sources of variation in reproduction: Prairie lizards in a food rich environment. *Oikos* 48:325–335.

Jones, S. M., S. R. Waldschmidt, and M. A. Potvin. 1987b. An experimental manipulation of food and water: Growth and time-space utilization of hatchling lizards (*Sceloporus undulatus*). *Oecologia* 73:53–59.

Kahmann, H. 1939. Uber das Jacobsonsche Organ der Echsen. *Z. Vergl. Physiol.* 26:669–695.

Kaplan, R. H. 1992. Greater maternal investment can decrease offspring survival in the frog *Bombina orientalis*. *Ecology* 73:280–288.

Karges, J. P., and J. W. Wright. 1987. A new species of *Barisia* (Sauria,

Anguidae) from Oaxaca, Mexico. *Contr. Sci. Nat. Hist. Mus. Los Angeles County* 381:1–11.

Kiester, A. R. 1985. Sex-specific dynamics of aggregation and dispersal in reptiles and amphibians. In *Migration: Mechanisms and adaptive significance*, edited by M. A. Rankin, 425–434. Contr. Marine Sci. 27 suppl.

Kiester, A. R., and M. Slatkin. 1974. A strategy of movement and resource utilization. *Theoret. Pop. Biol.* 6:1–20.

Kingsbury, B. A. 1989. Factors influencing activity in *Coleonyx variegatus*. *J. Herpetol.* 23:399–404.

Kirkpatrick, M., and R. Lande. 1989. Selection, inheritance, and evolution of maternal characters. *Evolution* 43:485–503.

Kluge, A. G. 1987. Cladistic relationships in the Gekkonoidea (Squamata, Sauria). *Misc. Pub. Mus. Zool., Univ. Michigan* 173:1–54.

Kobayashi, D., W. J. Mautz, and K. A. Nagy. 1983. Evaporative water loss: Humidity acclimation in *Anolis carolinensis* lizards. *Copeia* 1983:701–704.

Koenig, W. D., F. A. Pitelka, W. J. Carmen, R. L. Mumme, and M. T. Stanback. 1992. The evolution of delayed dispersal in cooperative breeders. *Quart. Rev. Biol.* 67:111–150.

Kramer, G. 1951. Body proportions of mainland and island lizards. *Evolution* 5:193–206.

Krebs, C. J. 1988. The experimental approach to rodent population dynamics. *Oikos* 52:143–149.

———. 1992. The role of dispersal in cyclic rodent populations. In *Animal dispersal: Small mammals as a model*, edited by N. C. Stenseth and W. Z. Lidicker, Jr., 160–175. Chapman and Hall, London.

Krebs, J. R., and N. B. Davies. 1978. *Behavioural ecology: An evolutionary approach*. Blackwell Scientific Publications, Oxford.

Krekorian, C. O. 1989. Field and laboratory observations on chemoreception in the desert iguana, *Dipsosaurus dorsalis*. *J. Herpetol.* 23:267–273.

Lacey, E. P. 1988. Latitudinal variation in reproductive timing of a short-lived monocarp, *Daucus carota* (Apiaceae). *Ecology* 69:220–232.

Lack, D. 1954. *The natural regulation of animal numbers*. Clarendon Press, Oxford.

Laerm, J. 1974. A functional analysis of morphological variation and differential niche utilization in basilisk lizards. *Ecology* 55:404–411.

Lande, R. 1976. Natural selection and random genetic drift in phenotypic evolution. *Evolution* 30:314–334.

———. Statistical tests for natural selection on quantitative characters. *Evolution* 31:442–444.

Lang, M. 1991. Generic relationships within the Cordyliformes (Reptilia: Squamata). *Bull. Institut Roy. Sci. Nat. Belg., Biol.* 61:121–188.

Larsen, K. R., and W. W. Tanner. 1974. Numeric analysis of the lizard genus *Sceloporus* with special reference to cranial osteology. *Great Basin Nat.* 34:1–41.

———. 1975. Evolution of the sceloporine lizards (Iguanidae). *Great Basin Nat.* 35:1–20.

Lauder, G. V. 1982. Historical biology and the problem of design. *J. Theor. Biol.* 97:57–67.

Lauder, G. V. 1990. Functional morphology and systematics: Studying functional patterns in a historical context. *Ann. Rev. Ecol. Syst.* 20:317–340.

———. 1991. Biomechanics and evolution: Integrating physical and historical biology in the study of complex systems. In *Biomechanics in evolution*, edited by J.M.V. Rayner and R. J. Wooton, 1–19. Cambridge University Press, Cambridge, UK.

Lauder, G. V., and K. F. Liem. 1989. The role of historical factors in the evolution of complex organismal functions. In *Complex organismal functions: Integration and evolution in vertebrates*, edited by D. B. Wake and G. Roth, 63–78. John Wiley and Sons, Chichester, UK.

Lauder, G. V., M. L. Armand, and M. R. Rose. 1993. Adaptations and history. *Trends Ecol. Evol.* 8:294–297.

Laurie, W. A. 1989. Effects of the 1982–83 El Niño sea warming on marine iguana (*Amblyrhynchus cristatus* Bell, 1825) populations on Galapagos. In *Global ecological consequences of the 1982–83 El Niño: Southern oscillation*, edited by P. Glynn, 121–141. Elsevier, New York.

———. 1990. Population biology of marine iguanas (*Amblyrhynchus cristatus*). I. Changes in fecundity related to a population crash. *J. Anim. Ecol.* 59:515–528.

Laurie, W. A., and D. Brown. 1990. Population biology of marine iguanas (*Amblyrhynchus cristatus*). II. Changes in annual survival rates and the effects of size, sex, age and fecundity in a population crash. *J. Anim. Ecol.* 59:529–544.

Lebreton, J. D., and G. Gonzalez-Davila. 1993. An introduction to models of subdivided populations. *J. Biol. Sys.* (in press).

Lecomte, J. 1993. Rôle du comportement dans l'organisation et la régulation des populations de lézards vivipares. Ph.D. thesis, Ecole Normale Supérieure et Université of Paris XI.

Lecomte, J., J. Clobert, and M. Massot. 1993. Shift in behaviour related to pregnancy in *Lacerta vivipara*. *Rev. Ecol. (Terre et Vie)* 48:99–107.

Lee, D. S. 1974. The possible role of fire on population density of the Florida scrub lizard, *Sceloporus woodi* Stejneger. *Bull. Md. Herp. Soc.* 10:20–22.

Legendre, S., J. Clobert, and Ferrière, R. 1993. *ULM, unified life models: A user's guide*. Publ. Laboratoire d'Ecologie, Ecole Normale Supérieure, Paris.

Leigh, E. G., Jr., A. S. Rand, and D. M. Windsor, eds. 1982. *The ecology of a tropical forest: Seasonal rhythms and long-term changes.* Smithsonian Institution Press, Washington, D.C.

Levings, S. C. 1983. Seasonal, annual, and among-site variation in the ground ant community of a deciduous tropical forest: Some causes of patchy species distributions. *Ecol. Monogr.* 53:435–455.

Levings, S. C., and D. M. Windsor. 1982. Seasonal and annual variation in litter arthropod populations. In *Advances in herpetology and evolutionary biology: Essays in honor of Ernest E. Williams,* edited by A. G. Rhodin and K. Miyata, 355–387. Smithsonian Institution Press, Washington, D.C.

Licht, P. 1970. Effects of mammalian gonadotropins (ovine FSH and LH) in female lizards. *Gen. Comp. Endocrinol.* 22:463–469.

———. 1974. Response of *Anolis* lizards to food supplementation in nature. *Copeia* 1974:215–221.

Lidicker, W. Z., Jr. 1962. Emigration as a possible mechanism permitting the regulation of population density below carrying capacity. *Amer. Nat.* 96:29–33.

———. 1975. The role of dispersal in the demography of small mammals. In *Small mammals: Their productivity and population dynamics,* edited by F. B. Goley, K. Petrusewicz, and L. Ryskowski, 103–128. Cambridge University Press, Cambridge, UK.

Lidicker, W. Z., Jr., and N. C. Stenseth. 1992. To disperse or not to disperse: Who does it and why? In *Animal dispersal: Small mammals as a model,* edited by N. C. Stenseth and W. Z. Lidicker, Jr., 21–34. Chapman and Hall, London.

Lieberman, S. S. 1986. Ecology of the leaf litter herpetofauna of a neotropical rainforest: La Selva, Costa Rica. *Acta Zool. Mex. new ser.* 15:1–71.

Lima, S. L., and L. M. Dill. 1990. Behavioral decisions made under the risk of predation: A review and prospectus. *Can. J. Zool.* 68:619–640.

Lindén, M., and A. P. Møller. 1989. Costs of reproduction and covariation of life history traits in birds. *Trends Ecol. Evol.* 4:367–371.

Loop, M. S., and S. A. Scoville. 1972. Response of newborn *Eumeces inexpectatus* to prey-object extracts. *Herpetologica* 28:254–256.

Lopez, P., and A. Salvador. 1992. The role of chemosensory cues in discrimination of prey odors by the amphisbaenian *Blanus cinereus*. *J. Chem. Ecol.* 18:87–93.

Losos, J. B. 1990a. Ecomorphology, performance capability, and scaling of West Indian *Anolis* lizards: An evolutionary analysis. *Ecol. Monogr.* 60:369–388.

———. 1990b. The evolution of form and function: Morphology and locomotor performance in West Indian *Anolis* lizards. *Evolution* 44:1189–1203.

————. 1990c. A phylogenetic analysis of character displacement in Caribbean *Anolis* lizards. *Evolution* 44:558–569.

————. 1992. The evolution of convergent structure in Caribbean *Anolis* communities. *Syst. Biol.* 41:403–420.

Losos, J. B., and D. B. Miles. 1994. Adaptation, constraint, and the comparative method: Phylogenetic issues and methods. In *Ecological morphology: Integrative organismal biology*, edited by P. C. Wainwright and S. Reilly. University Chicago Press, Chicago (in press).

Losos, J. B., D. J. Irschick, and T. W. Schoener. In press. Adaptation and constraint in the evolution of specialization of Bahamian *Anolis* lizards. *Evolution*.

Losos, J. B., J. C. Marks, and T. W. Schoener. 1993. Habitat use and ecological interactions of an introduced and a native species of *Anolis* lizard on Grand Cayman, with a review of the outcomes of anole introductions. *Oecologia* 95:525–532.

Losos, J. B., T. J. Papenfuss, and J. R. Macey. 1989. Correlates of sprinting, jumping and parachuting performance in the butterfly lizard, *Leiolepis belliani. J. Zool. London* 217:559–568.

Lynch, M. 1990. The rate of morphological evolution in mammals from the standpoint of the neutral expectation. *Amer. Nat.* 136:727–741.

————. 1991a. Methods for the analysis of comparative data in evolutionary biology. *Evolution* 45:1065–1080.

————. 1991b. The genetic interpretation of inbreeding depression and outbreeding depression. *Evolution* 45:622–629.

Lynch, M., and W. G. Hill. 1986. Phenotypic evolution by neutral mutation. *Evolution* 40:915–935.

M'Closkey, R. T., K. A. Baia, and R. W. Russel. 1987a. Defense of mates: A territory departure rule for male tree lizards following sex-ratio manipulation. *Oecologia* 73:28–31.

————. 1987b. Tree lizard (*Urosaurus ornatus*) territories: Experimental perturbation of the sex ratio. *Ecology* 68:2059–2062.

MacArthur, R. H. 1965. Patterns of species diversity. *Biol. Rev.* 40:510–533.

MacArthur, R. H., and E. O. Wilson. 1967. *The theory of island biogeography*. Princeton University Press, Princeton, New Jersey.

MacDonald, D. W., and H. Smith. 1990. Dispersal, dispersion and conservation in the agricultural ecosystem. In *Species dispersal in agricultural habitats*, edited by R.G.H. Bunce and D. C. Howard, 18–64. Belhaven Press, London.

MacMillen, R. E., and D. S. Hinds. 1992. Standard, cold induced, and exercise-induced metabolism of rodents. In *Mammalian energetics: Interdisciplinary views of metabolism and reproduction*, edited by T. E. Tomasi and T. H. Horton, 16–33. Comstock Publishing Associates, Ithaca, New York.

Maddison, W. P. 1989. Reconstructing character evolution on polytomous cladograms. *Cladistics* 5:365–377

———. 1990. A method for testing the correlated evolution of two binary characters: Are gains or losses concentrated on certain branches of a phylogenetic tree? *Evolution* 44:539–557.

Maddison, W. P., and D. R. Maddison. 1992. *MacClade: Analysis of phylogeny and character evolution. V3.0.* Sinauer Associates, Sunderland, Mass.

Maddison, W. P., M. J. Donoghue, and D. R. Maddison. 1984. Outgroup analysis and parsimony. *Syst. Zool.* 33:83–130.

Magnusson. W. E., L. J. Paiva, R. M. Rocha, C. R. Franke, L. A. Kasper, and A. P. Lima. 1985. The correlates of foraging mode in a community of Brazilian lizards. *Herpetologica* 41:324–332.

Marks, J. S., and R. L. Redmond. 1987. Parent-offspring conflict and natal dispersal in birds and mammals: Comments on the *Oedipus* hypothesis. *Amer. Nat.* 129:158–164.

Marler, C. A., and M. C. Moore. 1988. Evolutionary costs of aggression revealed by testosterone manipulations in free-living male lizards. *Behav. Ecol. Sociobiol.* 23:21–26.

———. 1989. Time and energy costs of aggression in testosterone-implanted free-living male mountain spiny lizards (*Sceloporus jarrovi*). *Physiol. Zool.* 62:1334–1350.

———. 1991. Supplementary feeding compensates for testosterone-induced costs of aggression in male mountain spiny lizards. *Anim. Behav.* 42:209–219.

Marshall, T. J., and J. W. Holmes. 1979. *Soil physics*. Cambridge University Press, Cambridge, UK.

Martins, E. P. 1991. A field study of individual and sex differences in the push-up display of the Sagebrush Lizard, *Sceloporus graciosus*. *Anim. Behav.* 41:403–416.

———. 1992a. Contextual use of the visual "push-up" display of the Sagebrush Lizard, *Sceloporus graciosus*. *Anim. Behav.* 45:25–36.

———. 1992b. Structure, function and evolution of the *Sceloporus* push-up display. Ph.D. Diss., University of Wisconsin, Madison.

———. 1993. A comparative study of the evolution of *Sceloporus* push-up displays. *Amer. Nat.* (in press).

———. 1994. Estimating rates of character change. *Amer. Nat.* (in press).

Martins, E., and T. Garland, Jr. 1991. Phylogenetic analyses of the correlated evolution of continuous characters: A simulation study. *Evolution* 45:534–557.

Massot, M. 1992. Déterminisme de la dispersion chez le lézard vivipare. Ph.D. thesis, Ecole Normale Supérieure et Université of Paris XI.

Massot, M., J. Clobert, A. Chambon, and Y. Michalakis. 1994b. Vertebrate natal dispersal: The problem of non independence of siblings. *Oikos* (in press).

Massot, M., J. Clobert, J. Lecomte, and R. Barbault. 1994a. Incumbent advantage in common lizards and their colonizing ability. *J. Anim. Ecol.* (in press).

Massot, M., J. Clobert, T. Pilorge, J. Lecomte, and R. Barbault. 1992. Density dependence in the common lizard: Demographic consequences of a density manipulation. *Ecology* 73:1742–1756.

Mautz, W. J., C. B. Daniels, and A. F. Bennett. 1992. Thermal dependence of locomotion and aggression in a xantusiid lizard. *Herpetologica* 48:271–279.

May, R. M. 1981. Models for single populations. In *Theoretical ecology: Principles and applications*, edited by R. M. May, 4–25. Sinauer Associates Inc., Sunderland, Mass.

Mayhew, W. W. 1963. Biology of the granite spiny lizard, *Sceloporus orcutti. Amer. Midl. Nat.* 69:310–327.

Maynard Smith, J. 1978. Optimization theory in evolution. *Ann. Rev. Ecol. Syst.* 9:31–56.

Mayr, E. 1963. *Animal species and evolution.* Belknap Press, Cambridge, Mass.

McCleery, R. H., and J. Clobert. 1990. Differences in recruitment of young by immigrant and resident great tits in Wytham wood. In *Population studies of passerine birds: An integrated approach*, edited by J. Blondel, A.G. Gosler, J.-D. Lebreton, and R. H. McCleery, 423–440. Springer Verlag, Berlin.

McDade, L. A., K. S. Bawa, H. A. Hespenheide, and G. S. Hartshorn. In press. *La Selva: Ecology and natural history of a neotropical rainforest.* University of Chicago Press, Chicago.

Mcdowell, S. B. 1972. The evolution of the tongue of snakes, and its bearing on snake origins. In *Evolutionary biology, vol. 6*, edited by T. Dobzhansky, M. K. Hecht, and W. C. Steere, 191–273. Appleton-Century-Crofts, New York.

McLaughlin, J. F., and J. Roughgarden. 1989. Avian predation on *Anolis* lizards in the northeastern Caribbean: An inter-island contrast. *Ecology* 70:617–628.

McLaughlin, R. L. 1989. Search modes of birds and lizards: Evidence for alternative movement patterns. *Amer. Nat.* 133:654–670.

McShea, W. J. 1990. Social tolerance and proximate mechanisms of dispersal among winter groups of meadow voles, *Microtus pennsylvanicus. Anim. Behav.* 39:346–351.

Miles, D. B., and A. E. Dunham. 1992. Comparative analyses of phylogenetic effects in the life-history patterns of iguanid reptiles. *Amer. Nat.* 139:848–869.

———. 1993. Historical perspectives in ecology and evolutionary biology: The use of phylogenetic comparative analyses. *Ann. Rev. Ecol. Syst.* 24:587–619.

Miles, D. B., and R. E. Ricklefs. 1984. The correlation between ecology and morphology in deciduous forest passerine birds. *Ecology* 65:1629–1640.

Miles, D. B., R. E. Ricklefs, and J. Travis. 1987. Concordance of ecomorphological relationships in three assemblages of passerine birds. *Amer. Nat.* 129:347–364.

Miller, K., G. C. Packard, T. J. Boardman, G. L. Paukstis, and M. J. Packard. 1987. Hydric conditions during incubation influence locomotor performance of hatchling snapping turtles. *J. Exp. Biol.* 127:401–412.

Milstead, W. W. 1970. Late summer behavior of the lizards *Sceloporus merriami* and *Urosaurus ornatus* in the field. *Herpetologica* 26:343–354.

Milstead, W. W., ed. 1967. *Lizard ecology: A symposium.* University of Missouri Press, Columbia.

Mindell, D. P., J. W. Sites, Jr., and D. Gaur. 1989. Speciational evolution: A phylogenetic test with allozymes in *Sceloporus* (Reptilia). *Cladistics* 5:49–62.

Mitchell-Olds, T., and R. G. Shaw. 1987. Regression analysis of natural selection: Statistical and biological interpretation. *Evolution* 41:1149–1161.

Moermond, T. C. 1979a. Habitat constraints on the behavior, morphology, and community structure of *Anolis* lizards. *Ecology* 60:152–164.

———. 1979b. The influence of habitat structure on *Anolis* foraging behavior. *Behaviour* 70:147–167.

Montanucci, R. R. 1987. A phylogenetic study of the horned lizards, genus *Phrynosoma*, based on skeletal and external morphology. *Contr. Sci. Nat. Hist. Mus. Los Angeles County* No. 390:1–36.

Moore, M. C. 1986. Elevated testosterone levels during non-breeding season territoriality in a fall-breeding lizard, *Sceloporus jarrovi*. *J. Comp. Physiol.* 158:159–163.

———. 1987. Castration affects territorial and sexual behaviour of free-living male lizards, *Sceloporus jarrovi*. *Anim. Behav.* 35:1193–1999.

———. 1988. Testosterone control of territorial behavior: Tonic-release implants fully restore seasonal and short-term aggressive responses in free-living castrated lizards. *Gen. Comp. Endocrinol.* 70:450–459.

Moore, M. C., and C. A. Marler. 1987. Effects of testosterone manipulations on non-breeding season territorial aggression in free-living male lizards, *Sceloporus jarrovi*. *Gen. Comp. Endocrinol.* 65:225–232.

Morris, D. W. 1982. Age-specific dispersal strategies in iteroparous species: Who leaves when? *Evol. Theory* 6:53–65.

Morris, K. A., G. C. Packard, T. J. Boardman, G. L. Paukstis, and M. J. Packard. 1983. Effect of the hydric environment on growth of embryonic snapping turtles (*Chelydra serpentina*). *Herpetologica* 39:272–285.

Morton, S. R., and C. D. James. 1988. The diversity and abundance of lizards in arid Australia: A new hypothesis. *Amer. Nat.* 132:237–256.

Moyle, P. B., and B. Vondracek. 1985. Persistence and structure of the fish assemblage in a small California stream. *Ecology* 66:1–13.

Murphy, R. W., W. E. Cooper, Jr., and W. S. Richardson. 1983. Phylogenetic relationships of the North American five-lined skinks, genus *Eumeces* (Sauria: Scincidae). *Herpetologica* 39:200–211.

Muth, A. 1977. Eggs and hatchlings of captive *Dipsosaurus dorsalis*. *Copeia* 1977:189–190.

———. 1980. Physiological ecology of the desert iguana (*Dipsosaurus dorsalis*) eggs: Temperature and water relations. *Ecology* 61:1335–1343.

Nagy, K. A. 1983. Ecological energetics. In *Lizard ecology: Studies of a model organism*, edited by R. B. Huey, E. R. Pianka, and T. W. Schoener, 24–54. Harvard University Press, Cambridge, Mass.

Newman, R. A. 1988. Adaptive plasticity in development of *Scaphiopus couchii* tadpoles in desert ponds. *Evolution* 42:774–783.

———. 1989. Developmental plasticity of *Scaphiopus couchii* tadpoles in an unpredictable environment. *Ecology* 70:1775–1787.

———. 1992. Adaptive plasticity in amphibian metamorphosis. *Bioscience* 42:671–678.

Nicoletto, P. F. 1985. The roles of vision and the chemical senses in predatory behavior of the skink, *Scincella lateralis*. *J. Herpetol.* 19:487–491.

Niewiarowski, P. H. 1992. Ecological and evolutionary sources of geographic variation in individual growth rates of the eastern fence lizard, *Sceloporus undulatus* (Iguanidae). Ph. D. Diss., University of Pennsylvania, Philadelphia.

Niewiarowski, P. H., and W. M. Roosenburg. 1993. Reciprocal transplant reveals sources of variation in growth rates of the lizard *Sceloporus undulatus*. *Ecology* 74:1992–2002.

Noble, G. K. 1939. The role of dominance in the social life of birds. *Auk* 56:263–273.

Noble, G. K., and K. F. Kumpf. 1936. The function of Jacobson's organ in lizards. *J. Genet. Psychol.* 48:371–382.

Norusis, M. J. 1986a. *SPSS/PC+ for the IBM PC/XT/AT.* SPSS, Chicago.

———. 1986b. *Advanced statistics SPSS/PC+ for the IBM PC/XT/AT.* SPSS, Chicago.

———. 1992. *SPSS/PC+ base system user's guide version 5.0.* SPSS Inc., Chicago.

Nussbaum, R. A. 1981. Seasonal shifts in clutch size and egg size in the side-blotched lizard, *Uta stansburiana* Baird and Girard. *Oecologia* 49:8–13.

O'Connell, B., R. Greenlee, and J. Bacon. 1985. Strike-induced chemosensory searching in elapid snakes (cobras, taipans, tiger snakes, and death adders) at San Diego Zoo. *Psychol. Rec.* 35:431–436.

Ogden, J. C., and J. P. Ebersole. 1981. Scale and community structure of coral reef fishes: A long-term study of a large artificial reef. *Marine Ecol. Prog. Ser.* 4:97–103.

Opdam, P. 1990. Dispersal in fragmented populations: The key to survival. In *Species dispersal in agricultural habitats,* edited by R.G.H. Bunce and D. C. Howard, 3–17. Belhaven Press, London.

Pacala, S., and J. Roughgarden. 1985. Population experiments with the *Anolis* lizards of St. Maarten and St. Eustatius. *Ecology* 66:129–141.

Packard, G. C., and M. J. Packard. 1987. Water relations and nitrogen excretion in embryos of the oviparous snake (*Coluber constrictor*). *Copeia* 1987:395–406.

Packard, G. C., and M. J. Packard. 1988. The physiological ecology of reptilian eggs and embryos. In *Biology of the Reptilia. Vol. 16. Ecology B. Defense and life history,* edited by C. Gans and R. B. Huey, 523–606. A. R. Liss, New York.

Packard, G. C., C. R. Tracy, and J. J. Roth. 1977. The physiological ecology of reptilian eggs and embryos, and the evolution of viviparity within the class Reptilia. *Biol. Rev.* 52:71–105.

Packard, G. C., M. J. Packard, K. Miller, and T. J. Boardman. 1987. Influence of moisture, temperature, and substrate on snapping turtle eggs and embryos. *Ecology* 68:983–993.

Packard, M. J., G. C. Packard, and T. J. Boardman. 1980. Water balance of eggs of a desert lizard (*Callisaurus draconoides*). *Can J. Zool.* 58:2051–2058.

———. 1982. Structure of eggshells and water relations of reptilian eggs. *Herpetologica* 38:136–155.

Packard, M. J., J. A. Phillips, and G. C. Packard. 1992. Sources of mineral for green iguanas (*Iguana iguana*) developing in eggs exposed to different hydric environments. *Copeia* 1992:851–858.

Pagel, M. D. 1993. Seeking the evolutionary regression coefficient: An analysis of what comparative methods measure. *J. Theor. Biol.* 164:191–205.

Paine, R. T. 1974. Intertidal community structure: Experimental studies on the relationship between a dominant competitor and its principal predator. *Oecologia* 15:93–120.

Parker, W. S., and E. R. Pianka. 1973. Notes on the ecology of the iguanid lizard *Sceloporus magister. Herpetologica* 29:143–152.

Partridge, L. 1992. Measuring reproductive costs. *Trends Ecol. Evol.* 7:99–100.

Partridge, L., and P. H. Harvey. 1985. Costs of reproduction. *Nature* 316:20–21.

Pasitschniak-Arts, M., and J. F. Bendel. 1990. Behavioural differences between locally recruiting and dispersing gray squirrels, *Sciurus carolinensis. Can. J. Zool.* 68:935–941.

Pease, C. M., and J. J. Bull. 1988. A critique of methods for measuring life history trade-offs. *J. Evol. Biol.* 1:293–303.

Pechmann, J.H.K., D. E. Scott, R. D. Semlitsch, J. P. Caldwell, L. J. Vitt, and J. W. Gibbons. 1991. Declining amphibian populations: The problem of separating human impacts from natural fluctuations. *Science* 253:892–895.

Perry, G., I. Lampl, A. Lerner, D. Rothenstein, E. Shani, N. Sivan, and Y. L. Werner. 1990. Foraging mode in lacertid lizards: Variation and correlates. *Amphibia-Reptilia* 11:373–384.

Peters, J. A., and R. Donoso Barros. 1970. *Catalogue of the neotropical Squamata: Part II. Lizards and amphisbaenians.* Smithsonian Institution Press, Washington, D.C.

Peterson, J. A. 1984. The locomotion of *Chamaeleo* (Reptilia: Sauria) with particular reference to the forelimb. *J. Zool. London* 202:1–42.

Petney, T. N., and C. M. Bull. 1984. Microhabitat selection by two reptile ticks at their parapatric boundary. *Austral. J. Ecol.* 9:233–239.

Phillips, J. A., A. Garel, G. C. Packard, and M. J. Packard. 1990. Influence of moisture and temperature on eggs and embryos of green iguanas (*Iguana iguana*). *Herpetologica* 46:238–245.

Pianka, E. R. 1966. Convexity, desert lizards, and spatial heterogeneity. *Ecology* 47:1055–1059.

———. 1967. On lizard species diversity: North American flatland deserts. *Ecology* 48:333–351.

———. 1969a. Habitat specificity, speciation, and species density in Australian desert lizards. *Ecology* 50:498–502.

———. 1969b. Sympatry of desert lizards (*Ctenotus*) in Western Australia. *Ecology* 50:1012–1030.

———. 1970. Comparative autecology of the lizard *Cnemidophorus tigris* in different parts of its geographic range. *Ecology* 51:703–720.

———. 1971. Lizard species density in the Kalahari Desert. *Ecology* 52:1024–1029.

———. 1972. Zoogeography and speciation of Australian desert lizards: An ecological perspective. *Copeia* 1972:127–145.

———. 1973. The structure of lizard communities. *Ann. Rev. Ecol. Syst.* 4:53–74.

———. 1976. Natural selection of optimal reproductive tactics. *Amer. Zool.* 16:775–784.

———. 1986. *Ecology and natural history of desert lizards: Analyses of the ecological niche and community structure.* Princeton University Press, Princeton, New Jersey.

————. 1989. Desert lizard diversity: Additional comments and some data. *Amer. Nat.* 134:344–364.

Pianka, E. R., and W. S. Parker. 1975. Age-specific reproductive tactics. *Amer. Nat.* 109:453–464.

Pianka, E. R., R. B. Huey, and L. R. Lawlor. 1979. Niche segregation in desert lizards. In *Analysis of ecological systems,* edited by D. J. Horn, G. R. Stairs, and R. D. Mitchell, 67–115. Ohio State University Press, Columbus.

Pietruszka, R. D. 1986. Search tactics of desert lizards: How polarized are they? *Anim. Behav.* 34:1742–1758.

Pilorge, T. 1982. Stratégie adaptative d'une population de montagne de *Lacerta vivipara. Oikos* 39:206–212.

————. 1987. Density, size structure, and reproductive characteristics of three populations of *Lacerta vivipara* (Sauria: Lacertidae). *Herpetologica* 43:345–356.

Pilorge, T., F. Xavier, and R. Barbault, 1983. Variations in litter size and reproductive effort within and between some populations of *Lacerta vivipara. Holartic Ecol.* 6:381–386.

Pilorge, T., J. Clobert, and M. Massot. 1987. Life history variation according to sex and age in *Lacerta vivipara.* In *Proceedings of the 4th ordinary general meeting of the Societas Europaea Herpetologica,* edited by J. J. vanGelder, H. Strijbosch, and P.J.M. Bergers, 311–315. Faculty of Sciences, Nijmegen.

Pimm, S. L. 1991. *The balance of nature?* The University of Chicago Press, Chicago.

Piperno, D. R. 1990. Fitolitos, arqueología y cambios prehistóricos de la vegetación en un lote de cincuenta hectáreas de la isla Barro Colorado. In *Ecologiá de un bosque tropical: Ciclos estacionales y cambios a largo plazo,* edited by E. G. Leigh, Jr., A. S. Rand, and D. M. Windsor, 153–156. Smithsonian Institution Press, Washington, D.C.

Pitelka, F. A. 1959. Numbers, breeding schedule, and territoriality in pectoral sandpipers of northern Alaska. *Condor* 61:233–264.

Plummer, M. V., and H. L. Snell. 1988. Nest site selection and water relations of eggs in the snake *Opheodrys aestivus. Copeia* 1988:58–64.

Pollock, K. H., J. D. Nichols, C. Brownie, and J. E. Hines. 1990. Statistical inference for capture-recapture experiments. *Wildl. Monogr.* 107:1–97.

Pounds, J. A. 1989. Ecomorphology, locomotion, and microhabitat structure: Patterns in a tropical mainland *Anolis* community. *Ecol. Monogr.* 58:299–320.

Pounds, J. A., J. F. Jackson, and S. H. Shively. 1983. Allometric growth of the hindlimb of some terrestrial iguanid lizards. *Amer. Midl. Nat.* 110:201–207.

Presch, W. 1974. Evolutionary relationships and biogeography of the macroteiid lizards (family Teiidae; subfamily Teiinae). *Bull. So. Cal. Acad. Sci.* 73:23–32.

———. 1983. The lizard family Teiidae: Is it a monophyletic group? *Zool. J. Linn. Soc.* 77:189–197.

Pulliam, H. R. 1988. Sources, sinks, and population regulation. *Amer. Nat.* 132:652–661.

Purvis, A., and T. Garland, Jr. 1993. Polytomies in comparative analyses of continuous characters. *Syst. Biol.* 42:569–575.

Rand, A. S. 1967. The adaptive significance of territoriality in iguanid lizards. In *Lizard ecology, a symposium.* edited by W. W. Milstead, 106–116. University of Missouri Press, Columbia.

———. 1972. Temperatures of iguana nests and their relation to incubation optima and to nesting sites and season. *Herpetologica* 28:252–253.

Rand, A. S., and H. W. Greene. 1982. Latitude and climate in the phenology of reproduction in the green iguana *Iguana iguana*. In *Iguanas of the world*, edited by G. M. Burghardt and A. S. Rand, 142–149. Noyes Publishing Co., Park Ridge, New Jersey.

Rand, A. S., M. J. Ryan, and K. Troyer. 1983. A population explosion in a tropical tree frog: *Hyla rufitela* on Barro Colorado Island, Panama. *Biotropica* 15:72–73.

Raup, D. M. 1987. Major features of the fossil record and their implications for evolutionary rate studies. In *Rates of evolution,* edited by K.S.W. Campbell and M. F. Day, 1–14. Allen and Unwin Press, London.

Regal, P. J. 1978. Behavioral differences between reptiles and mammals: An analysis of activity and mental capabilities. In *Behavior and neurology of lizards*, edited by B. Greenberg and P. D. MacLean, 183–202. United States Department of Health, Education, and Welfare, Washington, D.C.

———. 1983. The adaptive zone and behavior of lizards. In *Lizard ecology: Studies of a model organism,* edited by R. B. Huey, E. R. Pianka, and T. W. Schoener, 105–118. Harvard University Press, Cambridge, Mass.

Reilly, S. M., and G. V. Lauder. 1992. Morphology, behavior, and evolution: Comparative kinematics of aquatic feeding in salamanders. *Brain, Behav. Evol.* 40:182–196.

Rewcastle, S. C. 1980. Form and function in lacertilian knee and mesotarsal joints: A contribution to the analysis of sprawling locomotion. *J. Zool. London* 191:147–170.

———. 1983. Fundamental adaptations in the lacertilian hind limb: A partial analysis of the sprawling limb posture and gait. *Copeia* 1983: 476–487.

Reznick, D. N. 1983. The structure of guppy life histories: The tradeoff between growth and reproduction. *Ecology* 64:862–873.

————. 1985. Costs of reproduction: An evaluation of the empirical evidence. *Oikos* 44:257–267.

————. 1990. Plasticity in age and size at maturity in male guppies (*Poecilia reticulata*): An experimental evaluation of alternative models of development. *J. Evol. Biol.* 3:185–203.

————. 1992. Measuring the costs of reproduction. *Trends Ecol. Evol.* 7:42–45.

Reznick, D. N., and H. Bryga. 1987. Life history evolution in guppies (*Poecilia reticulata*): 1. Phenotypic and genetic changes in an introduction experiment. *Evolution* 41:1370–1385.

Reznick, D. N., H. Bryga, and J. A. Endler. 1990. Experimentally induced life history evolution in a natural population. *Nature* 346:357–359.

Reznick, D. N., and J. A. Endler. 1982. The impact of predation on life history evolution in Trinidadian guppies (*Poecilia reticulata*). *Evolution* 36:160–177.

Rice, W. R. 1989. Analyzing tables of statistical tests. *Evolution* 43: 223–225.

Ricklefs, R. E. 1987. Community diversity: Relative roles of local and regional processes. *Science* 235:167–171.

Ricklefs, R. E., and D. B. Miles 1994. Ecological and evolutionary inferences from morphology: An ecological perspective. In *Ecological morphology: Integrative organismal biology,* edited by P. C. Wainwright and S. M. Reilly. Univerity of Chicago Press, Chicago (in press).

Ricklefs, R. E., and J. Cullen. 1973. Embryonic growth of the green iguana (*Iguana iguana*). *Copeia* 1973:296–305.

Ricklefs, R. E., D. Cochran, and E. R. Pianka. 1981. A morphological analysis of the structure of communities of lizards in desert habitats. *Ecology* 62:1474–1483.

Roff, D. A. 1992. *The evolution of life histories.* Routledge, Chapman and Hall, Inc., New York.

Roitberg, B. D. 1989. The cost of reproduction in rosehip flies, *Rhagoletis basiola*: Eggs are time. *Evol. Ecol.* 3:183–188.

Root, R. B., and N. Cappuccino. 1992. Patterns in population change and the organization of the insect community associated with goldenrod. *Ecol. Monogr.* 62:393–420.

Rose, B. 1982. Food intake and reproduction in *Anolis acutus. Copeia* 1982:322–330.

Rose, B. R. 1976. Habitat and prey selection of *Sceloporus occidentalis* and *Sceloporus graciosus. Ecology* 57:531–541.

————. 1981. Factors affecting activity in *Sceloporus virgatus. Ecology* 62:706–716.

————. 1982. Lizard home ranges: Methodology and functions. *J. Herpetol.* 16:253–270.

Rose, M. R., and B. Charlesworth. 1981a. Genetics of life history in *Drosophila melanogaster.* I. Sib analysis of adult females. *Genetics* 97:173–186.

Rose, M. R., and B. Charlesworth. 1981b. Genetics of life history in *Drosophila melanogaster.* II. Exploratory selection experiments. *Genetics* 97:187–196.

Roughgarden, J., and S. Pacala. 1989. Taxon cycle among *Anolis* lizard populations: Review of the evidence. In *Speciation and its consequences,* edited by D. Otte and J. Endler, 403–432. Sinauer Associates Inc., Sunderland, Mass.

Ruby, D. E. 1977. Winter activity in Yarrow's spiny lizard, *Sceloporus jarrovi. Herpetologica* 33:322–333.

———. 1978. Seasonal changes in the territorial behavior of the iguanid lizard *Sceloporus jarrovi. Copeia* 1978:430–438.

———. 1981. Phenotypic correlates of male reproductive success in the lizard, *Sceloporus jarrovi.* In *Natural selection and social behavior,* edited by R. D. Alexander and D. W. Tinkle, 96–107. Chiron Press, New York.

———. 1986. Selection of home range site in the female of the lizard, *Sceloporus jarrovi. J. Herpetol.* 20:432–435.

Ruby, D. E., and A. E. Dunham. 1984. A population analysis of the ovoviviparous lizard *Sceloporus jarrovi* in the Piñaleno mountains of southeastern Arizona. *Herpetologica* 40:425–436.

———. 1987a. Ecological factors affecting home range size in a desert lizard, *Sceloporus merriami. Oecologia* 71:473–480.

———. 1987b. Variation in home range size along an elevational gradient in the iguanid lizard *Sceloporus merriami. Oecologia* 71:473–480.

Ruby, D. E., and D. I. Baird. 1994. Effects of sex and size on agonistic encounters between juvenile and adult lizards, *Sceloporus jarrovi. J. Herpetol.* (in press).

Russell, M. P. 1991. Modern death assemblages and Pleistocene fossil assemblages in open coast high energy environments, San Nicolas Island, California. *Palaios* 6:179–191.

Sale, P. F. 1977. Maintenance of high diversity in coral reef fish communities. *Amer. Nat.* 111:337–359.

———. 1984. The structure of communities of fish on coral reefs and the merit of a hypothesis-testing, manipulative approach to ecology. In *Ecological communities: Conceptual issues and the evidence,* edited by D. R. Strong, D. Simberloff, L. G. Abele, and A. B. Thistle, 478–490. Princeton University Press, Princeton, New Jersey.

Sale, P. F., and R. Dybdahl. 1978. Determinants of community structure for coral reef fishes in isolated coral heads at lagoonal and reef slope sites. *Oecologia* 34:57–74.

Sale, P. F., P. J. Doherty, G. J. Eckert, W. A. Douglas, and D. J. Ferrell. 1984.

Large-scale spatial and temporal variation in recruitment to fish populations on coral reefs. *Oecologia* 64:191–198.

Sanderson, M. J. 1993. Reversibility in evolution: A maximum likelihood approach to character gain/loss bias in phylogenies. *Evolution* 47:236–252.

SAS Institute, Inc. 1982. *SAS User's guide: Statistics. 1982 edition.* SAS Institute, Cary, North Carolina.

———. 1985. *SAS User's guide: Statistics, version 5.* SAS Institute, Cary, North Carolina.

———. 1989. *SAS/STAT ® User's guide, version 6, fourth edition.* SAS Institute, Cary, North Carolina.

———. 1990. *SAS/STAT User's guide, release 6.03 edition.* SAS Institute, Cary, North Carolina.

Satrawaha, R., and C. M. Bull. 1981. The area occupied by an omnivorous lizard, *Trachydosaurus rugosus. Aust. Wildl. Res.* 8:435–442.

Savage, M. J., A. Cass, and J. M. de Jager. 1981. Calibration of thermocouple hygrometers. *Irrig. Sci.* 2:113–125.

Schall, J. J. 1983. Lizard malaria: Parasite-host ecology. In *Lizard ecology: Studies of a model organism,* R. B. Huey, E. R. Pianka, and T. W. Schoener, 84–100. Harvard University Press, Cambridge, Mass.

Schall, J. J., and E. R. Pianka. 1980. Evolution of escape behavior diversity. *Amer. Nat.* 115:551–566.

Scheibe, J. S. 1987. Climate, competition, and the structure of temperate zone lizard communities. *Ecology* 68:1424–1436.

Schlosser, I. J. 1982. Fish community structure and function along two habitat gradients in a headwater stream. *Ecol. Monogr.* 52:395–414.

Schluter, D. 1988. Estimating the form of natural selection on a quantitative trait. *Evolution* 42:849–861.

Schoener, T. W. 1968a. Sizes of feeding territories among birds. *Ecology* 49:123–141.

———. 1968b. The *Anolis* lizards of Bimini: Resource partitioning in a complex fauna. *Ecology* 49:704–726.

———. 1970. Size patterns in West Indian *Anolis* lizards. II. Correlations with the size of particular sympatric species—Displacement and convergence. *Amer. Nat.* 104:155–174.

———. 1974. Resource partitioning in ecological communities. *Science* 185:27–39.

———. 1985. Are lizard population sizes unusually constant through time? *Amer. Nat.* 126:633–641.

———. 1986. Resource partitioning. In *Community ecology: Pattern and process,* edited by J. Kikkawa and D. J. Anderson, 91–126. Blackwell Scientific Publ., Melbourne, Australia.

Schoener, T. W., and G. H. Adler. 1992. Greater resolution of distributional complementarities by controlling for habitat affinities: A study with Bahamian lizards and birds. *Amer. Nat.* 137:669–692.

Schoener, T. W., and A. Schoener. 1978. Estimating and interpreting body-size growth in some *Anolis* lizards. *Copeia* 1978:390–405.

———. 1983. The time to extinction of a colonizing propagule of lizards increases with island area. *Nature* 302:332–334.

Schroeder, M. A., and D. A. Boag. 1988. Dispersal in spruce grouse: Is inheritance involved? *Anim. Behav.* 36:305–307.

Schwarzkopf, L. 1992. Annual variation of litter size and offspring size in a viviparous skink. *Herpetologica* 48:390–395.

———. 1993. Costs of reproduction in water skinks. *Ecology* 74:1970–1981.

Schwarzkopf, L., and R. Shine. 1991. Thermal biology of reproduction in viviparous skinks, *Eulamprus tympanum*: Why do gravid females bask more? *Oecologia* 88:562–569.

———. 1992. Costs of reproduction in lizards: Escape tactics and susceptibility to predation. *Behav. Ecol. Sociobiol.* 31:17–25.

Schwenk, K. 1986. Morphology of the tongue in the tuatara, *Sphenodon punctatus* (Reptilia, Lepidosauria), with comments on function and phylogeny. *J. Morphol.* 188:129–156.

———. 1988. Comparative morphology of the lepidosaur tongue and its relevance to squamate phylogeny. In *Phylogenetic relationships of the lizard families: Essays commemorating Charles L. Camp*, edited by R. Estes and G. Pregill, 569–598. Stanford University Press, Stanford.

———. 1993. The evolution of chemoreception in squamate reptiles: A phylogenetic approach. *Brain, Behav. Evol.* 41:124–137.

Schwenk, K., and G. S. Throckmorton. 1989. Functional and evolutionary morphology of lingual feeding in squamate reptiles: Phylogenetics and kinematics. *J. Zool. London* 219:153–175.

Secor, S. M., B. C. Jayne, and A. F. Bennett. 1992. Locomotor performance and energetic cost of sidewinding by the snake *Crotalus cerastes*. *J. Exp. Biol.* 163:1–14.

Seigel, R. A., and N. B. Ford. 1987. Reproductive ecology. In *Snakes: Ecology and evolutionary biology*, R. A. Seigel, J. T. Collins, and S. S. Novak, 210–252. Macmillan Pub. Co., New York.

Sexton, O. J. 1967. Population changes in a tropical lizard *Anolis limifrons* on Barro Colorado Island, Panama Canal Zone. *Copeia* 1967:219–222.

Sexton, O. J., and H. Heatwole. 1968. An experimental investigation of habitat selection and water loss in some anoline lizards. *Ecology* 49:762–769.

Sexton, O. J., and K. R. Marion. 1974. Duration of incubation of *Sceloporus undulatus* eggs at constant temperature. *Physiol. Zool.* 47:91–98.

Sexton, O. J., E. P. Ortleb, L. M. Hathaway, R. E. Ballinger, and P. Licht. 1971. Reproductive cycles of three species of anoline lizards from the

Isthmus of Panama. *Ecology* 52:201–215.

Sexton, O. J., G. M. Veith, and D. M. Phillips. 1979. Ultrastructure of the eggshell of two species of anoline lizards. *J. Exp. Zool.* 207:227–236.

Sexton, O. J., J. Bauman, and E. Ortleb. 1972. Seasonal food habits of *Anolis limifrons*. *Ecology* 53:182–186.

Shea, G. M. 1990. The genera *Tiliqua* and *Cyclodomorphus* (Lacertilia: Scincidae): Generic diagnoses and systematic relationships. *Mem. Queensland Museum* 29:495–519.

Shenbrot, G. I., K. A. Rogovin, and A. V. Surov. 1991. Comparative analysis of spatial organization of desert lizard communities in middle Asia and Mexico. *Oikos* 61:157–168.

Shields, W. M. 1983. Optimal inbreeding and the evolution of philopatry. In *The ecology of animal movement,* edited by I. R. Swingland and P. J. Greenwood, 132–159. Clarendon Press, Oxford.

———. 1987. Dispersal and mating systems: Investigating their causal connections. In *Mammalian dispersal patterns: The effect of social structure on population genetics,* edited by B. D. Chepko-Sade and Z. T. Halpin, 3–24. University of Chicago Press, Chicago.

Shine, R. 1980. "Costs" of reproduction in reptiles. *Oecologia* 46:92–100.

———. 1983. Reptilian reproductive modes: The oviparity-viviparity continuum. *Herpetologica* 39:1–8.

———. 1986. Evolutionary advantages of limblessness: Evidence from the pygopodid lizards. *Copeia* 1986:525–529.

———. 1988. Parental care in reptiles. In *Biology of the Reptilia. Vol. 16. Ecology B. Defense and life history,* edited by C. Gans and R. B. Huey, 275–330. Alan R. Liss, New York.

———. 1992. Relative clutch mass and body shape in lizards and snakes: Is reproductive investment constrained or optimized? *Evolution* 46:828–833.

Shine, R., and A. E. Greer. 1991. Why are clutch sizes more variable in some species than in others? *Evolution* 45:1696–1706.

Shine, R., and L. Schwarzkopf. 1992. The evolution of reproductive effort in lizards and snakes. *Evolution* 46:62–75.

Simon, C. A. 1975. The influence of food abundance on territory size in the iguanid lizard *Sceloporus jarrovi. Ecology* 56:993–998.

Simon, C. A., and G. A. Middendorf. 1976. Resource partitioning by an iguanid lizard: Temporal and microhabitat aspects. *Ecology* 57:1317–1320.

———. 1980. Spacing in juvenile lizards (*Sceloporus jarrovi*). *Copeia* 1980:141–146.

Simon, C. A., K. Gravelle, B. E. Bissinger, I. Eiss, and R. Ruibal. 1981. The role of chemoreception in the iguanid lizard *Sceloporus jarrovi. Anim. Behav.* 29:46–54.

Sinclair, A.R.E. 1989. Population regulation in animals. In *Ecological concepts: The contribution of ecology to an understanding of the natural world*, edited by J. M. Cherrett, 197–241. Blackwell, Oxford.

Sinervo, B. 1990a. The evolution of maternal investment in lizards: An experimental and comparative analysis of egg size and its effects on offspring performance. *Evolution* 44:279–294.

———. 1990b. The evolution of thermal physiology and growth rate between populations of the western fence lizard (*Sceloporus occidentalis*). *Oecologia* 83:228–237.

———. 1993. The effect of offspring size on physiology and life history: An experimental analysis using allometric engineering. *Bioscience* 43:210–218.

Sinervo, B., and J. B. Losos. 1991. Walking the tight rope: Arboreal sprint performance among *Sceloporus occidentalis* lizard populations. *Ecology* 72:1225–1233.

Sinervo, B., and P. Licht. 1991a. The physiological and hormonal control of clutch size, egg size, and egg shape in *Uta stansburiana*: Constraints on the evolution of lizard life histories. *J. Exp. Zool.* 257:252–264.

———. 1991b. Proximate constraints on the evolution of egg size, egg number, and total clutch mass in lizards. *Science* 252:1300–1302.

Sinervo, B., and R. B. Huey. 1990. Allometric engineering: An experimental test of the causes of interpopulational differences in performance. *Science* 248:1106–1109.

Sinervo, B., and S. C. Adolph. 1989. Thermal sensitivity of growth rate in hatchling *Sceloporus* lizards: Environmental, behavioral and genetic aspects. *Oecologia* 78:411–419.

Sinervo, B., P. Doughty, R. B. Huey, and K. Zamudio. 1992. Allometric engineering: A causal analysis of natural selection on offspring size. *Science* 258:1927–1930.

Sinervo, B., R. Hedges, and S. C. Adolph. 1991. Decreased sprint speed as a cost of reproduction in the lizard *Sceloporus occidentalis*: Variation among populations. *J. Exp. Biol.* 155:323–336.

Sites, J. W., J. W. Archie, and O. F. Villela. 1992. A review of phylogenetic hypotheses for the lizard genus *Sceloporus* (Phrynosomatidae): Implications for ecological and evolutionary studies. *Bull. Amer. Mus. Nat. Hist.* 213:1–110.

Sjögren, P. 1991. Extinction and isolation gradients in metapopulations: The case of the pool frog (*Rana lessonae*). *Biol. J. Linn. Soc.* 42:135–147.

Slatyer, R. O. 1962. Climate of the Alice Springs area. *Austr. CSIRO Land Res. Ser.* 6:109–128.

Smith, C. C., and S. D. Fretwell. 1974. The optimal balance between size and number of offspring. *Amer. Nat.* 108:499–506.

Smith, D. C. 1981. Competitive interactions of the striped plateau lizard (*Sceloporus virgatus*) and the tree lizard (*Urosaurus ornatus*). *Ecology* 62:679–687.

———. 1985. Home range and territory in the striped plateau lizard (*Sceloporus virgatus*). *Anim. Behav.* 33:417–427.

Smith, H. M., G. Sinelnik, J. D. Fawcett, and R. E. Jones. 1973. A survey of the chronology of ovulation in anoline lizard genera. *Trans. Kans. Acad. Sci.* 75:107–120.

Smythe, N. 1982. The seasonal abundance of night-flying insects in a Neotropical forest. In *The ecology of a tropical forest: Seasonal rhythms and long-term changes*, edited by E. G. Leigh, Jr., A. S. Rand, and D. M. Windsor, 309–318. Smithsonian Institution Press, Washington, D.C.

Snedecor, G. W., and W. G. Cochran. 1956. *Statistical methods.* Iowa State University Press, Ames.

Snell, H. L., and C. R. Tracy. 1985. Behavioral and morphological adaptations by Galapagos land iguanas (*Conolophus subcristatus*) to water and energy requirements of eggs and neonates. *Amer. Zool.* 25:1009–1018.

Snell, H. L., R. D. Jennings, H. M. Snell, and S. Harcourt. 1988. Intrapopulation variation in predator-avoidance performance of Galapagos lava lizards: The interaction of sexual and natural selection. *Evol. Ecol.* 2:353–369.

Snyder, R. C. 1954. The anatomy and function of the pelvic girdle and hindlimb in lizard locomotion. *Amer. J. Anat.* 95:1–41.

Sokal, R. R., and F. J. Rohlf. 1981. *Biometry.* W. H. Freeman and Co., San Francisco.

Sorci, G., M. Massot, and J. Clobert. 1994. Maternal parasite load increases sprint speed and philopatry in female offspring of the common lizard. *Amer. Nat.* (in press).

Soulé, M. E. 1986. *Conservation biology.* Sinauer Associates, Inc., Sunderland, Mass.

———., ed. 1987. *Viable populations for conservation.* Cambridge University Press, Cambridge, UK.

Sousa, W. P. 1979. Disturbance in marine intertidal boulder fields: The nonequilibrial maintenance of species diversity. *Ecology* 60:1225–1239.

Stafford Smith, D. M., and S. R. Morton. 1990. A framework for the ecology of arid Australia. *J. Arid Environ.* 18:255–278.

Stamps, J. A. 1977. Social behavior and spacing patterns in lizards. In *Biology of the Reptilia. Vol. 7. Ecology and behaviour,* edited by C. Gans and D. W. Tinkle, 265–334. Academic Press, New York.

———. 1983. Sexual selection, sexual dimorphism and territoriality. In *Lizard ecology: Studies of a model organism*, edited by R. B. Huey, E. R. Pianka, and T. W. Schoener, 169–204. Harvard University Press, Cambridge, Mass.

————. 1987. Conspecifics as cues to territory quality: A preference of juvenile lizards (*Anolis aeneus*) for previously used territories. *Amer. Nat.* 129:629–642.

————. 1988. Conspecific attraction and aggregation in territorial species. *Amer. Nat.* 131:329–347.

Stamps, J. A., and S. K. Tanaka. 1981. The relationship between food and social behavior in juvenile lizards (*Anolis aeneus*). *Copeia* 1981:422–434.

————. 1981. The influence of food and water on growth rates in a tropical lizard (*Anolis aeneus*). *Ecology* 62:33–40.

Stamps, J. A., M. Buechner, and V. V. Krishnan. 1987. The effect of edge permeability and habitat geometry on emigration from patches of habitat. *Amer. Nat.* 129:533–552.

Stearns, S. C. 1976. Life-history tactics: A review of the ideas. *Quart. Rev. Biol.* 51:3–47.

————. 1977. The evolution of life history traits: A critique of the theory and a review of the data. *Ann. Rev. Ecol. Syst.* 8:145–171.

————. 1980. A new view of life history evolution. *Oikos* 35:266–281.

————. 1989a. Trade-offs in life-history evolution. *Funct. Ecol.* 3:259–268.

————. 1989b. The evolutionary significance of phenotypic plasticity. *Bioscience* 39:436–445.

————. 1992. *The evolution of life histories*. Oxford University Press, Oxford.

Stearns, S. C., and J. C. Koella. 1986. The evolution of phenotypic plasticity in life-history traits: Predictions of reaction norms for age and size at maturity. *Evolution* 40:893–913.

Stearns, S. C., and R. E. Crandall. 1981. Quantitative predictions of delayed maturity. *Evolution* 35:455–463.

Stebbins, R. C. 1944. Field notes on a lizard, the mountain swift, with special reference to territorial behavior. *Ecology* 25:233–245.

————. 1985. *A field guide to western reptiles and amphibians, 2nd ed., revised.* Houghton Mifflin Co., Boston, Mass.

Steel, R.G.D., and J. H. Torrie. 1980. *Principles and procedures of statistics.* McGraw-Hill, New York.

Stenseth, N. C., and W. Z. Lidicker, Jr., eds. 1992. *Animal dispersal: Small mammals as a model.* Chapman and Hall, London.

Stewart, J. R. 1985. Growth and survivorship in a California population of *Gerrhonotus coeruleus*, with comments on intraspecific variation in female size. *Amer. Midl. Nat.* 113:30–44.

Storr, G. M., L. A. Smith, and R. E. Johnstone. 1981. *Lizards of Western Australia. I. Skinks.* University of Western Australia Press, Nedlands.

————. 1983. *Lizards of Western Australia. II. Dragons and monitors.*

Western Australian Museum, Perth.

———. 1990. *Lizards of Western Australia. III. Geckos and pygopods*. Western Australian Museum, Perth.

Strijbosch, H., and R.C.M. Creemers. 1988. Comparative demography of sympatric populations of *Lacerta vivipara* and *Lacerta agilis*. *Oecologia* 76:20–26.

Strong, D.R.J. 1983. Natural variability and the manifold mechanisms of ecological communities. *Amer. Nat.* 122:636–660.

Sukhanov, V. B. 1968. *General system of symmetrical locomotion of terrestrial vertebrates and some features of movement of lower tetrapods*. Nauka Publ., Leningrad; translated to English, Amerinde Publishing Co., New Dehli.

Swain, T. A. 1977. The autecology of two Venezuelan lizard species. Ph. D. Diss., University of Colorado, Boulder.

Swingland, I. R. 1983. Intraspecific differences in movement. In *The ecology of animal movement*, edited by I. R. Swingland and P. J. Greenwood, 105–115. Clarendon Press, Oxford.

Swingland, I. R., and P. J. Greenwood, eds. 1983. *The ecology of animal movement*. Clarendon Press, Oxford.

Swingland, I. R., P. M. North, A. Dennis, and M. J. Parker. 1989. Movements patterns and morphometrics in giant tortoises. *J. Anim. Ecol.* 58:971–985.

SYSTAT. 1992. *SYSTAT: Statistics*. SYSTAT, Inc., Evanston, Illinois.

Tanner, W. W., and J. E. Krogh. 1973. Ecology of *Sceloporus magister* at the Nevada test site, Nye County, Nevada. *Great Basin Nat.* 33:133–146.

Tanner, W. W., and J. M. Hopkin. 1972. Ecology of *Sceloporus occidentalis longipes* Baird and *Uta stansburiana stansburiana* Baird and Girard on Rainier Mesa, Nevada Test Site, Nye County, Nevada. *Brigham Young Univ. Sci. Bull., Biol. Ser.* 3:1–31.

Taper, M. L., and T. J. Case. 1992. Coevolution among competitors. *Oxford Surveys Evol. Biol.* 8:63–109.

Taylor, C. R., N. Heglund, and G.M.O. Maloiy. 1982. Energetics and mechanics of terrestrial locomotion. I. Metabolic energy consumption as a function of speed and body size in birds and mammals. *J. Exp. Biol.* 97:1–21.

Templeton, A. R. 1986. Relation of humans to African apes: A statistical appraisal of diverse types of data. In *Evolutionary processes and theory*, edited by S. Karlin and E. Nevo, 218–234. Academic Press, Orlando, Florida.

Terborgh, J., and S. Robinson. 1986. Guilds and their utility in ecology. In *Community ecology: Pattern and process*, edited by J. Kikkawa and D. J. Anderson, 65–90. Blackwell Scientific, Melbourne, Australia.

Thompson, S. D. 1985. Bipedal hopping and seed-dispersion selection by heteromyid rodents: The role of locomotion energetics. *Ecology* 66:220–229.

Tinkle, D. W. 1967. Home range, density, dynamics, and structure of a Texas population of the lizard *Uta stansburiana*. In *Lizard ecology: A symposium*, edited by W. W. Milstead, 5–29. University of Missouri Press, Columbia.

———. 1969. The concept of reproductive effort and its relation to the evolution of life histories in lizards. *Amer. Nat.* 103:501–516.

———. 1972. The dynamics of a Utah population of *Sceloporus undulatus*. *Herpetologica* 28:351–359.

———. 1973. A population analysis of the sagebrush lizard *Sceloporus graciosus* in southern Utah. *Copeia* 1973:284–296.

———. 1976. Comparative data on the population ecology of the desert spiny lizard, *Sceloporus magister*. *Herpetologica* 32:1–6.

Tinkle, D. W., and Dunham, A. E. 1986. Comparative life histories of two syntopic sceloporine lizards. *Copeia* 1986:1–18.

Tinkle, D. W., and J. W. Gibbons. 1977. The distribution and evolution of viviparity in reptiles. *Misc. Publ. Univ. Michigan Mus. Zool.* 154:1–55.

Tinkle, D. W., and N. F. Hadley. 1973. Reproductive effort and winter activity in the viviparous montane lizard *Sceloporus jarrovi*. *Copeia* 1973:272–277.

Tinkle, D. W., and R. E. Ballinger. 1972. *Sceloporus undulatus:* A study of the intraspecific comparative demography of a lizard. *Ecology* 53:570–584.

Tinkle, D. W., H. M. Wilbur, and S. G. Tilley. 1970. Evolutionary strategies in lizard reproduction. *Evolution* 24:55–74.

Tinkle, D. W., J. D. Congdon, and P. C. Rosen. 1981. Nesting frequency and success: Implications for the demography of painted turtles. *Ecology* 62:1426–1432.

Tokarz, R. R. 1992. Male mating preference for unfamiliar females in the lizard *Anolis sagrei*. *Anim. Behav.* 44:843–849.

Tracy, C. R. 1980. On the water relations of parchment-shelled lizard (*Sceloporus undulatus*) eggs. *Copeia* 1980:478–482.

Tracy, C. R., and H. L. Snell. 1985. Interrelations among water and energy relations of relations of reptilian eggs, embryos, and hatchlings. *Amer. Zool.* 25:999–1008.

Trivers, R. L. 1972. Parental investment and sexual selection. In *Sexual selection and descent of man*, edited by B. Campbell, 136–179. Aldine-Atherton, Chicago.

Troyer, K. 1983. Post-hatchling yolk energy in a lizard: Utilization pattern and interclutch variation. *Oecologia* 58:340–343.

Tsuji, J. S., R. B. Huey, F. H. van Berkum, T. Garland, Jr., and R. G. Shaw. 1989. Locomotor performance of hatchling fence lizards (*Sceloporus occidentalis*): Quantitative genetics and morphometric correlates. *Funct. Ecol.* 3:240–252.

Tucker, V. A. 1967. The role of the cardiovascular system in oxygen transport and thermoregulation in lizards. In *Lizard ecology: A symposium*, edited by W. W. Milstead, 258–269. University of Missouri Press, Columbia.

Tuomi, J., T. Hakala, and E. Haukioja. 1983. Alternative concepts of reproductive effort, costs of reproduction, and selection in life-history evolution. *Amer. Zool.* 23:25–34.

Turelli, M., J. H. Gillespie, and R. Lande. 1988. Rate tests for selection on quantitative characters during macroevolution and microevolution. *Evolution* 42:1085–1089.

Turner, F. B. 1977. The dynamics of populations of squamates, crocodilians and rhynchocephalians. In *Biology of the Reptilia. Vol. 7. Ecology and behaviour*, edited by C. Gans and D. W. Tinkle, 157–264. Academic Press, New York.

Turner, F. B., G. A. Hoddenbach, P. A. Medica, and J. R. Lannom. 1970. The demography of the lizard, *Uta stansburiana* Baird and Girard, in southern Nevada. *J. Anim. Ecol.* 39:505–519.

Turner, F. B., R. I. Jennrich, and J. D. Weintraub. 1969. Home ranges and body size of lizards. *Ecology* 50:1076–1081

Van Berkum, F. H. 1986. Evolutionary patterns of the thermal sensitivity of sprint speed in *Anolis* lizards. *Evolution* 40:594–604.

Van Damme, R., D. Bauwens, and R. F. Verheyen. 1989. Effect of relative clutch mass on sprint speed in the lizard *Lacerta vivipara*. *J. Herpetol.* 23:459–461.

Van Damme, R., D. Bauwens, F. Braña, and R. F. Verheyen. 1992. Incubation temperature differentially affects hatching time, egg survival, and hatchling performance in the lizard *Podarcis muralis*. *Herpetologica* 48:220–228.

van Noordwijk, A. J., and G. de Jong. 1986. Acquisition and allocation of resources: Their influence on variation in life-history tactics. *Amer. Nat.* 128:137–142.

Vanzolini, P. E., and E. E. Williams. 1970. South American anoles: The geographic differentiation and evolution of the *Anolis chrysolepis* species group (Sauria, Iguanidae). *Arq. Zool. S. Paulo* 19:1–289.

Vermeij, G. J. 1982. Unsuccessful predation and evolution. *Amer. Nat.* 120:701–720.

Vernet, R., C. Grenot, and S. Nouira. 1988. Flux hydriques et métabolisme énergétiques dans un peuplement de lacertidés des îles Kerkennah (Tunisie). *Can. J. Zool.* 66:555–561.

Via, S., and R. Lande. 1985. Genotype-environment interaction and the evolution of phenotypic plasticity. *Evolution* 39:505–522.

Vince, M. A., and S. Chinn. 1971. Effect of accelerated hatching on initiation of standing and walking in Japanese Quail. *Anim. Behav.* 19:62–66.

Vinegar, M. B. 1975a. Comparative aggression in *Sceloporus virgatus*, *S. undulatus consobrinus*, and *S. u. tristichus* (Sauria: Iguanidae). *Anim. Behav.* 23:279–286.

———. 1975b. Demography of the striped plateau lizard, *Sceloporus virgatus*. *Ecology* 56:172–182.

———. 1975c. Life history phenomena in two populations of the lizard *Sceloporus undulatus* in southwestern New Mexico. *Amer. Midl. Nat.* 93:388–402.

Vitt, L. J. 1981. Lizard reproduction: Habitat specificity and constraints on relative clutch mass. *Amer. Nat.* 117:506–514.

———. 1986. Reproductive tactics of sympatric gekkonid lizards with a comment on the evolutionary and ecological consequences of invariant clutch size. *Copeia* 1986:773–786.

———. 1990. The influence of foraging mode and phylogeny on seasonality of tropical lizard reproduction. *Pap. Avul. Zool. (S. Paulo)* 37:107–123.

———. 1992. Diversity of reproductive strategies among Brazilian lizards and snakes: The significance of lineage and adaptation. In *Reproductive biology of South American vertebrates*, edited by W. C. Hamlett, 135–149. Springer-Verlag, New York.

Vitt, L. J., and C. M. de Carvalho. 1992. Life in the trees: The ecology and life history of *Kentropyx striatus* (Teiidae) in the lavrado area of Roraima, Brazil, with comments on the life histories of tropical teiid lizards. *Can. J. Zool.* 70:1995–2006.

Vitt, L. J., and G. L. Breitenbach. 1993. Life histories and reproductive tactics among lizards in the genus *Cnemidophorus* (Sauria: Teiidae). In *Biology of whiptail lizards (genus Cnemidophorus)*, edited by J. W. Wright and L. J. Vitt, 211–243. Oklahoma Mus. Nat. Hist., Norman.

Vitt, L. J., and H. J. Price. 1982. Ecological and evolutionary determinants of relative clutch mass in lizards. *Herpetologica* 38:237–255.

Vitt, L. J., and J. D. Congdon. 1978. Body shape, reproductive effort, and relative clutch mass in lizards: Resolution of a paradox. *Amer. Nat.* 112:595–608.

Vitt, L. J., and R. A. Seigel. 1985. Life history traits of lizards and snakes. *Amer. Nat.* 125:480–484.

Vitt, L. J., and T. E. Lacher, Jr. 1981. Behavior, habitat, diet, and reproduction of the iguanid lizard *Polychrus acutirostris* in the caatinga of northeastern Brazil. *Herpetologica* 37:53–63.

Vitt, L. J., and W. E. Cooper, Jr. 1985. The relationship between reproduction and lipid cycling in the skink *Eumeces laticeps* with comments on brooding ecology. *Herpetologica* 41:419–432.

———. 1986. Foraging and diet of a diurnal predator (*Eumeces laticeps*) feeding on hidden prey. *J. Herpetol.* 20:408–415.

Von Achen, P. H., and J. L. Rakestraw. 1984. The role of chemoreception in the prey selection of neonate reptiles. In *Vertebrate ecology and systematics: A tribute to Henry S. Fitch*, edited by R. A. Seigel, L. E. Hunt, J. L. Knight, L. Malaret, and N. L. Zuschlag, 163–172. Mus. Nat. Hist. University of Kansas, Lawrence.

Wade, M. J., and S. Kalisz. 1990. The causes of natural selection. *Evolution* 44:1947–1955.

Walls, G. Y. 1981. Feeding ecology of the tuatara (*Sphenodon punctatus*) on Stephens Island, Cook Strait. *New Zeal. J. Ecol.* 4:89–97.

Walls, S. C., and R. Altig. 1986. Female reproductive biology and larval life history of *Ambystoma* salamanders: A comparison of egg size, hatchling size, and larval growth. *Herpetologica* 42:334–345.

Walton, M., B. C. Jayne, and A. F. Bennett. 1990. The energetic cost of limbless locomotion. *Science* 249:524–527.

Warheit, K. I. 1992. A review of the fossil seabirds from the Tertiary of the North Pacific: Plate tectonics, paleoceanography, and faunal change. *Paleobiology* 18:401–424.

Webb, G.J.W., and H. Cooper-Preston. 1989. Effects of incubation temperature on crocodiles and the evolution of reptilian oviparity. *Amer. Zool.* 29:953–971.

Webb, G.J.W., S. C. Manolis, P. J. Whitehead, and K. Dempsey. 1987. The possible relationship between embryo orientation, opaque banding, and the dehydration of albumin in crocodile eggs. *Copeia* 1987:252–257.

Werner, D. I. 1988. The effect of varying water potential on body weight, yolk and fat bodies in neonate green iguana. *Copeia* 1988:406–411.

Werner, Y. L. 1989. Egg size and egg shape in near-eastern gekkonid lizards. *Israel J. Zool.* 35:199–213.

White, G. C., D. R. Anderson, K. P. Burnham, and D. L. Otis. 1982. *Capture-recapture and removal methods for sampling closed populations*. Los Alamos National Laboratory, LA 8787–NERP. Los Alamos, New Mexico.

Whitford, W. G., and F. M. Creusere. 1977. Seasonal and yearly fluctuations in Chihuahuan Desert lizard communities. *Herpetologica* 33:54–65.

Wiens, J. A. 1977. On competition and variable environments. *Amer. Sci.* 65:590–597.

———. 1984. On understanding a non-equilibrial world: Myth and reality in community patterns and processes. In *Ecological communities: Conceptual issues and the evidence*, edited by D.R.J. Strong, D. Simberloff, L. G. Abele, and A. B. Thistle, 439–457. Princeton University Press, Princeton, New Jersey.

————. 1989a. *The ecology of bird communities. Vol. 2. Processes and variations.* Cambridge University Press, Cambridge, UK.

————. 1989b. Spatial scaling in ecology. *Funct. Ecol.* 3:385–397.

Wiens, J. J. 1993. Phylogenetic relationships of phrynosomatid lizards and monophyly of the *Sceloporus* group. *Copeia* 1993:287–299.

Wilcox, B. A. 1978. Supersaturated island faunas: A species-age relationship for lizards on post-Pleistocene land-bridge islands. *Science* 199:996–998.

Wiley, E. O., D. Siegel-Causey, D. R. Brooks, and V. A. Funk. 1991. *The compleat cladist: A primer of phylogenetic procedures.* Mus. Nat. Hist., University of Kansas, Lawrence.

Williams, D. M., and P. F. Sale. 1981. Spatial and temporal patterns of recruitment of juvenile coral reef fishes to coral habitats within One Tree Lagoon, Great Barrier Reef. *Marine Biol.* 65:245–253.

Williams, E. E. 1972. The origin of faunas: Evolution of lizard congeners in a complex island fauna: A trial analysis. *Evol. Biol.* 6: 47–89.

————. 1983. Ecomorphs, faunas, island size, and diverse end points in island radiations of *Anolis*. In *Lizard ecology: Studies of a model organism*, edited by R. B. Huey, E. R. Pianka, and T. W. Schoener, 326–370. Harvard University Press, Cambridge, Mass.

————. 1989. A critique of Guyer and Savage (1986): Cladistic relationships among anoles (Sauria: Iguanidae); are the data available to reclassify the anoles? In *Biogeography of the West Indies*, edited by C. A. Woods, 433–478. Sandhill Crane Press, Gainesville, Florida.

Williams, G. C. 1966a. Natural selection, the costs of reproduction, and a refinement of Lack's principle. *Amer. Nat.* 100:687–690.

————. 1966b. *Adaptation and natural selection: A critique of some current evolutionary thought.* Princeton University Press, New Jersey.

————. 1992. *Natural selection: Domains, levels, and challenges.* Oxford University Press, Oxford.

Wilson, B. S., and J. C. Wingfield. 1992. Correlation between female reproductive condition and plasma corticosterone in the lizard *Uta stansburiana*. *Copeia* 1992:691–697.

Wilson, L. D., and L. Porras. 1983. *The ecological impact of man on the south Florida herpetofauna.* University of Kansas Press, Lawrence.

Wolda, H., and R. Foster. 1978. *Zunacetha annulata* (Lepidoptera: Dioptidae), an outbreak insect in a neotropical forest. *Geo-Eco-Trop* 2:443–454.

Wolda, H., and S. J. Wright. 1992. Artificial dry season rain and its effects on tropical insect abundance and seasonality. *Proc. Kon. Ned. Akad. v. Wetensch.* 95:535–548.

Wright, J. W., and L. J. Vitt, eds. 1993. *Biology of whiptail lizards (genus Cnemidophorus).* Oklahoma Mus. Nat. Hist., Norman.

Wright, S. J. 1991. Seasonal drought and the phenology of understory shrubs in a tropical moist forest. *Ecology* 72:1643–1657.

Wright, S. J., and F. H. Cornejo. 1990. Seasonal drought and leaf fall in a tropical forest. *Ecology* 71:1165–1175.

Wright, S. J., R. Kimsey, and C. J. Campbell. 1984. Mortality rates of insular *Anolis* lizards: A systematic effect of island area? *Amer. Nat.* 123:134–142.

Wyles, J. S., and V. M. Sarich. 1983. Are the Galapagos iguanas older than the Galapagos? In *Patterns of evolution in Galapagos organisms*, edited by R. I. Bowman, M. Berson, and A. E. Leviton, 177–186. Pacific Division, Amer. Assoc. Adv. Sci., San Francisco.

Yanoskey, A. A., D. E. Iriart, and M. R. Perrin. 1993. Predatory behavior in *Tupinambis teguixin* (Sauria, Teiidae). I. Tongue-flicking responses to chemical food stimuli. *J. Chem. Ecol.* 19:291–299.

Yant, P. R., J. R. Karr, and P. L. Angermeier. 1984. Stochasticity in stream fish communities: An alternate interpretation. *Amer. Nat.* 124:573–582.

Yaron, Z., and L. Widzer. 1978. The control of vitellogenesis by ovarian hormones in the lizard *Xantusia vigilis*. *Comp. Biochem. Physiol.* 60:279–284.

Yeatman, E. M. 1988. Resource partitioning by three congeneric species of skink (*Tiliqua*) in sympatry in South Australia. Ph. D. Diss., Flinders University of South Australia, Adelaide.

Zug, G. R. 1992. *Herpetology: An introductory biology of amphibians and reptiles*. Academic Press, San Diego.

AUTHOR INDEX

SPECIES INDEX

Lightning Source UK Ltd.
Milton Keynes UK
UKHW021558031222
413284UK00009B/448

9 780691 601960